Neighbourhood Effects Research: New Perspectives

Maarten van Ham • David Manley • Nick Bailey
Ludi Simpson • Duncan Maclennan
Editors

Neighbourhood Effects Research: New Perspectives

Springer

Editors
Maarten van Ham
OTB Research Institute for the Built Environment
Delft University of Technology
P.O. Box 5030
2600 GA Delft
The Netherlands
m.vanham@tudelft.nl

Nick Bailey
Urban Studies
University of Glasgow
Glasgow G12 8RS
United Kingdom
nick.bailey@glasgow.ac.uk

Duncan Maclennan
Centre for Housing Research
School of Geography and Geosciences
University of St. Andrews
Fife KY16 9AL
United Kingdom
dm103@st-andrews.ac.uk

David Manley
Centre for Housing Research
School of Geography and Geosciences
University of St. Andrews
Fife KY16 9AL
United Kingdom
d.manley@st-andrews.ac.uk

Ludi Simpson
The Cathie Marsh Centre for Census and Survey Research
University of Manchester
Manchester M13 9PL
United Kingdom
ludi.simpson@manchester.ac.uk

ISBN 978-94-007-2308-5 e-ISBN 978-94-007-2309-2
DOI 10.1007/978-94-007-2309-2
Springer Dordrecht Heidelberg London New York

Library of Congress Control Number: 2011942227

Printed on acid-free paper

Springer is part of Springer Science+Business Media (www.springer.com)

Preface

There is little doubt that neighbourhood effects exist, but after decades of research we seem no closer to knowing how important they are. Neighbourhood effects research is academically intriguing, but also has high policy relevance. Area-based policies, and especially mixed communities policies, are a direct response to the idea that the neighbourhood where you live can have a negative effect on your well-being. It is therefore important to establish how influential such effects really are, what causal mechanisms produce them, and under which circumstances and in which places neighbourhood effects are most significant. Answering these questions helps to develop more effective policy interventions.

The research that is reported in the chapters of this book addresses many of the key issues in the neighbourhood effects debate. The book reviews theories about how neighbourhoods might shape individual lives, exploring the potential causal pathways between neighbourhood context and individual outcomes. Given that one of the main challenges in neighbourhood effects research is the identification of true causal neighbourhood effects, special attention is paid to causality. The book also presents new empirical research on neighbourhood effects, highlighting various methodological problems associated with investigating these effects. Finally, the book increases our understanding of data and methods suitable to analyse neighbourhood effects.

Collectively, the chapters in this book offer new perspectives on this field of research, and refocus the academic debate. It enriches the neighbourhood effects literature with insights from a wide range of disciplines and countries. The introduction of the book summarises seven ways forward for neighbourhood effects research: development of clear hypotheses; empirically testing explicit hypotheses; investigating neighbourhood selection; integrate models of neighbourhood selection and models of neighbourhood effects; investigate various spatial scales; development of better longitudinal data; and the use of mixed methods research.

Many of the contributions in this book were presented at the seminar *Neighbourhood effects: theory and evidence* on 4 and 5 February 2010 at the University of St. Andrews. The seminar was part of a wider ESRC Seminar Series: *Challenges in neighbourhood effects research: does it really matter where you live and what are the implications for policy* (RES-451-26-0704). The seminar series was a collaboration between the Centre for Housing Research (CHR) at the University of St Andrews (lead), Urban Studies at the University of Glasgow, and the Cathie Marsh Centre for Census and Survey Research (CCSR) at the University of Manchester.

Delft, NL	Maarten van Ham
St. Andrews, UK	David Manley
Glasgow, UK	Nick Bailey
Manchester, UK	Ludi Simpson
St. Andrews, UK	Duncan Maclennan

Acknowledgements

The editors would like to acknowledge the financial support from the Economic and Social Research Council in the form of an ESRC Seminar Series grant (RES-451-26-0704). The editors also wish to acknowledge the generous financial and staff time support of the Centre for Housing Research (CHR) at the University of St. Andrews. The editors would like to express their thanks to all the authors for submitting their initial manuscripts on schedule, responding positively to comments and suggestions from the editors and delivering final versions of chapters with minimum delay. We are also grateful to Graeme Sandeman, the cartographer of the School of Geography and Geosciences at the University of St Andrews, who has redrawn the graphs and maps. Maarten van Ham contributed to this volume while working at CHR, University of St Andrews.

Contents

Contributors

Kathy Arthurson Faculty of Health Sciences, Southgate Institute for Health, Society and Equity, Flinders University, Bedford Park, SA, Australia, kathy.arthurson@flinders.edu.au

Nick Bailey Urban Studies, School of Social and Political Sciences, University of Glasgow, Glasgow, Scotland, UK, nick.bailey@glasgow.ac.uk

Venla Bernelius Department of Geosciences and Geography, University of Helsinki, Helsinki, Finland, venla.bernelius@helsinki.fi

Paul Cheshire Department of Geography and Environment, London School of Economics, London, UK, p.cheshire@lse.ac.uk

Michael Darcy School of Social Sciences, University of Western Sydney, Penrith, NSW, Australia, m.darcy@uws.edu.au

Stefanie DeLuca Johns Hopkins University, Baltimore, MD, USA, sdeluca@jhu.edu

Greg J. Duncan Department of Education, University of California, Irvine, CA, USA, gduncan@uci.edu

Jessica Feldman Department of Sociology, The University of Chicago, Chicago, IL, USA, jessicaf@uchicago.edu

George C. Galster Department of Urban Studies and Planning, Wayne State University, Detroit, MI, USA, aa3571@wayne.edu

Gabrielle Gwyther School of Social Sciences, University of Western Sydney, Penrith, NSW, Australia, g.gwyther@bigpond.com

Lina Hedman Institute for Housing and Urban Research, Uppsala University, Gävle, Sweden, lina.hedman@ibf.uu.se

Timo M. Kauppinen Department of Social Research, University of Turku, Turku, Finland, timo.kauppinen@utu.fi

Micere Keels Department of Comparative Human Development, University of Chicago, Chicago, IL, USA, micere@uchicago.edu

Dylan Kneale International Longevity Centre, London, UK

Centre for Longitudinal Studies, Institute of Education, London, UK, dylankneale@ilcuk.org.uk

Ruth Lupton Centre for Analysis of Social Exclusion (CASE), London School of Economics and Political Science, London, UK, r.lupton@lse.ac.uk

Duncan Maclennan Centre for Housing Research, School of Geography and Geosciences, University of St Andrews, Fife, Scotland, UK, dm103@st-andrews.ac.uk

David Manley Centre for Housing Research, School of Geography and Geosciences, University of St Andrews, Fife, Scotland, UK, d.manley@st-andrews.ac.uk

Ruby Mendenhall African American Studies and Sociology, University of Illinois, Urbana, IL, USA, rubymen@illinois.edu

Ludi Simpson The Cathie Marsh Centre for Census and Survey Research (CCSR), School of Social Sciences, University of Manchester, Manchester, UK, Ludi.Simpson@manchester.ac.uk

Mario Luis Small Department of Sociology, The University of Chicago, Chicago, IL, USA, mariosmall@uchicago.edu

Gindo Tampubolon Institute for Social Change, The University of Manchester, Manchester, UK, tampubolon@manchester.ac.uk

Maarten van Ham OTB Research Institute for the Built Environment, Delft University of Technology, Delft, The Netherlands, m.vanham@tudelft.nl

Chapter 1
Neighbourhood Effects Research: New Perspectives

Maarten van Ham, David Manley, Nick Bailey, Ludi Simpson, and Duncan Maclennan

Introduction

Over the last 25 years a vast body of literature has been published on neighbourhood effects: the idea that living in deprived neighbourhoods has a negative effect on residents' life chances over and above the effect of their individual characteristics (van Ham and Manley 2010). Neighbourhood effects have been reported on outcomes such as educational achievement, school dropout rates, deviant behaviour, social exclusion, health, transition rates from welfare to work, and social and occupational mobility (see for a review Ellen and Turner 1997; Galster 2002; Dietz 2002; Durlauf 2004). The concept of neighbourhood effects – as an independent residential and social environment effect – is academically intriguing, but has also been embraced by policy makers. Area-based policies aimed at socially mixing neighbourhood populations through mixed tenure policies are seen as a solution to create a more diverse socio-economic mix in neighbourhoods, removing the

M. van Ham (✉)
OTB Research Institute for the Built Environment,
Delft University of Technology, Delft, The Netherlands
e-mail: m.vanham@tudelft.nl

D. Manley • D. Maclennan
Centre for Housing Research, School of Geography and Geosciences,
University of St Andrews, Fife, Scotland, UK
e-mail: d.manley@st-andrews.ac.uk; dm103@st-andrews.ac.uk

N. Bailey
Urban Studies, School of Social and Political Sciences,
University of Glasgow, Glasgow, Scotland, UK
e-mail: nick.bailey@glasgow.ac.uk

L. Simpson
The Cathie Marsh Centre for Census and Survey Research,
School of Social Sciences, University of Manchester, Manchester, UK
e-mail: Ludi.Simpson@manchester.ac.uk

M. van Ham et al. (eds.), *Neighbourhood Effects Research: New Perspectives*,
DOI 10.1007/978-94-007-2309-2_1,

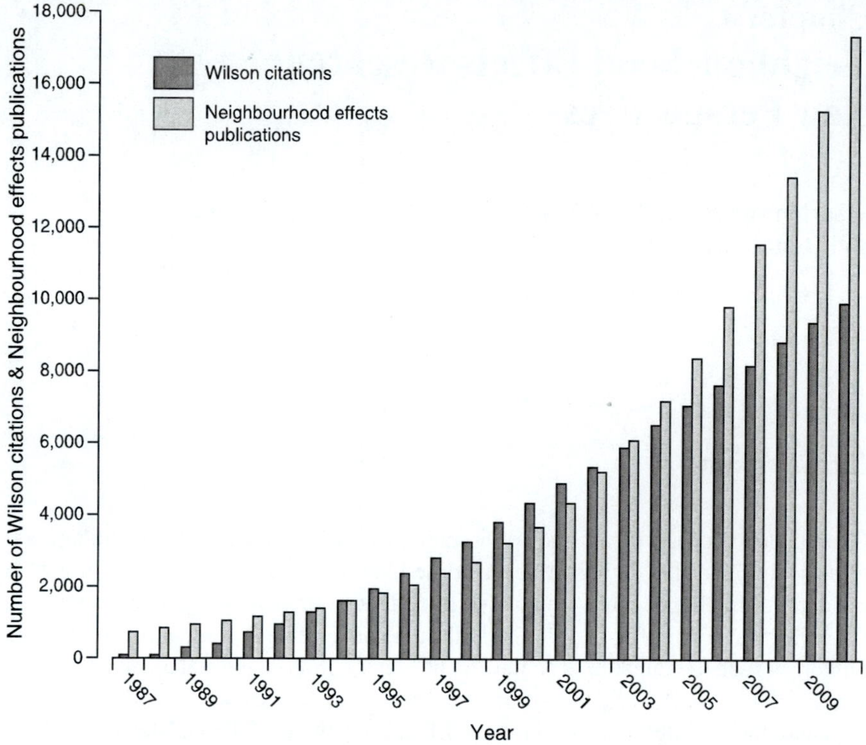

Fig. 1.1 Number of hits in Google Scholar including the words "neighbo(u)rhood effects" by year and the number of hits including the words "The Truly Disadvantaged" combined with "Wilson"

potential of negative neighbourhood effects (Musterd and Andersson 2005). Mixed housing strategies are stated explicitly by many governments including those in the Netherlands, the United Kingdom, Germany, France, Finland and Sweden (Atkinson and Kintrea 2002; Kearns 2002; Musterd 2002).

To illustrate the popularity of the neighbourhood effects discourse we performed a simple Google search on the term "neighbourhood effects" [1] which yielded 203,100 hits (on 24 Feb. 2011). To get more insight in the academic literature we performed a similar search in Google Scholar (on 24 Feb. 2011). Figure 1.1 gives a breakdown of these Google Scholar hits since 1987.[2] In the first year, Google Scholar returned

[1] In our search we used both the UK and US spelling of "neighbo(u)rhood effect", excluding the pluaral "neighbo(u)rhood effects" to avoid double counting documents which mention both singular and plural forms (we found a total of 27,500 hits on "neighbo(u)rhood effect"). Counting both hits in UK and US spelling will potentially also result in some double counts as both spellings can occur in the same document as reference lists typically use the original spelling of a title, regardless the spelling of the document.

[2] 1987 was chosen because this was the year Julius Wilson published his famous book The Truly Disadvantaged. Google Scholar also returned publications containing the words "neighbo(u)rhood effects" from before 1987, since Wilson's book was by no means the starting point of the debate.

772 documents (books, journal articles and reports) that included the words "neighbo(u)rhood effects". In 2010, 23 years later, the number had increased to 17,420 documents. Figure 1.1 clearly shows that measured in publications, the interest in neighbourhood effects has accelerated over the years and is still growing fast.

Although the neighbourhood effects literature can be traced back to the work of the American sociologist Herbert Gans (1968) in the 1960s, the current popularity of the concept is largely driven by the work of William Julius Wilson and his book "The Truly Disadvantaged: The Inner City, the Underclass and Public Policy" (Wilson 1987). In this book Wilson used ethnographic research to provide an in-depth and detailed account of the effects of living in concentrations of poverty in Chicago and he concluded that the "local conditions and the social practices of residents of poor areas cannot be understood independently of the macro social and economic forces which shape them" (Darcy and Gwyther 2011). The popularity of the book by Wilson can be illustrated by a Google Scholar search on a combination of "The truly disadvantaged" and "Wilson". A breakdown by year shows a steady and growing interest in the book, starting with 33 references in 1987, and growing to 9,880 references (on 24 Feb. 2011) in 2010. The very high volume of references to Wilson's book illustrates the impact of the work on the neighbourhood effects discourse. It is interesting to note that since 2004 the volume of work on neighbourhood effects is growing faster than the number of references to Wilson's book, which suggests that many more recent publications on these effects are moving away from Wilson' original work.

The large volume of work on neighbourhood effects not only reflects the interest in the topic, but possibly also reflects the fact that we are still a long way from answering the question how important these effects actually are (see also Small and Feldman 2011, in this volume). Sampson and colleagues have described the search for neighbourhood effects as the "cottage industry in the social sciences" (Sampson, et al. 2002 p. 444). There is little doubt that these effects exist, but we do not know enough about the causal mechanisms which produce them, their relative importance compared to individual characteristics such as education, and under which circumstances and where these effects are important.

One of the main challenges in neighbourhood effects research is the identification of true causal effects (Durlauf 2004) and many existing studies fail to do this convincingly. This leaves the impression that neighbourhood effects are important, while in reality many studies just show correlations between individual outcomes and neighbourhood characteristics (Cheshire 2007; van Ham and Manley 2010). Critics[3] have even stated that "there is surprisingly little evidence that living in poor

[3] Some go a lot further in criticizing the neighbourhood effects literature and reject the whole concept of neighbourhood effects by suggesting that they are the product of an ideological discourse. Bauder (2002) presents a strong critique of the neighbourhood effects literature, and notes that "neighbourhood effects are implicit in the culture-of-poverty and underclass concepts" (2002, p.88) through the pathologising of unwed pregnancies, high school dropouts, number of female headed households as de facto societal ills. Bauder argues that "the idea of neighbourhood effects can be interpreted as yet another episode in the on-going discourse of inner-city marginality that blames marginal communities for their own misery" (ibid). Bauder accuses those who research neighbourhood effects of reproducing the very notions of marginality that they seek to understand.

neighbourhoods makes people poorer and erodes their life chances, independently of those factors that contribute to their poverty in the first place" (Cheshire 2007, p. 2). It is in the interest of those involved in the neighbourhood effects discourse and in policy development, to develop a better understanding of the current evidence base for neighbourhood effects, the problems associated with the empirical testing of the neighbourhood effects hypothesis, and to explore new directions for future research.

This book is specifically designed to offer new perspectives on neighbourhood effects research with the aim to further the academic debate on neighbourhood effects and to aid the development of effective policies. It will enrich the neighbourhood effects literature with views from various disciplines and countries. The book will address three key issues. First, it will review theories about how neighbourhoods might shape individual lives, exploring the potential causal pathways between neighbourhood context and individual behaviour. Specific attention will be paid to the issue of causality. Surprisingly, given the awareness of (self) selection processes, the neighbourhood effects literature pays scant attention to the literature on selective residential mobility into and out of neighbourhoods. Second, the book will provide new empirical research on neighbourhood effects. Subsequent chapters will explore various problems associated with investigating neighbourhood effects. Third, the book will increase our understanding of data and methods suitable to analyse neighbourhood effects – free of bias – and the limitations of these methods.

Identifying Causal Neighbourhood Effects

There is a substantial divide in the neighbourhood effects literature between evidence from studies that use qualitative methodologies and the evidence from those studies using quantitative techniques. Studies using qualitative methods, which focus on the experiences and perceptions of residents, have tended to report stronger and more consistent evidence of neighbourhood effects than those that use quantitative methodologies. For instance, using qualitative techniques, neighbourhood effects of poor reputations of neighbourhoods have been repeatedly identified on employment outcomes (see Atkinson and Kintrea 2001), and on social processes, including social networks, acting on other socio-economic outcomes of residents living in deprived neighbourhoods (Pinkster 2009).

This is in stark contrast to the quantitative literature where there has been much less clarity in outcomes. Taking as an example work on labour market outcomes and the effects of the neighbourhood context, there are some papers that claim they have identified causal neighbourhood effects (see for instance Musterd and Andersson 2005; Galster et al. 2008; Overman 2002), while other studies conclude that there may be other mechanisms (such as neighbourhood selection) driving the apparent correlations between poor individual labour market outcomes and neighbourhood context variables (see for instance Oreopoulos 2003; Bolster et al. 2007;

Van Ham and Manley 2010). This critical literature argues that policies designed to tackle poverty should target individuals rather than the areas within which they live (see also Cheshire 2007), without dismissing the importance of area-based policies to direct funding to those individuals who most need it.

The divide in evidence between methodologies is not overly surprising given the epistemological differences between qualitative and quantitative approaches. Qualitative work explicitly draws on real world experiences, while quantitative studies require "abstractions of the world [...] and thus are inherently once-removed from empirical reality" (Small 2008, p. 170). Ideally quantitative research aims to identify independent generalizable causal mechanisms, although it has to be acknowledged that many quantitative studies operate within a 'black box' approach without explicitly identifying specific causal mechanisms. Qualitative studies ideally aim to identify plausible causal mechanism (from residents' perceptions or from other sources) and then investigate what evidence there is that they are operating. Many qualitative studies seek to give voice to individuals and their perceptions, and if an individual *perceives* that they have experienced negative outcomes because of their neighbourhood situation, such as difficulty getting employment because of neighbourhood stigma, then this is of interest in itself.

The qualitative literature emphasises richness and depth of participants' life courses and can use theoretical constructions to move from the one to the many and to generalise findings. Conversely, generalizability in the quantitative literature is gained through the use of samples that reflect the structures of the wider population. In this literature (ideally), proof of a causal neighbourhood effect can only be accepted once a set of analytical and econometric principles have been met and all other possibilities have been controlled for in the modelling approach. Small and Feldman (2011, in this volume) argue that for neighbourhood effects research to move on, qualitative and quantitative methods should meet within one and the same research design (see also Galster 2011, in this volume). Deluca and colleagues (2011, in this volume) offer an empirical exemplar using such a mixed method approach. They use qualitative methods to help to understand some of the unexpected findings of quantitative work from the Moving to Opportunity program.

The main challenge in the quantitative literature is the econometric identification of real causal neighbourhood effects (Moffitt 1998; Durlauf 2004). Sceptics could argue that using quantitative methods it is not possible to identify real causal effects, as there will always be the potential of omitted variable bias and selection bias. A lot can be done to reduce such biases, but many studies do not make an effort to do so. Maybe even more importantly, many studies do not discuss how bias can affect their modelling results, and what the potential implications of bias are for the interpretation of the study outcomes. As a result, many studies which claim to have found causal neighbourhood effects cannot rule out the possibility that some or all of the results are actually a consequence of omitted variables or selection effects. The most obvious examples of quantitative studies which cannot make any claims about causality are ecological studies (see Graham et al. 2009). Such studies can only show correlations between area characteristics and have the potential to fall foul of the ecological fallacy (Robinson 1950). Also quantitative studies using individual level data suffer from a number of problems which may inhibit the identification of causal effects. The most noticeable are omitted (context) variable bias, the simultaneity problem and the endogenous membership problem (Moffitt 1998; Durlauf 2004).

Omitted variable bias occurs when a key explanatory variable is not available in the data used and other variables in the model, which serve as statistical proxies for the missing variable, pick up the effect. A well-known example of an omitted context variable problem is the racial proxy hypothesis (Harris 1999), where race serves as a proxy for economic deprivation (see van Ham and Feijten 2008; Feijten and van Ham 2009). One way to avoid omitted variable bias is to decide which data to collect on the basis of explicit theory and hypotheses (see Galster 2011 in this volume for an example), although it has to be acknowledged that there will always be relevant factors not included in data. The simultaneity problem (also known as the reflection problem, see Manski 1993) is concerned with the fact that measures of neighbourhood characteristics are not independent from the individuals living in neighbourhoods. When testing the hypothesis that the level of unemployment in a neighbourhood has a negative effect on individual unemployment, the individuals in the model should not simultaneously be included in the neighbourhood level measure. An empirical solution is to use longitudinal data and to associate neighbourhood characteristics from a previous point in time to current outcomes. The endogenous group membership problem mainly refers to the issue that households do not select their neighbourhood at random. This is a problem in neighbourhood effects research when the selection mechanism is related to the outcome under study, which is often the case. Pinkster (2009) argues that selection bias is less of a problem in qualitative studies as such investigations focus on the mechanisms through which the neighbourhood context may mediate individual outcomes. Neighbourhood selection is highly structured by demographic and socio-economic characteristics of household, and characteristics of the local housing market (see Hedman et al. 2011). As a result, many quantitative studies of neighbourhood effects suffer from selection bias. The literature offers several econometric techniques aimed at overcoming selection bias, but it is probably realistic to say that selection bias can never be fully ruled out in observational studies.

An approach which can potentially overcome the problem of selection bias is the use of experimental data instead of observational data. Prime examples of such an approach are derived from the poverty deconcentration programs in the US including the Gautreaux project in Chicago and the Moving to Opportunity (MTO) and HOPE VI programs (see Deluca et al. 2011 in this volume). However, although the experimental research design is often seen as the gold standard within the social sciences, in reality many experimental settings still suffer from selection bias. Participation in the deconcentration programs was never completely random as households had to nominate themselves for inclusion in the programs. Often strict selection criteria were used, and there is also some evidence, especially in the Gautreaux project, that some of the allocations were based on judgements of whether or not households were considered as deserving (Rosebaum 1995).

Interestingly, the outcomes of the experimental data analyses are as mixed as those from the observational data. Durlauf (2004) reports that quasi-experimental studies, such as Gautreaux and the Moving to Opportunity program (Rosebaum 1995; Ludwig et al. 2001; Goering et al. 2002) or randomised education studies (see Leventhal and Brooks-Gunn 2004) find little impact on adults' outcomes. Conversely, work by Popkin and Cunningham (2009) reported that, following the HOPE VI program, there were dramatic improvements in social wellbeing for residents who had been moved into neighbourhoods with lower levels of poverty. Clark (2008) reported that many of the studies that had reported an advantage for movers were poorly conceived or failed to take into account the appropriate populations for comparison. Clark concluded that the gains attributed to the deconcentration programs were more likely to be the result of structural improvements, for instance through economic conditions improving, rather than effects directly relating to changes in the neighbourhood and the social environment.

Neighbourhood Effects Research at a Crossroads?

According to Small and Feldman (2011, this volume), neighbourhood effects research is at a crossroads since current empirical and theoretical approaches to the topic do not seem to be moving the debate forward. The body of research is increasing at such a rate that it has become impossible for anyone to gain an overview of the whole literature, and to systematically assess where and under which circumstances neighbourhood effects are important or not, and how important they are compared to individual characteristics. Many studies suffer from a lack of clarity about causality and fail to set out clear hypotheses on the causal mechanisms under investigation.

One of the problems in the quantitative neighbourhood effects literature is that progress has almost exclusively focussed on statistical techniques to overcome selection bias. While these techniques are important, they will never be able to overcome these, or other econometric problems, completely. Moreover, as observed by Rubin (2008), there are potentially greater gains in terms of casual inference to be made through good study design rather than through complex statistical modelling techniques. The emphasis on statistical techniques has also hampered our understanding of why certain households move to certain neighbourhoods and how this is related to neighbourhood effects. More importantly, the emphasis on technical solutions to solve selection bias has distracted us from a much more important issue: the theoretical and empirical identification of potential causal pathways which may lead to neighbourhood effects (see both Galster 2011 and Small and Feldman 2011 in this volume). Many studies simply search for correlations between neighbourhood characteristics and individual outcomes, control for a range of econometric problems (if at all) and, when some correlation remains, conclude that they have (most likely) found a neighbourhood effect. There is a lack of research that starts from a clear theoretical framework, and searches for robust and defensible causal mechanisms.

Jencks and Mayer (1990) concluded that in many studies neighbourhood effects are essentially treated as a "black-box" term identifying a set of unexplained relationship(s) to be further investigated rather than an entity that can be used to explain a set of outcomes. Over 20 years later, that criticism is still relevant.

It is apparent that there is a real need for a re-evaluation of the way in which we research neighbourhood effects. The chapters in this book offer multiple ways forward. First of all, future work should concentrate on deriving and testing clear hypotheses on causal neighbourhood effect mechanisms. Small and Feldman (2011) in this volume identify a need to integrate ethnography more effectively in neighbourhood effects research to generate explicit, testable hypotheses that guide quantitative research. Second, studies should explicitly investigate the relationship between neighbourhood context and individual outcomes. Are there duration effects? Are there thresholds? (see Galster 2011 in this volume). Third, future work should also concentrate on understanding mechanisms behind neighbourhood selection. Simply controlling for selection is not enough as selection is at the heart of understanding why certain households end up in certain neighbourhoods (Hedman and van Ham 2011 in this volume). Fourth, instead of treating neighbourhood selection as a nuisance which needs to be controlled away, future work should attempt to incorporate models of neighbourhood selection in models of neighbourhood effects (Manley and van Ham 2011 in this volume). Fifth, future work should acknowledge that neighbourhood effects might operate at various spatial scales and include multiple scales in the empirical investigation of neighbourhood effects (Lupton and Kneale 2011. in this volume). A specification of scale should be incorporated in the hypotheses set out. Sixth, better data are needed to test neighbourhood effects hypotheses. Longitudinal data are crucial in investigating causal mechanisms, but such data should also contain a richer array of individual level and spatial context variables than is now the case. The seventh and final way forward as identified in this volume is mixed methods research. Ethnographic research is crucial in exploring and identifying potential causal mechanisms. Quantitative analysis of large scale longitudinal data enriched with contextual data are crucial in testing the generalisability of causal mechanisms, but the combination of qualitative and quantitative work is very powerful when it comes to understanding the unexpected (see Deluca et al. 2011 in this volume).

Book Structure and Contents

The remainder of this book is organised around 11 chapters by researchers from Australia, Finland, Sweden, the United Kingdom and the United States of America. The first three chapters by George Galster, Mario Small and Jessica Feldman, and Lina Hedman and Maarten van Ham offer theoretical contributions to the literature. The next five chapters by Kathy Arthurson, Ruth Lupton and Dylan Kneale, David Manley and Maarten van Ham, Gindo Tampubolon, and Stefanie DeLuca, Greg Duncan, Micere Keels, and Ruby Mendenhall report empirical work using case studies from five different national contexts. In the third part of the book, Venla Bernelius and Timo Kauppinen, and Michael Darcy and Gabrielle Gwyther present data collection proposals aimed at overcoming some of the challenges mentioned earlier in this introduction, from a quantitative and qualitative perspective. In the final chapter, Paul Cheshire provides a critique of mixed communities policies through analyzing the evidence base for neighbourhood effects.

There are several important links between chapters in different sections of the book. For example, both Galster and Small and Feldman call for more mixed methods research where qualitative techniques are used to interrogate the broad findings produced by quantitative neighbourhood effects studies. In Chap. 9, Deluca and colleagues provide an exemplar of how such research should be undertaken. In a similar vein, the chapters by Galster, Bernelius and Kauppinen, and Darcy and Gwyther all present designs of new data collection projects. The work of Darcy and Gwyther also has links with the work of Arthurson, as both highlight the lack of voices from individuals living in deprived neighbourhoods in the majority of neighbourhood effects work. Finally, the work of Hedman and van Ham points to the importance of considering selective mobility in neighbourhood effects research, a theme picked up again in the chapter by Manley and van Ham. The remainder of this introduction provides a detailed overview and summary of all the book chapters.

In **Chapter 2 George Galster** posits the idea that although there is now a large body of empirical research on neighbourhood effects, we know relatively little about the causal mechanisms responsible for relationships between neighbourhood attributes and individual outcomes. Without an in-depth understanding of these mechanisms and an understanding of the circumstances under which neighbourhood effects matter, scholarship on neighbourhood effects cannot advance, and public policy cannot be adequately directed (see Small and Feldman 2011 in this volume). Galster offers a list of 15 potential causal pathways which may lead to neighbourhood effects, grouped into four categories: social-interactive mechanisms, environmental mechanisms, geographical mechanisms, and institutional mechanisms. Social-interactive mechanisms refer to social processes endogenous to neighbourhoods, which are generally seen as the core of the neighbourhood effects argument (social contagion, collective socialisation, social networks, social cohesion and control, competition, relative deprivation, and parental mediation). Environmental mechanisms operate through natural and human-made attributes of neighbourhoods that may affect directly the mental and/or physical health of residents without affecting their behaviours (exposure to violence; physical surroundings; and toxic exposure). Geographical mechanisms refer to effects of the relative location of neighbourhoods (spatial mismatch of jobs and workers and a lack of quality public services). And finally institutional mechanisms which are related to the behaviour of actors external to neighbourhoods, who control the resources available and access to housing, services and markets for neighbourhood residents (stigmatisation, local institutional resources, and local market actors).

Galster continues his argument by stating that the ultimate goal of neighbourhood effects research is not only to identify which mechanisms are responsible for neighbourhood effects, but also to ascertain quantitatively their relative contributions to the outcome of interest. He uses the pharmacological metaphor of "dosage-response" to understand how the theoretical mechanisms could be causally linked to individual outcomes. He formulates 17 questions regarding the composition of the neighbourhood dosage, the administration of the neighbourhood dosage, and the neighbourhood dosage-response relationship which need to be answered to fully understand how the neighbourhood context affects residents. Neighbourhood residents can be exposed to a certain composition of mechanisms, over a certain time, with a certain frequency, and intensity. The relationship between the "dosage" of neighbourhood to an individual and certain outcomes may be nonlinear (thresholds), be temporary or long-lasting, take time to have an effect, and only have an effect in combination with other factors.

Existing qualitative and quantitative studies have not been able to adequately answer the 17 questions and uncover the dominant neighbourhood effect mechanisms at work. There is no definitive, comprehensive study of neighbourhood effect mechanisms. No study examines more than one or two of the 17 questions for an array of potential causal mechanisms and many of the questions have not been addressed explicitly in the theoretical or empirical literature. Field studies have yielded important insights on potential mechanisms, but are often limited in their ability to discern the relative contributions of alternative causes. Multivariate statistical studies often look for average effects (see also Small and Feldman 2011 in this volume) and are very limited in their ability to distinguish multiple mechanisms and dosage-response relationships for a variety of cities, neighbourhoods and groups of individuals.

Galster concludes by stating that, despite the ever growing literature on neighbourhood effects, there is far too little scholarship to make many claims about which causal links dominate for which outcomes for which people in which national contexts and any conclusions on the existence of neighbourhood effects should be treated as provisional at best. Galster calls for more, but especially different research (see also Small and Feldman 2011). Mixed method strategies should be embedded within the same study design; studies should explore residential histories; studies should consider a wider range of neighbourhood conditions and characteristics; and studies should collect more data on social interactions and mobility within neighbourhoods. Those developing public policy on health, employment and housing are urged to be careful when basing public policy responses on neighbourhood effects research as the causal pathways are not yet not clear.

In Chapter 3 Small and Feldman begin with the observation that research on neighbourhood effects is at a crossroads. After decades of qualitative and quantitative empirical studies (including Moving To Opportunity) aiming to ascertain how much neighbourhoods affect life chances, we seem nowhere near a coherent answer. They identify three concerns from the literature on neighbourhood effects.

The first concern is that most quantitative empirical studies into neighbourhood effects most likely suffer from selection bias (see also the chapters by Hedman and van Ham 2011; Cheshire 2011; and Manley and van Ham 2011 in this volume). The second is that much of the neighbourhood effects literature is searching for average effects: a single answer to the question whether neighbourhood effects exist, for any given outcome, regardless of location, context, or other conditions. They argue that "an entire generation of researchers concerned themselves with answering either a yes-or-no question (do neighbourhoods matter?) or a question of degree (how much do they matter?)—rather than a conditional question (under what circumstances do they matter?)." The third concern is that it is unclear how much progress has been made on the question of which mechanisms potentially causing neighbourhood effects matter (see also Galster 2011 in this volume). Small and Feldman use the three concerns to evaluate the Moving to Opportunity (MTO) randomized trials, which are generally seen as a turning point in neighbourhood effects studies. Despite MTO's claim of providing a solution to the selection bias problems, it failed to do so convincingly. Small and Feldman draw two important lessons from the MTO work, which will guide future work on neighbourhood effects.

The first is that it is often assumed that neighbourhood effects operate homogeneously across subpopulations and across treatment settings. Small and Feldman argue that future work on neighbourhood effects should move away from a perspective focused on average effects to one that expects and explains heterogeneity: whether neighbourhoods matter is conditional on the characteristics of individuals, neighbourhoods, and cities. To illustrate this point they test the de-institutionalization hypotheses that concentrated poverty undermines organizational density. This hypothesis is derived from work primarily done in Chicago, which is generally seen as a laboratory where phenomena occurring in the average large city can be observed. Small and Feldman show that Chicago's poor neighbourhoods are substantially less organizationally dense than not only the average poor neighbourhood in U.S. cities and but also the average for Rustbelt cities. So Chicago cannot be seen as a representative city and hypotheses derived from Chicago might be place-specific rather than general. This is not to say that neighbourhoods do not matter, but that whether and how they matter may depend on the context.

The second lesson from the MTO work is that future work should better integrate ethnographic research into the quantitative empirical research program. Ethnographic research has the capacity to help explain the often contradictory results of previous neighbourhood effect studies, and to generate hypotheses for future studies. Many findings from previous work cannot be understood without talking to residents of poor neighbourhoods to find out how they make decisions under their circumstances. Neighbourhood effects research would benefit from ethnographic research specifically designed to generate explicit, testable hypotheses that guide quantitative research. Such research should study neighbourhood effects for different cities, neighbourhoods, and types of individuals to explain heterogeneity. Study sites should be selected in cities other than the conventional locations, particularly Chicago.

To conclude, Small and Feldman call for integrating ethnography more effectively in neighbourhood effects research, accompanied by a reorientation of practical and theoretical assumptions behind the work, and a reorientation from homogeneity and average effects toward heterogeneity and conditional relationships.

Chapter 4 by Hedman and van Ham argues that the most severe problem in the identification of causal neighbourhood effects is selection bias as a result of selective sorting into neighbourhoods. People sort themselves into and out of neighbourhoods and selection bias occurs when the selection mechanism into neighbourhoods is not independent from the outcome studied. Many studies do not control their models of neighbourhood effects for selection bias. As a result it is impossible to say whether correlations between neighbourhood characteristics and individual outcomes are causal effects, or the result of neighbourhood selection. For example, unemployed people are more likely to move into deprived neighbourhoods than employed people. If this selection mechanism is not adequately controlled for in modelling the effect of living in a deprived neighbourhood on unemployment, a correlation between unemployment and neighbourhood deprivation might be mistaken for a neighbourhood effect. The chapter argues that to better understand mechanisms behind neighbourhood effects, more knowledge is needed about residential mobility and the selective sorting into and out of neighbourhoods.

Using data from three neighbourhoods in Stockholm, Sweden, Hedman and van Ham show that selective mobility of neighbourhood residents can either change the neighbourhood population or reproduce existing patterns. If, in a neighbourhood with relatively low employment levels, those who get a job leave the neighbourhood, and are replaced by others without a job, it is not the neighbourhood which causes unemployment, but the neighbourhood housing stock which attracts unemployed people who cannot afford to live elsewhere. This is not the same as concluding that neighbourhood effects do not exist. Instead, the conclusion is that the selection mechanisms outlined above must be accounted for in empirical models. The chapter proposes a conceptual model linking neighbourhood choices made by individuals and households with individual level outcomes. Both real causal effects and selection effects are featured in the model.

The chapter continues to argue that in order to further our understanding of neighbourhood effects we should incorporate neighbourhood sorting into our models of neighbourhood effects. Many approaches to deal with selection bias treat neighbourhood sorting as a statistical nuisance and reveal nothing about the processes behind the potential bias. Neighbourhood sorting is of interest in its own right and surprisingly few studies focus on why certain households 'choose' certain neighbourhoods. A better understanding of neighbourhood sorting is also central in understanding residential segregation and the production and reproduction of neighbourhoods of different characteristics and status. Neighbourhood effect studies are thus in the situation where the processes behind one of its key methodological problems (selection bias) are also critical to fully understand the neighbourhood context itself.

Moving the focus towards empirical investigations, **Chapter 5 by Kathy Arthurson** explores some of the debates about poor reputations and stigmatisation of neighbourhoods in which social housing is concentrated. She argues that living in a neighbourhood with a poor reputation can have a negative effect on individual outcomes, over and above other neighbourhood characteristics. For example, employers may discriminate against neighbourhood residents based on the postcode area in which they live. Residents of neighbourhoods with a poor reputation can also adopt self-defeating behaviours linked to the place in which they live. The reputation of a neighbourhood is not necessarily based on current attributes, but can be rooted in the history of a place. Neighbourhood regeneration programs often have as one of their aims changing the reputation of a neighbourhood. Despite the debates about the potentially harmful effects of living in a neighbourhood with a poor reputation, in-depth knowledge and understandings of the dynamics of stigma and whether the situations are improved post-neighbourhood regeneration with changes to social mix are limited, especially from residents' perspectives. Arthurson's chapter aims to get more insight in how neighbourhood residents see their neighbourhood and how they think others see their neighbourhood. Data are collected from three neighbourhoods across the city of Adelaide, Australia, using a questionnaire and in-depth interviews. Results are presented on four neighbourhood dimensions – house condition, attractiveness, safety and density. Overall, when analysing the differences between internal and external ratings within four housing tenure groups, on all four measures respondents' internal (self) ratings from their own perspectives, were more favourable than their judgements of how they felt that people from outside the area would view the neighbourhoods. It is hypothesised that this negative external perception might influence the behaviour of neighbourhood residents. The interview results show that interviewees overall expressed the view that, post-regeneration, their neighbourhoods were more attractive and the condition of housing was much improved. In general, the findings support those of other studies, which suggest that introducing homeowners onto social housing estates as part of regeneration initiatives to some extent improves the external reputation of the neighbourhoods.

In **Chapter 6, Ruth Lupton and Dylan Kneale** investigate neighbourhood and place effects on the likelihood of becoming a teenage parent in England. They argue that government policies to reduce teenage parenthood are in part informed by a persistent belief in neighbourhood effects. They also identify that current evidence for neighbourhood effects on teenage parenthood is remarkably weak. The chapter is designed to make a dual contribution to this volume. First, it highlights some of the conceptual problems in much existing neighbourhood effects research around the role of place and the importance of geography. Lupton and Kneale critique the lack of theoretical basis to much of the existing literature on neighbourhood effects. Their critique closely matches some of the points made in the chapters by Galster (2011) and Small and Feldman (2011): many studies search for more general evidence of neighbourhood effects without formulating specific hypotheses on causal mechanisms, and often without detailed knowledge of the dependent variable under study. Lupton and Kneale also critique the lack of attention to what is the most appropriate spatial scale to study specific neighbourhood effects (see also the chapter by Manley and van Ham 2011). Many studies use geographical units without any particular logic or theoretical justification, simply because a certain level of geography is available in the data. They call for a much closer and also a more critical collaboration between quantitative and qualitative researchers so that qualitative understandings of place are better reflected in quantitative models (see also the chapter by Small and Feldman 2011)

Second, the chapter offers an empirical investigation into neighbourhood effects and adds to the evidence base on teenage parenthood. They use data from the British Cohort Study (BCS70), a longitudinal study of people born in 1970, with unique postcode geo-coding of neighbourhood characteristics. In many studies of neighbourhood effects it is usual that only one neighbourhood geography is tested. To extend their analysis Lupton and Kneale test several geographies (see also chapter by Manley and van Ham 2011 for a study using multiple geographies). They use the standard geographies available in the data in combination with bespoke geographies designed to more closely represent the spatial scales over which they believe the relevant mechanisms operate. The bespoke geographies are based on newly created spatial units, for example around clusters of contiguous similar areas, and on considerations of the characteristics of neighbouring units. They found some evidence of value-related place effects at the neighbourhood level and labour market structural effects at the sub-regional level. The results suggested that place effects on values around fertility operate at a relatively fine spatial scale. The study emphasises the limitation that it did not take into account selection of people into neighbourhoods, which is likely to have lead them to overestimate the propensity to experience a teenage birth in certain types of neighbourhoods. The overall conclusion is that although in principle a theory-driven approach that identifies and tests specific mechanisms is the right one, in practice it may be impossible to separate the social processes leading to early parenthood from one another using quantitative methods and data. A second conclusion is that neighbourhood effects research should move towards more explicit and transparent considerations of geography in order to make a stronger contribution to knowledge of place effects.

In **Chapter 7, Manley and van Ham** explore labour market outcomes for individuals living in concentrations of unemployment using data from the Scottish Longitudinal Study (SLS). They highlight a number of serious shortcomings in much of the existing literature on neighbourhood effects, which leads them to question the current evidence base for neighbourhood effects. Many existing studies suffer from selection bias in their results as they are not able to control for selective mobility into deprived neighbourhoods. As a result, they are likely to show correlations between individual outcomes and neighbourhood characteristics, instead of real causal effects. They pay special attention to the outcomes of (quasi)-experimental studies, which should (in theory) be able to overcome the selection bias issue.

The empirical section of the chapter investigates whether the level of unemployment in a neighbourhood is related to the employment outcomes of residents. Using logistic regression models they estimate the probability that an unemployed person in 1991 has a job in 2001, and the probability than an employed person in 1991 still has a job in 2001. The models control for a wide variety of individual and household contexts and clearly show a correlation between neighbourhood characteristics and individual employment outcomes. The results suggest that living in a concentration of unemployment is harmful for getting or keeping a job.

Most studies of neighbourhood effects would stop at this point and claim to have found evidence for neighbourhood effects. Manley and van Ham argue that at this point it is important to further explore the data and run models for sub-populations (such as age groups, gender, housing tenure). The only sub-populations to yield interesting results were separate models by housing tenure: the models showed clear "neighbourhood effects" for homeowners, but not for social renters. Manley and van Ham argue that this can be explained by selection bias for homeowners, which was largely absent for social renters. In 1991 most social renters were allocated a dwelling and neighbourhood by housing officers. Although this allocation process was not entirely random, it approximated a random assignment of neighbourhoods to households. Owner-occupiers on the other hand where "free" to choose where to live. For them, neighbourhood selection was closely associated with their earnings and earning potential, affected their ability to get a mortgage. Those with a low income, or without job security selected themselves into the most deprived neighbourhoods, where cheap (affordable) housing could be found. These were also the workers who were most of risk of losing their job.

The main substantive conclusion of the chapter is that (self-) selection should be more fully explored in studies of neighbourhood effects. Wherever possible, models investigating the impact of neighbourhood contexts on individual outcomes should take into account the different routes through which households enter neighbourhoods.

Chapter 8 by Tampubolon is an example of a formal econometric approach to neighbourhood effects research, which uses complex econometric solutions in an attempt to identify causal neighbourhood effects. He identifies a recent and strong interest in neighbourhood effects from within the literature on public health and social epidemiology, which focuses on neighbourhood effects on individual health outcomes such as obesity, mental health, physical health and health-related quality of life. In his chapter Tampubolon focuses on the relationship between neighbourhood social capital and individual mental health. The current empirical evidence on this relationship is divided.

Based on the literature, the chapter identifies four mechanisms linking neighbourhood social capital and individual health. First, more cohesive neighbourhoods are better equipped to disseminate information and mobilize collective action. Second, more cohesive neighbourhoods are better equipped to enforce and maintain social norms. Third, collective efficacy and informal control in preventing crime and violence reduce environmental stresses suffered by residents in their day to day activities. Fourth, high levels of neighbourhood social capital enable communities to be more responsive to national and local organisations that seek involvement and engagement at the local level.

Tampubolon contributes to the literature on neighbourhood effects and health outcomes by proposing an extension of the influential Grossman model of health with the explicit inclusion of interactions within the neighbourhood context. He draws upon the Blume-Brock-Durlauf social interaction model to study the effect of neighbourhood social capital on mental health, using data from the Welsh Health Survey 2007 (WHS) and the Living in Wales 2007 (LiW) survey. He proposes various instrumental variables to identify causal effects, uses objective measures of neighbourhood social capital for small geographies, and uses a measure of mental health derived from the SF36 (Short Form Heath Survey). Using his approach, and contrary to some other studies, Tampubolon concludes that neighbourhood social capital is generally being benificial to individual mental health.

Chapter 9 by DeLuca, Duncan, Keels, and Mendenhall provides a unique contribution to the neighbourhood effects literature by demonstrating that data from in-depth interviews is capable of revealing some of the mechanisms behind unexpected quantitative findings of how the Moving to Opportunity (MTO) program did and did not affect outcomes for individuals. Such a mixed methods approach is regarded a major step forward in neighbourhood effects research (see the chapters by Galster 2011 and Small and Feldman 2011 in this volume who call for such a mixed methods approach). The study by DeLuca and colleagues was triggered by the observation that whereas the earlier Gautreaux residential mobility program documented dramatic improvements in the lives of people placed in more affluent neighbourhoods, the results of the MTO program were not nearly as positive.

The chapter begins with a review of the process model behind the MTO experiment, which assumes that program participants make rational choices and that neighbourhood improvement would be a sufficient condition to enhance outcomes for children and their families. The MTO program was designed to understand the long-term effects of moving poor families out of subsidized housing in high-poverty communities and into low-poverty neighbourhoods in five cities: Baltimore, Boston, Chicago, Low Angeles, and New York. Families were randomly assigned to three groups to minimize the effects of selection bias.

DeLuca and colleagues highlight how MTO researchers encountered a mixed bag of program effects (using mainly quantitative analysis) and use evidence from mixed methods studies and their own data collection to understand some of the program's outcomes. They subsequently describe and attempt to explain unexpected findings (mental health improvements which were not originally anticipated in the MTO program); a weak 'treatment' effect for many families (initial and subsequent moves to segregated, economically declining areas instead of higher opportunity neighbourhoods); "null" findings where large effects on individual outcomes were expected instead (MTO was primarily designed to enhance the employment prospects of adults and to improve the educational outcomes of children, but no effects on employment and education were found); and a set of conflicting findings (moves to low poverty neighbourhoods were found to be beneficial to girls, but harmful for boys).

The mixed-method approach adopted by DeLuca and colleagues enabled them to extend MTO's original process model to a broader model which is better capable of understanding how individual actions and (historical) social conditions reinforce or limit the effects of neighbourhood interventions on individual outcomes. They conclude that it is too early to label MTO-based policy approaches as ineffective, and that neighbourhood interventions are more likely to be one part of a wider solution for solving the problems of poor families, rather than the ultimate solution *per se*. The use of mixed methods has allowed DeLuca and colleagues to show how the potential of MTO-based policy approaches is limited by structural barriers, and the dynamics of poor families' beliefs, backgrounds and constraints. They showed that some of the assumptions underlying the original MTO process model were off base and that many families are not able to relocate to higher opportunity neighbourhoods, or to utilise the higher quality services in those communities.

Chapter 10 by Bernelius and Kauppinen investigates neighbourhood effects on educational outcomes in Finland. They critique the common perception that Finland is a country with equal opportunities for education. Studies consistently show that the Finnish educational system is one of the best in the world with only small variations in educational outcomes between pupils and schools. They argue that these observations at the country level hide variation in equality within the country: when educational outcomes are studied for the Helsinki Metropolitan area, large variations can be found between neighbourhoods, schools, and individuals. Recent research suggests that the differences between neighbourhoods and schools are growing, which makes the Helsinki Metropolitan area an attractive "urban laboratory" for research as neighbourhood effects are generally assumed to intensify as socio-spatial segregation increases. The aim of the chapter is to explore the possibility of neighbourhood effects in the Finnish context. The chapter starts a presentation of results from a study on neighbourhood effects and educational outcomes, using data for Finland. It then highlights some of the weaknesses of this study. The chapter ends with the presentation of the design of a new research project funded by the Finnish National Research Council, and the Academy of Finland, which should be able to overcome some of the shortcomings of previous research. The study will collect longitudinal data on a large sample of pupils with detailed information about individuals, households, schools and neighbourhoods. This design will allow the use of multilevel models to estimate neighbourhood effects.

Chapter 11 by Darcy and Gwyther also presents a new approach and research design to study neighbourhood effects, but from a completely different methodological and epistemological angle than the previous chapter. Although the language of the chapter is very different to the language used in many of the other chapters in this book, one of the messages is surprisingly similar: current neighbourhood effects research falls short on delivering convincing evidence of causal neighbourhood effects. They argue that most studies simply show unsurprising correlations between neighbourhood characteristics and individual outcomes, without adding to our understanding of the mechanisms behind these correlations. Many of the mechanisms are assumed rather than discovered. In essence, this argument is similar to the ones made by Galster (2011) and Small and Feldman (2011).

Darcy and Gwyther go one step further and critique what they see as the dominant discourses of place and disadvantage as well as the epistemology underlying this discourse. They see the current attention given to neighbourhood effects as part of a larger 'spatial turn' in social science, which attempts to explain the disadvantage of poor households concentrated in poor neighbourhoods. They critique the 'culture of poverty explanation' of disadvantage and the associated policy response of de-concentrating poverty through the creation of mixed income neighbourhoods. If there is little evidence of neighbourhood effects in the first place, then creating mixed neighbourhoods will lead to little benefit for the neighbourhood residents, a large proportion of who will be displaced as a result of the policy. This argument is very similar to the one made in the chapters by Manley and van Ham (2011) and Cheshire (2011) in this volume.

Darcy and Gwyther further critique quantitative research for ignoring the voice and perspectives of neighbourhood residents. They state that conventional positivist epistemology systematically excludes important aspects of community life as experienced by those most affected. Not everyone will agree with this viewpoint, and many researchers will not see such a "black and white" distinction between epistemologies. It is true however, that most neighbourhood effects research searches for 'effects' without an understanding of the (assumed) underlying mechanisms. This is why Small and Feldman (2011) call for integrating ethnography more effectively in neighbourhood effects research to identify the mechanisms underlying causal associations and generating the hypotheses that should inform future (quantitative) studies. Darcy and Gwyther distance themselves completely from positivist epistemology and propose an alternative approach based on phenomenological epistemology and participatory action research. Their research is based on a 'collaborative university – community research' design to understand residents' perspectives of their neighbourhood and concentrated public housing and the policy proposals for mixed housing in Australia.

Finally, **Chapter 12 by Cheshire** assesses the evidence base underlying mixed communities policies, which are now firmly established in the national policies of most OECD countries. Mixed communities policies are partly based on a firm belief in neighbourhood effects. The idea behind mixed communities policies is that creating neighbourhoods in which populations are mixed will take away these negative effects. Cheshire argues that such policies are essentially faith-based since there is still scant evidence that making communities more mixed significantly improves the life chances of the poor.

The main challenge in neighbourhood effects research is the identification of causal neighbourhood effects. As Cheshire puts it "do poor neighbourhoods make residents poorer, or do poor people simply live in poor neighbourhoods because living in affluent ones costs too much?" There is overwhelming evidence that the attributes which make neighbourhoods attractive are capitalised into house prices/rents. The result is that poor people cannot afford to buy into nicer neighbourhoods and have to concentrate into deprived neighbourhoods where housing is cheap. Cheshire extensively reviews the evidence relating to why social segregation develops and generates specialised neighbourhoods in cities and why this pattern is more obvious the larger a city is and the more unequal a country's society is.

The question is whether living in a poor neighbourhood is a separate, independent cause of poverty? Cheshire reviews the evidence from a range of neighbourhood effects studies designed to overcome problems with selection bias: studies based on quasi-experimental data (the US Moving to Opportunity experiment) and longitudinal individual level data from a variety of countries. He concludes that the evidence supporting the significance, even the existence of neighbourhood effects is remarkably thin when subjected to rigorous evaluation (see the chapter by DeLuca and colleagues, 2011 for a more nuanced view on the MTO results and a mixed methods approach to understanding why MTO did not find clear evidence of neighbourhood effects).

Based on this review of evidence Cheshire argues that policies for mixed neighbourhoods treat the symptoms rather than the causes of poverty and that efforts to improve the lives of the poor would be more effectively directed towards people themselves rather than moving people around to mix neighbourhoods. Moreover, he also finds that there are real welfare and productivity benefits of living in specialised neighbourhoods. Mixed neighbourhood policies run the risk of destroying these and these possible losses need to be balanced against any potential benefits of reduced negative 'neighbourhood effects'.

Cheshire ends with the remark that his chapter does not argue that neighbourhood effects do not exist and that he is open to the idea that living in the most deprived neighbourhoods can have negative effects on individuals. But up to now there is not enough convincing empirical evidence to justify spending substantial resources to use policy to force neighbourhoods to be mixed. However, the lack of evidence for neighbourhood effects does not imply that it is not useful to target neighbourhoods with people-targeted policies aimed at reducing societal inequality. It is in the poorest neighbourhoods that those who most need help are concentrated.

References

Atkinson, R., & Kintrea, K. (2001). Disentangling area effects: Evidence from deprived and non-deprived neighbourhoods. *Urban Studies, 38*(11), 2277–2298.

Atkinson, R., & Kintrea, K. (2002). Area effects: What do they mean for British housing and regeneration policy? *European Journal of Housing Research, 2*(2), 147–166.

Bauder, H. (2002). Neighbourhood effects and cultural exclusion. *Urban Studies, 39*(1), 85–93.

Bolster, A., Burgess, S., Johnston, R., Jones, K., Propper, C., & Sarker, R. (2007). Neighbourhoods, households and income dynamics: A semi-parametric investigation of neighbourhood effects. *Journal of Economic Geography, 7*(1), 1–38.

Cheshire, P. (2007). *Are mixed communities the answer to segregaiton and poverty?* York: Joseph Rowntree Foundation.

Cheshire, P. (2011). Are mixed community policies evidence based? A review of the research on neighbourhood effects. In M. van Ham, D. Manley, N. Bailey, L. Simpson, & D. Maclennan (Eds.), *Neighbourhood effects research: New perspectives*. Dordrecht: Springer.

Clark, W. A. V. (2008). Reexamining the moving to opportunity study and its contribution to changing the distribution of poverty and ethnic concentration. *Demography, 45*(3), 515–535.

Darcy, M., & Gwyther, G. (2011). Recasting research on 'neighbourhood effects': A collaborative, participatory, trans-national approach. In M. van Ham, D. Manley, N. Bailey, L. Simpson, & D. Maclennan (Eds.), *Neighbourhood effects research: New perspectives*. Dordrecht: Springer.

Deluca, S., Duncan, G. J., Keels, M., & Mendenhall, R. (2011). The notable and the null: Using mixed methods to understand the diverse impacts of residential mobility programs. In M. van Ham, D. Manley, N. Bailey, L. Simpson, & D. Maclennan (Eds.), *Neighbourhood effects research: New perspectives*. Dordrecht: Springer.

Dietz, R. D. (2002). The estimation of neighborhood effects in the social sciences: An interdisciplinary approach. *Social Science Research, 31*(4), 539–575.

Durlauf, S. N. (2004). Neighbourhood effects. In J. V. Henderson & J. F. Thisse (Eds.), *Handbook of regional and urban economics. Volume 4 Cities and geography*. Amsterdam: Elsevier.

Ellen, I. G., & Turner, M. A. (1997). Does neighbourhood matter? Assessing recent evidence. *Housing Policy Debate, 8*(4), 833–866.

Feijten, P. M., & van Ham, M. (2009). Neighbourhood change... reason to leave? *Urban Studies, 46*(10), 2103–2122.

Galster, G. (2002). An economic efficiency analysis of deconcentrating poverty populations. *Journal of Housing Economics, 11*(4), 303–329.

Galster, G. (2011). The mechanism(s) of neighbourhood effects: Theory, evidence, and policy implications. In M. van Ham, D. Manley, N. Bailey, L. Simpson, & D. Maclennan (Eds.), *Neighbourhood effects research: New perspectives*. Dordrecht: Springer.

Galster, G., Andersson, R., Musterd, S., & Kauppinen, T. (2008). Does neighbourhood income mix affect earnings of adults? New evidence from Sweden. *Journal of Urban Economics, 63*(3), 858–870.

Gans, H. J. (1968). Culture and class in the study of poverty: An approach to antipoverty research. In D. P. Moynihan (Ed.), *Understanding poverty: Perspectives from the social sciences* (pp. 229–259). New York: Basic Books.

Goering, J., Feins, J. D., & Richardson, T. M. (2002). A cross-site analysis of initial moving to opportunity demonstration results. *Journal of Housing Research, 13*(1), 1–30.

Graham, E., Manley, D., Hiscock, R., Boyle, P., & Doherty, J. (2009). Mixing housing tenures: Is it good for social well-being? *Urban Studies, 46*(1), 139–165.

Harris, D. R. (1999). Property values drop when blacks move in, because...: Racial and socioeconomic determinants of neighbourhood desirability. *American Sociological Review, 64*(3), 461–479.

Hedman, L., van Ham, M., & Manley, D. (2011). Neighbourhood choice and neighbourhood reproduction. *Environment and Planning A, 43*(6), 1381–1399.

Hedman, L., & van Ham, M. (2011). Understanding neighbourhood effects: Selection bias and residential mobility. In M. van Ham, D. Manley, N. Bailey, L. Simpson, & D. Maclennan (Eds.), *Neighbourhood effects research: New perspectives*. Dordrecht: Springer.

Jencks, C., & Mayer, S. E. (1990). The social consequences of growing up in a poor neighbourhood. In M. G. H. McGeary & E. L. Lawrence (Eds.), *Inner city poverty in the United States* (pp. 111–186). Washington, DC: National Academic Press.

Kearns, A. (2002). Response: From residential disadvantage to opportunity? Reflections on British and European policy and research. *Housing Studies, 17*(1), 145–150.

Leventhal, T., & Brooks-Gunn, J. (2004). A randomized study of neighborhood effects on low-income children's educational outcomes. *Developmental Psychology, 40*(4), 488–507.

Ludwig, J., Duncan, G. J., & Hirschfield, P. (2001). Urban poverty and juvenile crime: Evidence from a randomized housing-mobility experiment. *Quarterly Journal of Economics, 116*(2), 655–680.

Lupton, R., & Kneale, D. (2011). Theorising and measuring place in neighbourhood effects research: The example of teenage parenthood in England. In M. van Ham, D. Manley, N. Bailey, L. Simpson, & D. Maclennan (Eds.), *Neighbourhood effects research: New perspectives*. Dordrecht: Springer.

Manley, D., & van Ham, M. (2011). Neighbourhood effects, housing tenure and individual employment outcomes. In M. van Ham, D. Manley, N. Bailey, L. Simpson, & D. Maclennan (Eds.), *Neighbourhood effects research: New perspectives*. Dordrecht: Springer.

Manski, C. (1993). Identification of endogenous social effects: The reflection problem. *The Review of Economic Studies, 60*(3), 531–542.

Moffitt, R. (1998). *Policy interventions, low-level equilibria, and social interactions*. Working Paper: 432. Baltimore, MD: Department of Economics, Johns Hopkins University.

Musterd, S. (2002). Response: Mixed housing policy: A European (Dutch) perspective. *Housing Studies, 17*(1), 139–143.

Musterd, S., & Andersson, R. (2005). Housing mix, social mix and social opportunities. *Urban Affairs Review, 40*(6), 761–790.

Oreopoulos, P. (2003). The long-run consequences of living in a poor neighborhood. *Quarterly Journal of Economics, 118*(4), 1533–1575.

Overman, H. G. (2002). Neighbourhood effects in large and small neighbourhoods. *Urban Studies, 39*(1), 117–130.

Pinkster, F. M. (2009). *Living in concentrated poverty*. Amsterdam: University of Amsterdam.

Popkin, S. J., & Cunningham, M. K. (2009). Has HOPE VI transformed residents' lives? In H. G. Cisneros & L. Engdahl (Eds.), *Despair to hope: HOPE VI and the new promise of public housing in America's cities* (pp. 191–204). Washington, DC: Brookings Institution Press.

Robinson, W. S. (1950). Ecological correlations and the behavior of individuals. *American Sociological Review, 15*(3), 351–357.

Rosebaum, J. (1995). Changing the geography of opportunity by expanding residential choice: Lessons from the gatreaux program. *Housing Policy Debate, 6*(1), 231–269.

Rubin, D. B. (2008). For objective causal inference, design trumps analysis. *Annals of Applied Statistics, 2*(3), 808–840.

Sampson, R., Morenoff, J. D., & Gannon-Rowley, T. (2002). Assessing "neighbourhood effects": Social processes and new directions. *Annual Sociological Review, 28*, 443–478.

Small, M. L. (2008). Lost in translation: How not to make qualitative research more scientific. In M. Lamont & P. White (Eds.), *Workshop on interdisciplinary standards for systematic qualitative research*. Washington, DC: National Science Foundation.

Small, M. L., & Feldman, J. (2011). Ethnographic evidence, heterogeneity, and neighbourhood effects after moving to opportunity. In M. van Ham, D. Manley, N. Bailey, L. Simpson, & D. Maclennan (Eds.), *Neighbourhood effects research: New perspectives*. Dordrecht: Springer.

van Ham, M., & Feijten, P. M. (2008). Who wants to leave the neighbourhood? The effect of being different from the neighbourhood population on wishes to move. *Environment and Planning A, 40*(5), 1151–1170.

van Ham, M., & Manley, D. (2010). The effect of neighbourhood housing tenure mix on labour market outcomes: A longitudinal investigation of neighbourhood effects. *Journal of Economic Geography, 10*(2), 257–282.

Wilson, W. J. (1987). *The truly disadvantaged: the inner city, the underclass and public policy*. Chicago: University of Chicago Press.

Chapter 2
The Mechanism(s) of Neighbourhood Effects: Theory, Evidence, and Policy Implications

George C. Galster

Introduction

Although there has been a burgeoning literature on quantifying the relationship between various aspects of the residential environment and numerous outcomes for individual adults and children residing in that environment, comparably less attention has been given to uncovering empirically the causal mechanisms that yield these relationships. There have been many discussions of the potential causal connections between neighbourhood context and individual behavioural and health outcomes; see especially Jencks and Mayer (1990), Gephart (1997), Ellen and Turner (1997), Wandersman and Nation (1998), Friedrichs (1998), Green and Ottoson (1999), Atkinson and Kintrea (2001), Booth and Crouter (2001), Small and Newman (2001), Sampson (2001), Ellen et al. (2001), Haurin et al. (2002), Sampson et al. (2002), Ellen and Turner (2003), Ioannides and Loury (2004), Pinkster (2008), and Phibbs (2009). Though often in these works the listings of potential mechanisms differ in labelling and categorizations, there is a broad consensus about how the underlying causal paths are thought to operate in theory. Unfortunately, there are few tentative conclusions, let alone any consensus, about which mechanisms demonstrate the strongest empirical support. The following quotes are illustrative. "In general research findings…are too scant to draw any firm conclusions about the potential pathways through which neighbourhood effects may be transmitted…" (Leventhal and Brooks-Gunn 2000, p. 322). "The causal pathways that underlie hypotheses about the effects of neighbourhood social factors are often not explicit…This clearly is an important next step for understanding the relationship between neighbourhood and health" (Pickett and Pearl 2001,

G.C. Galster (✉)
Department of Urban Studies and Planning, Wayne State University, Detroit, MI, USA
e-mail: aa3571@wayne.edu

M. van Ham et al. (eds.), *Neighbourhood Effects Research: New Perspectives*,
DOI 10.1007/978-94-007-2309-2_2, © Springer Science+Business Media B.V. 2012

p. 117–120). "One important question is *how* a less advantaged neighbourhood increases the risk of low birth weight and of children developing behavioural problems…Understanding of the causal chains in both of these areas is…incomplete" (Sellström and Bremberg 2006, p. 553).

Given this lack of scholarly consensus, my purposes in this paper are four-fold: (1) offer a comprehensive listing of 15 potential causal pathways between neighbourhood context and individual behavioural and health outcomes, which synthesizes both sociological and epidemiological perspectives; (2) provide a new conceptualization of dimensions of neighbourhood effect mechanisms that uses a pharmacological analogy to clarify the empirical challenges of this field of enquiry; (3) provide an updated, international review of empirical studies related to neighbourhood effect mechanisms; and (4) draw provisional conclusions about the dominant neighbourhood effect mechanisms operating and implications from this review for scholarship and public policy. The paper is organized as follows. I begin with an overview of the numerous possible neighbourhood effect mechanisms that have been hypothesized, and group them into four categories. Next I will examine a variety of issues that render the identification of neighbourhood causal mechanisms particularly challenging for social scientists, and the two methodological approaches that have been adopted thus far. These issues are brought into clear relief by use of a pharmacological metaphor: *dosage-response*. I will then synthesize the international evidence related to the four categories of mechanisms, each in its own subsection, in an effort to assess the state of empirical scholarship and offer some provisional conclusions. Finally, I close the paper by drawing implications for future scholarship and public policy.

How Might Neighbourhood Effects Transpire?

Prior scholarly works addressing this question have been distinctly segregated, with social scientists focusing on behavioural outcomes and epidemiologists focusing on health outcomes. However, within each subset there is broad theoretical agreement about potential causal pathways of neighbourhood effects. I therefore will list these mechanisms and describe them only briefly here. My synthesis of these disparate literatures suggests that fifteen (15) distinctive linkages have been advanced. I think it is most useful to group these 15 mechanisms of neighbourhood effects under four broad rubrics: social interactive; environmental; geographical; and institutional.[1]

[1] By contrast, Manski (1995) groups them into "endogenous," "exogenous," and "correlated" categories. Ellen and Turner (1997) group them into five categories: concentration, location, socialization, physical, and services. Leventhal and Brooks-Gunn (2000) use the rubrics "institutional resources," "relationships," and "norms/collective efficacy."

Social-Interactive Mechanisms

This set of mechanisms refers to social processes endogenous to neighbourhoods. These processes include:

- *Social Contagion*: Behaviours, aspirations, and attitudes may be changed by contact with peers who are neighbours. Under certain conditions these changes can take on contagion dynamics that are akin to "epidemics."
- *Collective Socialization*: Individuals may be encouraged to conform to local social norms conveyed by neighbourhood role models and other social pressures. This socialization effect is characterized by a minimum threshold or critical mass being achieved before a norm can produce noticeable consequences for others in the neighbourhood.
- *Social Networks*: Individuals may be influenced by the interpersonal communication of information and resources of various kinds transmitted through neighbours. These networks can involve either "strong ties" and/or "weak ties."
- *Social cohesion and control*: The degree of neighbourhood social disorder and its converse, "collective efficacy" (Sampson et al. 1999), may influence a variety of behaviours and psychological reactions of residents.
- *Competition*: Under the premise that certain local resources are limited and not pure public goods, this mechanism posits that groups within the neighbourhood will compete for these resources amongst themselves. Because the outcome is a zero-sum game, residents' access to these resources (and their resulting opportunities) may be influenced by the ultimate success of their group in "winning" this competition.
- *Relative Deprivation*: This mechanism suggests that residents who have achieved some socioeconomic success will be a source of disamenities for their less-well off neighbours. The latter, it is argued, will view the successful with envy and/or will make them perceive their own relative inferiority as a source of dissatisfaction.
- *Parental Mediation*: The neighbourhood may affect (through any of the mechanisms listed under all categories here) parents' physical and mental health, stress, coping skills, sense of efficacy, behaviours, and material resources. All of these, in turn, may affect the home environment in which children are raised.

Environmental Mechanisms

Environmental mechanisms refer to natural and human-made attributes of the local space that may affect directly the mental and/or physical health of residents without affecting their behaviours. As in the case of social-interactive mechanism, the environmental category can also assume distinct forms:

- *Exposure to Violence*: If people sense that their property or person is in danger they may suffer psychological and physical responses that may impair their functioning or sensed well-being. These consequences are likely to be even more pronounced if the person has been victimized.

- *Physical Surroundings*: Decayed physical conditions of the built environment (e.g., deteriorated structures and public infrastructure, litter, graffiti) may impart psychological effects on residents, such as a sense of powerlessness. Noise may create stress and inhibit decision-making through a process of "environmental overload" (Bell et al. 1996).
- *Toxic Exposure*: People may be exposed to unhealthy levels of air-, soil-, and/or water-borne pollutants because of the current and historical land uses and other ecological conditions in the neighbourhood.

Geographical Mechanisms

Geographic mechanisms refer to aspects of spaces that may affect residents' life courses yet do not arise within the neighbourhood but rather purely because of the neighbourhood's location relative to larger-scale political and economic forces such as:

- *Spatial Mismatch*: Certain neighbourhoods may have little accessibility (in either spatial proximity or as mediated by transportation networks) to job opportunities appropriate to the skills of their residents, thereby restricting their employment opportunities.
- *Public Services*: Some neighbourhoods may be located within local political jurisdictions that offer inferior public services and facilities because of their limited tax base resources, incompetence, corruption, or other operational challenges. These, in turn, may adversely affect the personal development and educational opportunities of residents.

Institutional Mechanisms

The last category of mechanisms involves actions by those typically not residing in the given neighbourhood who control important institutional resources located there and/or points of interface between neighbourhood residents and vital markets:

- *Stigmatization*: Neighbourhoods may be stigmatized on the basis of public stereotypes held by powerful institutional or private actors about its current residents. In other cases this may occur regardless of the neighbourhood's current population because of its history, environmental or topographical disamenities, style, scale and type of dwellings, or condition of their commercial districts and public spaces. Such stigma may reduce the opportunities and perceptions of residents of stigmatized areas in a variety of ways, such as job opportunities and self-esteem.
- *Local Institutional Resources*: Some neighbourhoods may have access to few and/or high-quality private, non-profit, or public institutions and organizations, such as benevolent charities, day care facilities, schools, and medical clinics. The lack of same may adversely affect the personal development opportunities of residents.

- *Local Market Actors*: There may be substantial spatial variations in the prevalence of certain private market actors that may encourage or discourage certain behaviours by neighbourhood residents, such as liquor stores, fresh food markets, fast food restaurants, and illegal drug markets.

Conceptual Issues in Uncovering and Measuring Mechanism(s) of Neighbourhood Effects

I begin this discussion with the premise that the ultimate goal of social science is to not only identify which mechanisms are responsible for creating a designated effect on residents but to ascertain *quantitatively their relative contributions* to the outcome of interest. For the purposes of this discussion it is useful to employ a pharmacological metaphor: "dosage-response." There is substantial empirical evidence that several sorts of variables measuring neighbourhood-level indicators are correlated with a variety of behavioural and health outcomes for children, youth, and adults; for reviews see: Haveman and Wolfe (1994), Duncan et al. (1997), Van Kempen (1997), Gephart (1997), Ellen and Turner (1997), Friedrichs (1998), Leventhal and Brooks-Gunn (2000), Booth and Crouter (2001), Atkinson and Kintrea (2001), Ellen et al. (2001), Pickett and Pearl (2001), Haurin et al. (2002), Dietz (2002), Sampson et al. (2002), Musterd (2002), Friedrichs et al. (2003), Kawachi and Berkman (2003), Galster (2005), Sellström and Bremberg (2006), and Schaefer-McDaniel et al. (2009). The question here is "Why?" I find it revealing to employ a pharmacological metaphor here and frame the questions as follows: What about this "dose of neighbourhood" might be *causing* the observed individual "response?" The challenges in answering this deceptively simple question are legion, and my purpose here is to present some of the major ones.[2] If we are to deeply understand why aspects of the neighbourhood context affect residents we ultimately must answer 17 questions arrayed under three overarching rubrics regarding the *composition, administration, and response to the neighbourhood dosage*.

The Composition of the Neighbourhood Dosage

- *What are the "active ingredients" that constitute the dosage?* What is it about this space in terms of internal social interactions, environmental conditions, geographic attributes, and reactions of external institutional drivers that is (are) the causal agent(s) and how can it (they) be measured precisely? If neighbourhood is a multi-dimensional package of causal attributes, as is likely, each part of the package will need to be identified and measured directly.

[2] Note that this discussion is related to but distinct from the question of how to accurately measure the magnitude of this dosage-response relationship, about which I wrote in Galster (2003; 2008).

The Administration of the Neighbourhood Dosage

- *Frequency*: *How often is the dosage administered?* For example, does a particular form of social interaction occur only rarely or (as in the case of air pollution) is the exposure occurring during each inhalation?
- *Duration*: *How long does the dosage continue, once begun?* Certain social interactions can vary dramatically in their length, whereas the dosage of unresponsive public services and non-existent facilities can be omnipresent.
- *Intensity*: *What is the size of the dosage?* How concentrated are the toxins? How weak are the local services? In the case of social interactive causes, the answers to the frequency, duration and intensity questions will be related to the amount of time that the individual spends in the neighbourhood and outside of the home in "routine activities."
- *Consistency*: *Is the same dosage being applied each time it is administered?* Do pollutants or the threat of victimization vary daily based on meteorological conditions or time of day?
- *Trajectory*: *Is the frequency, duration, and/or intensity of dosage growing, declining, or staying constant over time for the resident in question?* Do the individuals in a rising trajectory context evince fewer effects because they get more "immune" or evince more effects because their resistance is "weathered?"
- *Spatial Extent*: *Over what scale does the dosage remain constant?* How rapidly does the frequency, duration, intensity and/or consistency of dosage decay when the subject travels away from the residence? Do any of these gradients vary according to the direction of movement away from the residence?
- *Passivity*: *Does the dosage require any action by residents (cognitive or physical) to take effect?* I.e., do residents need to engage in any activities or behaviours, or even be cognizant of the forces operating upon them for the effect to transpire? In the case of endogenous local social interactions, the answer is probably yes, but not in the case of the other categories of mechanisms.
- *Mediation*: *Is the dosage received directly or indirectly by the resident in question?* For example, neighbourhood influences on children may be mediated by parents who are directly affected by the neighbourhood.

The Neighbourhood Dosage-Response Relationship

- *Thresholds*: *Is the relationship between variation in any dimension of dosage administration and the response nonlinear?* Are there critical points at which marginal changes in the dosage have non-marginal effects?
- *Timing*: *Does the response to the dosage occur immediately, after a substantial lag, or only after cumulative administration?* For example, you might become stigmatized as soon as you move into a certain neighbourhood, but eroded health due to lack of local recreational facilities may not show up until an extended period has elapsed.

- *Durability*: *Does the response to the dosage persist indefinitely or decay over time slowly or quickly?* The developmental damage done by lead poisoning is, for illustration, indelible.
- *Generality*: *Are there many predictable responses to the given dosage administration, or only one?* Peers may influence a wide variety of adolescent behaviours, whereas certain environmental toxins may have rather narrowly defined health impacts.
- *Universality*: *Is the relationship between variation in any dimension of dosage administration and the particular response similar across children's developmental stages, demographic groups, or socioeconomic groups*? The same dosage of neighbourhood may yield different responses depending on the developmental or socioeconomic status of those exposed.
- *Interactions*: *Are dosages of other intra- or extra-neighbourhood treatments also being administered that intensify the given dosage's expected response?* Different dimensions of neighbourhood may not be additive but multiplicative.
- *Antidotes*: *Are dosages of other intra- or extra-neighbourhood treatments also being administered that counteract the given dosage's expected response?* For example, efforts to improve residents' health by building new clinics and outreach facilities in the neighbourhood may founder if environmental pollution in the area gets worse.
- *Buffers*: *Are people, their families, and/or their communities responding to the dosage in ways that counteract its expected response?* Because residents individually and collectively potentially have agency they may engage in compensatory behaviours that offset negative neighbourhood effects, such as when parents keep their children in the home when certain violent youngsters are using the local playground.

Past Investigative Responses and Their Limitations

There are two broad sorts of approaches that social scientists have employed in an attempt to answer the above questions and uncover the dominant neighbourhood effect mechanisms at work: (1) field-interview studies of people's social relations and networks within neighbourhoods and non-residents' opinions about neighbourhoods, involving both quantitative and qualitative analyses of the data collected thereby; and (2) multivariate statistical studies estimating models of how various neighbourhood indicators are correlated with a variety of individual outcomes for children, youth, and adults. Field-interview studies try directly to observe potential mechanisms. In this vein, there have been numerous sociological and anthropological investigations, but they are often limited in their ability to discern the relative contributions of alternative causes because of their qualitative nature and their typical focus on only one set of mechanisms to the exclusion of others. Nevertheless, several have been revealing and remarkably consistent in their findings that allow us to rule out certain potential causes. Moreover, this style of investigation is more appropriate for probing many of the questions noted above, such as active ingredients, passivity, mediation, and buffering of dosages.

The multivariate statistical approach tries to draw inferences about neighbourhood effect mechanisms from the statistical patterns observed. It has its own challenges, akin to a physician making a differential diagnosis on the basis of a patient's symptoms and only a partial, poorly measured medical history. One inferential notion that has been used is that if particular sorts of descriptors of a neighbourhood prove more statistically and economically significant predictors of resident outcomes they may hint at which underlying process is dominant. For example, if the variable "percentage of poor residents in the neighbourhood" was not to prove significant but the variable "percentage of affluent residents in the neighbourhood" was to in a regression predicting outcomes for low-income residents, it would suggest that a positive social externality from the affluent group like role modelling, not a negative social externality from the poor group like peer effects, was present. Unfortunately, an overview of the research record typically does not produce such unambiguous results for coefficients. Moreover, most of this statistical literature is of little help to us here because it does not disaggregate findings by economic or demographic group (though see Galster et al. 2010). For example, how is one to interpret the finding from a regression model estimated over youth from all income groups that there is a negative correlation between the percentage of poor households in the neighbourhood and an individual's chances of dropping out from high school? One cannot make the deduction that non-poor youth are positively influencing poor youth through role modelling. A second inferential notion often employed draws upon the assumption that different types of neighbourhood social externalities yield distinctive functional forms for the relationship between the percentage of disadvantaged and/or advantaged residents in a neighbourhood and the amount of externality being generated. For example, collective social norms and social control likely come into play only after a threshold scale of the population group thought to be generating this effect has been achieved in the neighbourhood. This logic can be used to draw out implications for underlying mechanisms of neighbourhood effects if the statistical procedures used to investigate the relationship between neighbourhood indicator(s) and individual outcome permit the estimation of non-linear relationships. Unfortunately, few extant empirical studies test for non-linear relationships between neighbourhood indicators and various individual outcomes. Moreover, even if thresholds and other distinctive non-linearities are observed, it need not uniquely identify only one causal mechanism.

In the review that follows I will organize the presentation in subsections corresponding to the foregoing mechanisms of neighbourhood interaction,[3] bringing to bear evidence from the two approaches as relevant. Before turning to this empirical evidence, however, I note as preface that no definitive, comprehensive study of neighbourhood effect mechanisms exists; none examines more than one or two of

[3]I combine the competition and relative deprivation mechanisms because, to my knowledge, there is little extant statistical evidence that can distinguish between them.

the above questions for an array of potential causal mechanisms. Indeed, most of the questions to my knowledge have not been addressed explicitly in the theoretical or empirical literature. Thus, most empirical conclusions regarding neighbourhood effect mechanisms should be treated as provisional at best.

Evidence on Social-Interactive Mechanisms of Neighbourhood Effects

Social Contagion and Collective Socialization

There have been numerous studies that have examined in detail the social relationships of youth from disadvantaged neighbourhoods. They have identified links between deviant peer group influences and adolescents' grade point average (Darling and Steinberg 1997), mental health (Simons et al. 1996), anti-social behaviour, school attainment, and substance abuse (Dubow et al. 1997). One of the most notable, because of its sophisticated efforts to avoid statistical bias, is Case and Katz's (1991) investigation of youth in low-income Boston neighbourhoods. They find that neighbourhood peer influences among low-income youth are strong predictors of a variety of negative behaviours, including crime, substance abuse, and lack of labour force participation. For more supportive evidence on the importance of role models and peer effects in disadvantaged neighbourhoods, see Sinclair et al. (1994), Briggs (1997a), South and Baumer (2000), Ginther et al. (2000), South (2001), and Oberwittler (2004). To me this body of (primarily) U.S. work suggests that negative social externalities are often being generated through peer effects and role models among disadvantaged young neighbours.[4]

However, the extent to which such negative socialization would be diminished, or replaced by positive socialization, were more higher-income youth to be present is unclear. Rosenbaum (1991, 1995) and Rosenbaum et al. (2002) have provided a series of studies related to black families living in public housing in concentrated poverty neighbourhoods who were assisted (with rental vouchers and counselling) in finding apartments in majority white-occupied neighbourhoods of Chicago and its suburbs as part of a court-ordered remedy for the *Gautreaux* public housing discrimination suit. Though he provides one of the most optimistic portraits of the benefits that such moves can provide to black adults and their children, he does not

[4]However, it is not definitive about the extent to which such negative socialization is general across races. Turley (2003) probes beyond her discovery of overall positive correlations between median family income of neighborhood and youths' behavioral and psychological test scores to see whether there were interaction effects with proxies for number of peer interactions and time spent in neighborhood. She found such strong interaction effects for white but not black youths in her sample, and concluded "differences in neighborhood socializing may explain why neighborhood income affects black and white children differently" (2003: 70).

find a great deal of social interchange or networking between these new in-movers and the original residents. Rosenbaum (1991) concludes by stressing instead the importance of role models and social norms in middle class suburban environments for generating positive outcomes for those participating in the *Gautreaux* Program. However, this optimistic conclusion has been challenged by recent qualitative case studies revealing limited role modelling between upper-income and lower-income blacks in gentrifying neighbourhoods (Boyd 2008; Freeman 2006; Hyra 2008).

The threshold notion embedded in both the social contagion and collective socialization (norm) mechanisms potentially allows them to be identified by regression-based studies that allow for non-linear relationships between the measure of neighbourhood. My review (see Galster 2002) of the U.S. literature (Vartanian 1999a, b; Weinberg et al. 2004) suggests that the independent impacts of neighbourhood poverty rates in encouraging negative outcomes for individuals like crime, school leaving, and duration of poverty spells appear to be nil unless the neighbourhood exceeds about 20% poverty, whereupon the externality effects grow rapidly until the neighbourhood reaches approximately 40% poverty; subsequent increases in the poverty population appear to have no marginal external effect. Analogously, the independent impacts of neighbourhood poverty rates in discouraging positive behaviours like working appear to be nil unless the neighbourhood exceeds about 15% poverty, whereupon the effects grow rapidly until the neighbourhood reaches roughly 30% poverty; subsequent increases in poverty appear to have no marginal effect. This evidence supports the social contagion and/or collective socialization processes.

As far as non-linear relationships between individual outcomes and neighbourhood percentages of affluent residents, the work of Crane (1991), Duncan et al. (1997), and Chase-Lansdale et al. (1997) is relevant. Unfortunately, though they all suggest the existence of a threshold of neighbourhood affluence they differ on where this occurs. Crane's (1991) analysis finds strong evidence of epidemic-like effects on both secondary school leaving and teenage childbearing of the share of affluent (professional-managerial occupation) neighbours. For whites and blacks there is a threshold at 5% affluent neighbours, below which dropout rates skyrocket; for blacks not living in large cities there is another threshold at 20%, above which affluent neighbours cease having a positive impact. These thresholds are more dramatic for black males than females. A similar threshold at low percent affluent neighbours is observed for both black and white teen women's childbearing, especially in large cities. Crane (1991, pp. 1234–1241) interpreted these findings as consistent with intra-neighbourhood social interactions, but was unable to distinguish whether the high-status neighbours created an endogenous effect (such as serving as positive role models) or a correlated effect (such as bringing resources that made local institutions and services better). Duncan et al. (1997) find a different sort of nonlinear neighbourhood effect for educational attainment and the percentage of affluent neighbours. Here the threshold does not seem to occur at a small percentage of affluent, as in Crane's study.[5] The positive effect of the latter becomes dramatically stronger when

[5]Duncan et al. (1997) did not explicitly test for a threshold at a below-average percentage of affluent, however.

the percentage exceeds the national mean for the neighbourhood (for black men and women, and white women). Chase-Lansdale et al. (1997) examine how the percentage of affluent neighbours relates to a variety of intellectual and behavioural development test scores for youth. They find, controlling for family influences, that the percentage of affluent neighbours is positively associated with higher intellectual functioning scores for black children and female children only when the percentage exceeds the 25th percentile and is less than the 75th percentile; for other children the effect is linear. Both the Duncan et al. (1997) and Chase-Lansdale et al. (1997) findings support the notion of collective social norms taking hold only after a substantial share of the affluent group is present in the neighbourhood.[6]

Most Western European evidence related to potential non-linear neighbourhood effects focuses on labour market outcomes as they relate to percentages of disadvantaged neighbours. Here the findings are inconsistent in the extreme. Several studies did not observe any strong nonlinear relationships. Ostendorf et al. (2001) compared "income-mixed" neighbourhoods in Amsterdam with "homogeneous" ones, to ascertain whether this aspect of neighbourhood was related to an individual's chances of living in poverty. Bolster et al. (2004) compared 1-, 5- and 10-year income growth trajectories of British individuals living at the beginning of the period amid different degrees of disadvantage (measured by a composite index). Finally, McCulloch's (2001) multi-level analysis of British Household Panel Study data also failed to identify any strong non-linearities between a ward-level index of disadvantage and such outcomes as employment status, current financial situation, financial expectations, health status, or receipt of social support. Musterd et al. (2003) related the proportion of neighbouring households on social benefits to the chances of Dutch individuals' being employed consistently or not during the 1990s. Over a vast variation in neighbourhoods they found no relationship. Though arguably some non-linearities were evinced at the extreme values of neighbourhood conditions, they involved few neighbourhoods.

Other studies detected non-linear relationships, but of highly inconsistent natures. Buck's (2001) analysis of British Household Panel Study data (but, unlike McCulloch, using unemployment rate as the neighbourhood characteristic) identified substantial non-linearities with the probability of not starting work and the probability of not escaping from poverty, which suggested that the worst results for individuals occurred when the share of neighbourhood residents unemployed exceeded 23–24% (i.e., the highest 5% of all wards). Diametrically opposed results were generated by Musterd and Andersson (2006), who analyzed the Swedish

[6]Turley (2003) analyzes behavioral and psychological test scores for youth as measured in a special supplement of the PSID. She relates these scores to the median family income of the census tract, so one cannot be certain whether the relationship is being generated by share of affluent or share of poor. She tests for non-linearities by employing a quadratic version of neighborhood income variable and finds that its coefficient is statistically significant and negative for the self-esteem outcome, implying that improving the economic environment of youth has a much greater psychological impact for those initially in disadvantaged neighborhood circumstances. Unfortunately, quadratic specifications are not precise in identifying thresholds.

national register database for the three largest metropolitan areas in Sweden to ascertain the relationship between the odds that an individual remained unemployed in both 1995 and 1999 and the percentage of unemployed residents in their neighbourhood in 1995. They (like Buck) found a strong positive relationship until the neighbourhood percentage unemployed exceeded 16%; thereafter there appeared to be no further marginal impact (instead of increasing marginal impact, as per Buck). Van der Klaauw and van Ours (2003) found, using data from Rotterdam (NL) administrative records, that the neighbourhood unemployment rate had no statistically significant negative impact on the probability of exiting welfare into work for Dutch job losers or school leavers until it surpassed 11%, though there were no neighbourhood effect for non-Dutch job losers.

Only two studies using Western European data have investigated the potential nonlinear effects of affluent neighbours. Kauppinen (2004) used categorical variables to delineate neighbourhood affluence in Helsinki and, like Duncan et al. (1997), found that only in neighbourhoods with above-average educational levels does the neighbourhood seem to make a difference in individuals' post-secondary level of educational attainment.

Galster et al. (2008) study the effects of both disadvantaged and advantaged neighbours on individual earnings of adults using Swedish urban data. In the case of men who were not employed full time, it was the neighbourhood with the highest possible share of *middle-income neighbours* that was most conducive to their earning more. The fact that even a few low-income neighbours eroded these benefits suggested to the authors that a negative role modelling or peer effect was transpiring here. Replacing middle-income with high-income neighbours also had negative impacts on these less-advantaged males, implying that the former provided positive role models and/or resource rich networks but the latter did not, perhaps because the social distance between the groups was too great for social interactions. The collective socialization model of interaction was not supported by their findings, because no minimum threshold of low-income neighbours was observed past which their negative impacts began and because such would imply no distinctions between shares of middle- and high-income neighbours under the assumption that both provided comparable norms and social controls.

In sum, this Western European evidence on non-linear neighbourhood effects is so inconsistent that no clear implications can be drawn regarding social contagion and collective socialization causal mechanisms. Nevertheless, it is fair to say that it does not appear to evince non-linear neighbourhood effects similar to those more consistently appearing in the U.S.-based research.

Social Networks

Tiggs et al. (1998) investigate the social networks of blacks in U.S. urban areas. They find that, controlling for personal income, those in areas of concentrated poverty typically are more isolated within their households; they have fewer close external ties, especially with those who are employed or well-educated. These findings replicate those of Fernandez and Harris (1992), who also found that the volume, breadth and depth of social relationships in poor neighbourhoods were especially attenuated for black females. Coupled with consistent evidence that job-seekers in U.S. high-poverty areas rely upon neighbours for potential employment information, and the situation appears ripe for neighbourhood effects in disadvantaged U.S. places working through resource-poor social networks.

Two statistical studies provide further support to the hypothesis that the "social network" mechanism of neighbourhood effect has veracity when it comes to finding employment in the U.S. Bertrand et al. (2000) consider the impact of local social networks on welfare participation. They find welfare participation was enhanced not only by geographic proximity to others on welfare, but especially if these proximate others on welfare spoke the individual's language. Bayer et al. (2004) examine the degree to which people who live on the same census block also tend to work on the same census block. They find that individuals indeed interact very locally when exchanging information about jobs, even when controlling for personal characteristics. However, given the typical high degree of class and race segregation in American neighbourhoods it is not clear how much of the observed local social networks span across groups. Indeed, consistent with sociological field evidence above, Bayer et al. (2004) find that interactions are stronger when pairs of individuals are more likely to interact because of common education.

Evidence also suggests that the social networks established in disadvantaged U.S. neighbourhoods may be so influential that they are difficult to break even after moving away. Briggs (1998) examined the social networks of black and Hispanic youth who participated in a court-ordered, scattered-site public housing desegregation program in Yonkers (NY) during the 1990s. He found few differences in the network diversity or types of aid provided by networks comparing youth who moved to developments in white, middle-class neighbourhoods in Yonkers and those who remained in traditional public housing in poor, segregated neighbourhoods. The former group did not leverage any benefits of living in more affluent and racially diverse areas, and their social ties typically remained within the common race-class confines of their scattered-site developments. Popkin et al. (2002) and Rosenbaum et al. (2003) found that families participating in the Moving To Opportunity demonstration in Chicago were likely to maintain close social ties with their former, poverty-stricken neighbourhoods even after they moved a considerable distance away to low-poverty neighbourhoods. More than half indicated that their social networks were located someplace other than their new neighbourhood.

A complementary view is provided by U.S.-based field studies, which consistently show that the social interaction among members of different economic groups is quite limited, even within the same neighbourhood or housing complex. Members of the lower-status group often do not take advantage of propinquity to broaden their "weak ties" and enhance the resource-producing potential of their networks, instead often restricting their networks to nearby members of their own group. Schill (1997) investigated relationships between different classes of residents living in a newly modernized public housing complex in Chicago that intentionally tried to mix employed, moderate-income households amid unemployed, poor households. Few social ties developed between the groups in the development. Similar conclusions were reached by Clampet-Lundquist (2004) in her study of residents displaced from a revitalized Philadelphia public housing development and Kleit (2001a, b, 2002, 2005; Kleit and Carnegie 2009) in a series of mixed-income housing developments in Maryland and Washington. Several European-based studies have probed this topic as part of restructuring of social housing estates (Atkinson and Kintrea 1998; Jupp 1999; Van Beckhoven and Van Kempen 2003; Duyvendak et al. 2000) or post-war neighbourhoods (Blokland-Potters 1998; Pinkster 2008) and reached similar conclusions.[7]

Several multivariate studies based on European data contribute as well to our understanding of neighbourhood networks. Buck (2001) uses British Household Panel Study data to ascertain a positive relationship between the probability that individuals have no close friends employed and neighbourhood unemployment rates or disadvantage index scores. When coupled with the aforementioned positive relationship Buck observed between these neighbourhood indicators and an individual's probability of not starting work and remaining in poverty, the totality of results are supportive of the importance of local job information networks as a mechanism of transmitting a neighbourhood effect. Farwick (2004) finds that Turkish individuals' contacts with native Germans decline rapidly once the percentage of Turks in the apartment complex exceeds 20%. In turn, this lack of contact increases Turks' chances of having an unstable employment history. Galster et al. (2008, 2009) show that Swedish individuals with a weaker labour market position apparently benefited more from middle-income than high-income neighbours, consistent with the view that the resource-enhanced job information networks provided by better-off neighbours was only influential if the class divide ("social distance") was not too extreme (see van Ham and Manley 2010 for a Scottish labour market study of neighbourhood effects). Pinkster's (2008) study of networks in deprived neighbourhoods in The Hague (in the Netherlands) discovered that localized social ties helped low-income residents in the short-term find jobs but over the longer-term locked them in to these dead-end options and adversely affected their work ethic and expectations. Pinkster suggested that one possible explanation for these effects was that processes of social control limited residents' ability and willingness to interact with residents in the other groups and to look for opportunities outside of the neighbourhood (Pinkster 2008).

Social Cohesion and Control

The importance of social control has been emphasized in a number of studies by Sampson and his colleagues (Sampson 1992; Sampson and Groves 1989; Sampson et al. 1997; Morenoff et al. 2001). To understand the effects of disadvantaged

[7]See review in Kleinhans (2004).

neighbourhoods, they argue, one must understand their degree of social organization, which entails the context of community norms, values and structures enveloping residents' behaviours (what he has labelled "collective efficacy"). Sampson's work has empirically demonstrated that disorder and lack of social cohesion are associated with greater incidence of mental distress and criminality in neighbourhoods (see the review in Sampson et al. 2002).

In this regard there is a good deal of trans-Atlantic commonality of findings related to crime outcomes. Hirschfield and Bowers (1997) identify a strong relationship between neighbourhood social control and assault and robbery in their study using Merseyside (England) data. Veysey and Messner (1999) examine British Crime Survey data and find that unsupervised peer groups and weak organizational participation in the neighbourhood was associated with greater victimization. Markowitz et al. (2001) analysis of British Crime Survey data showed that neighbourhood cohesion mediated some, though not all, of the neighbourhood socio-economic status effects on burglary.

There also has been suggestive work in both North American and Western European contexts demonstrating that social control and disorder potentially have affects on a wider array of outcomes. Aneshensel and Sucoff (1996) find that neighbourhood social cohesion explains a large portion of the relationship between neighbourhood socioeconomic status and adolescent depression. Kohen et al. (2002) find in Canada that neighbourhood disorder is negatively related and neighbourhood cohesion is positively related to children's verbal ability, and that neighbourhood cohesion (though not disorder) is negatively associated with child behavioural problems. Steptoe and Feldman (2001) surveyed London adults and found that the effect of neighbourhood socio-economic status on individual psychological distress was mediated by social cohesion and informal control. Blasius and Friedrichs (2004) also found in Koln (Germany) that collective efficacy was a valid construct that was correlated with several individual outcomes.

Finally, Galster and Santiago (2006) provide a unique perspective on the issue by asking low-income parents what they thought the main mechanisms of neighbourhood effects upon their children were. The dominant plurality (24%) cited lack of norms and collective efficacy. By contrast, peers (12%), exposure to violence (11%), and institutional resources (3%) were cited much less often. Of interest, one-third reported that their neighbourhoods had no effect either because their children were too young or that they were able to buffer the impacts.

Competition and Relative Deprivation

Though the U.S statistical evidence (already cited) overwhelmingly suggests that affluent residents convey positive externalities to their less-well of neighbours, there is at least one dissenting study: Ginther et al. (2000) analyze U.S. high school graduation probabilities and total years of education attained. For the white subsample (only) they found that a larger percentage of high-income neighbours was negatively related to graduation probabilities, while a larger percentage of low-income families was positively related to educational attainment. The qualitative evidence from the U.S. is less clear, with some case studies indicating that upper-income gentrifiers can sometimes mobilize and compete in ways that can work to the detriment of the original, lower-income residents; cf: Hyra (2008), Boyd (2008) and Freeman (2006). The importance of these effects is, of course, impossible to quantify from these qualitative works.

The statistical evidence on the effect of affluent neighbours on less-fortunate ones is decidedly more mixed in Europe, with a non-trivial literature indicating that effects are negative. A hint of a social conflict-type of neighbourhood mechanism is embodied in the finding by Sampson and Groves (1989) in Britain that neighbourhood ethnic heterogeneity was associated with more unsupervised peer groups and lack of participation in local organizations. McCulloch's (2001) analysis of British data finds that disadvantaged women are more likely to experience a variety of negative outcomes if they live in affluent areas, indicative of relative deprivation or competition mechanisms. This is consistent with two other British studies that found that health issues for poor individuals were more problematic when they lived in more affluent areas (Duncan and Jones 1995; Shouls et al. 1996). Finally, Oberwittler (2007) observed that German adolescents living in households receiving welfare recipients scored substantially higher on an index of relative deprivation when they resided in neighbourhoods with the lowest overall welfare receipt rates. Finally, I note the Atkinson and Kintrea (2004) qualitative study of key informant opinions in Glasgow, in which some espoused the relative deprivation consequence of extreme social mixing within neighbourhoods.

It is less clear whether this potential relative deprivation effect in Europe extends to outcomes related to education. Kauppinen (2004) reports a strongly positive influence of affluent neighbours on educational achievement of individuals in Helsinki. Gibbons (2002) used the British National Child Development Study to examine the relationship between educational levels of neighbourhood experienced during adolescence and educational attainments by age 33. He found that, controlling for parental and school characteristics, the neighbourhood percentage of highly educated adults was strongly positively correlated with the probability that the children would be highly educated as young adults, and negatively correlated with the probability that they would fail to obtain any credentials, and that these relationships persisted similarly for various groups of children stratified by early childhood test scores. However, the marginal gains from more educated neighbours clearly attenuated within the highest quartile of neighbourhoods. Indeed, for children living in social housing the probability of not gaining any credentials was slightly *greater* in the most-educated quartile of neighbourhoods than in more modestly educated ones.

Parental Mediation

Few would argue that parents' mental and physical health, coping skills, sensed efficacy, irritability, parenting styles, and socio-psycho-economic resources loom large in how children develop. Thus, if any of the above elements are seriously affected by the neighbourhood (by whatever causal path), child outcomes are likely to be affected, though in this case the neighbourhood effect for children is indirect (Klebanov et al. 1997; Spencer 2001). For example, as I will explore in the following section, certain neighbourhoods generate much higher exposures to stress-generating events for parents that, in turn, has been shown to adversely affect children (Elder et al. 1995; Linares et al. 2001). Such neighbourhoods may also vary, however, in their degrees of social support networks that might serve to defuse the negative effects of stress. As another example, parenting styles related to responsiveness/warmth and harshness/control have been observed to vary across aspects of neighbourhood disadvantage (Klebanov et al. 1994; Earls et al. 1994; Simons et al. 1996; Briggs 1997a). Such variations, in turn, have been related to, among other outcomes, adolescent boys' psychological distress (Simons et al. 1996). Finally, riskier neighbourhoods have been linked to lower-quality home learning environments on many dimensions, resulting in lower reading abilities, verbal skills, and internalizing behaviour scores (Greenberg et al. 1999).

Evidence on Environmental Mechanisms of Neighbourhood Effects

In the U.S. it is clear that exposure to violence has reached epidemic proportions for low-income, minority youths (Martinez and Richters 1993; Richters and Martinez 1993; Aneshensel and Sucoff 1996). The Yonkers (NY) Family and Community Survey and Moving To Opportunity demonstration have provided strong support for the perceived importance of this factor, since safety concerns were cited as a prime reason for participating in these programs by most public housing families (Briggs 1997b; Goering and Feins 2003). One of the most significant results of the Moving To Opportunity demonstration was the substantial reductions in stress and other psychological benefits accrued by parents and children who moved from dangerous, high-poverty neighbourhoods to safer ones (Katz et al. 2001; Goering and Feins 2003). Other work also has demonstrated that youths and adults who have been exposed to violence as witnesses or victims suffer increased stress and declines in mental health (Aneshensel and Sucoff 1996; Martinez and Richters 1993; Ceballo et al. 2001; Hagan and Foster 2001). Exacerbated stress, in turn, can produce a variety of unhealthy stress-reduction behaviours such as smoking (Ganz 2000) and over the long term can reduce the efficacy of the body's immune system (Geronimus 1992). Exposure to violence has also been linked to higher risks of pregnancy (Linares et al. 2001), poorer pregnancy outcomes and low birth weight (Zapata et al. 1992), poorer educational outcomes (Hagan and Foster 2001; Lord and Mahoney 2007), more aggressive behaviours (Linares et al. 2001; Guerra et al. 2003), and reduced social cognition (Guerra et al. 2003), though some of these effects appear substantially mediated by the stress levels of parents (Linares et al. 2001).

Several aspects of the physical environment of the neighbourhood have been probed for their potential health impacts. A major proponent of the physical decay dimension is Ross et al. (2001), whose work suggests that prolonged exposure to a badly deteriorated environment weakens residents' sense of efficacy. A variant on this approach is the "broken windows" hypothesis in criminology, which suggests that physical symbols are strongly correlated with deviant and criminal behaviours in the neighbourhood (Kelling and Wilson 1982). It is less clear whether it is the decay that creates an effect in its own right, however, or whether it merely serves as proxy for lack of collective efficacy. Clearer links to health have been identified for another physical aspect of the environment: noise (Stansfeld et al. 2000; Schell and Denham 2003; Van Os 2004). Others have argued that the physical design of neighbourhoods (presence of sidewalks, local land use mixes, cul-de-sacs, etc.) can affect the amount of exercise that residents get, which in turn affects obesity rates and other health outcomes (Lopez and Hynes 2006), though the body of empirical evidence is small thus far. Results from the Moving To Opportunity demonstration found, however, that those moving from disadvantaged to low-poverty neighbourhoods had reduced rates of obesity, which supports the view that some (unspecified) physical feature(s) of the neighbourhood environment were at play (Goering and Feins 2003).

As for toxic exposure to environmental pollutants, there is a large body of U.S.-based literature that documents a common pattern whereby lower-income and minority-occupied neighbourhoods are exposed to higher concentrations of air-, water-, and soil-borne pollutants (Anderton et al. 1994; Bullard 1994; Hamilton 1995; Vrijheid 2000; Perlin et al. 2001; Kawachi and Berkman 2003; Ash and Fetter (2004); Litt et al. 2009; Saha 2009). In turn, air pollutants have been linked in many international epidemiological studies to lower life expectancy, higher infant and adult mortality risks, more hospital visits, poorer birth outcomes, and asthma (McConnochie et al. 1999; Brunekreef and Holgate 2002; Ritz et al. 2002; Clancy et al. 2002; McConnell et al. 2002; Kawachi and Berkman 2003; Chay and Greenstone 2003; Neidell 2004; Currie and Neidell 2005; Brook 2008; Hassing et al. 2009). Proximity to hazardous waste ("brownfield") sites has been linked to higher rates of mortality from cancer and other diseases (Litt et al. 2009). Potential physiological mechanisms by which pollution can create health risks have been elucidated by Holguin (2008) and Mills et al. (2009). All of these studies can be challenged on one or more methodological grounds, however (Vrijheid 2000). These include failure to control for many confounding personal factors, lack of precision in the local-area estimates of pollution concentrations, high sampling variability due to the small number of toxic waste sites, and potential selection bias where unobserved personal characteristics affect both their exposure to pollutants and their health outcomes. For fuller critical review, discussion and evaluation of this vast research literature on pollution and health, see Bernstein et al. (2004), Stillerman et al. (2008), Ren and Tong (2008), Chen et al. (2008), and Clougherty and Kubzansky (2009).

The one area where the health effects of exposure to environmental toxins seem incontrovertible is in the realm of lead poisoning. It has been shown that even small amounts of lead poisoning (typically produced by residue from deteriorated lead-based paint formerly used in homes) can produce harms to infants (Reyes 2005). Lead poisoning also harms the mental development, IQ, and behaviours of older children (Needleman and Gastsonis 1991; Pocock et al. 1994).

Evidence on Geographical Mechanisms of Neighbourhood Effects

Numerous rigorous empirical studies have investigated the issue of racial differentials in accessibility to work (the "spatial mismatch" hypothesis) in the U.S. context (for reviews see: Kain 1992). This literature generally suggests that mismatch can be an important aspect of spatial opportunity differentials in at least some metropolitan areas. Ethnographies (Sullivan 1989; Newman 1999) have shown that low-income youths can benefit greatly from part-time employment (by gaining resources, adult supervision, and routinized schedules), yet there neighbourhoods typically have few such jobs (Newman 1999; Wilson 1997). Evaluations of the Gautreaux program in Chicago showed that low-income black youths moving to the suburbs were more likely to hold jobs and earn more than their counterparts who stayed within the city (Rosenbaum 1995). Nevertheless, there is considerable statistical evidence that this spatial mismatch is of less importance to economic outcomes than the social-interactive dimensions of neighbourhoods (Weinberg et al. 2004; Dawkins et al. 2005). Spatial mismatch typically is not seen as major issue in Europe, perhaps because of lower levels of ethnic and income segregation, less concentration of work-sites, and more comprehensive public transportation systems (Gobillon et al. 2007). Nevertheless, the few studies have come to divergent conclusions (cf. Dujardin and Goffette-Nagot 2007; and Gobillon et al. 2010).

By contrast, what little evidence exists on the mechanism of neighbourhood stigmatization tends to be idiosyncratic, qualitative, and (with one exception) hard to evaluate or quantify. Nevertheless, considerable case study evidence suggests that place-based stigmatization is an often occurring process in Western Europe. The work of Wacquant (1993), Power (1997), Taylor (1998), Atkinson and Kintrea (1998), Dean and Hastings (2000), Hastings and Dean (2003), and Martin and Watkinson (2003) is noteworthy. This body of work does not, of course, help us to quantify the degree to which neighbourhood stigmatization diminishes the life-chances of residents or restricts the various public or private resources or institutions flowing into these areas. To my knowledge, only one study has attempted statistically to relate measured perceptions of key actors about neighbourhoods to socioeconomic or demographic indicators measured in those places. Permentier et al. (2007) asked households and real estate agents to evaluate on multiple grounds a variety of neighbourhoods in their city of Utrecht (NL) in which they did not live. They found that neighbourhood reputations were significantly correlated with their socio-economic characteristics, while their physical and functional features were of less importance. Unfortunately, these authors did not test for threshold points where the perceptions dramatically changed in response to neighbourhood social mix. Perhaps even more crucially, it is unclear the degree to which the reputation of a long-stigmatized neighbourhood can change as a consequence of more advantaged households being added to the social mix (Cole et al. 1997; Pawson et al. 2000; Beekman et al. 2001).

Evidence on Institutional Mechanisms of Neighbourhood Effects

Many U.S.-based studies have documented the vast differences in both public and private institutional resources serving different neighbourhoods (e.g., Condron and Roscigno 2003; Lankford et al. 2002). Though there has been considerable debate on this subject, the current consensus seems to be that measurable educational resources are strongly correlated with several aspects of student performance in both the U.S. (Jargowsky and El Komi 2009) and the U.K. (Bramley et al. 2005). Although the evidence linking these geographic differences to various outcomes for children has been subject to challenge (e.g., Morenoff et al. 2001; Popkin et al. 2002), there is increasing evidentiary prominence of some institutions, such as the public schools, serving as important mediators of neighbourhood context (Teitler and Weiss 1996). Moreover, it is clear that many parents believe that a paucity of local resources can adversely affect their children (Galster and Santiago 2006; Phibbs 2009) and often try to compensate for this lack by seeking them from outside of their neighbourhoods (Jarrett 1997).

There is also substantial evidence from the U.S. regarding the large spatial variations in many sorts of market actors whose proximity may affect health-related behaviours of neighbourhood residents. Several studies, for example, have documented distinctive race and class patterns in supermarket food store locations (Morland et al. 2002; Block et al. 2004; Zenk et al. 2005) and others have done the same for dietary habits (Diez-Roux et al. 1999). As another illustration, in his study of Latino and Black youth moving from concentrated poverty neighbourhoods in Yonkers, Briggs (1997b) finds that they had much less access to liquor stores in their non-poverty destinations and that their reported alcohol usage was lowered. Quantifying a convincing causal link between such contextual variations and individual's diets and consumption patterns and, ultimately, health, has proven more challenging, however; see Gallagher (2006, 2007) and Morland et al. (2002).

In the Western European context the effect of institutional and public service differentials across space are probably less severe, given that these welfare states have a more centralized funding mechanism and often try to provide compensatory services to disadvantaged neighbourhoods (Powell and Boyne 2002). However, Atkinson and Kintrea (2001), Buck (2001), and Hastings (2007, 2009b) offer several more subtle mechanisms about how such effects may be imparted nevertheless: (1) low expectations by residents of disadvantaged places create self-fulfilling prophecies; (2) inter-neighbourhood competition for scarce public services, skilled employees, and facilities; (3) "rationing" of public services in ways that are insufficient to equally meet needs in different locales; and (4) direct place-based discrimination by institutional actors controlling allocations of resources. Hastings (2009a) provides a comprehensive conceptual framework of a variety of endogenous relationships of relevance here.

The Western European evidence on these points is suggestive but hardly definitive. Lupton (2004) finds that schools in disadvantaged UK districts have a more difficult time attracting highly qualified, experienced teachers. Some studies have found that teachers in disadvantaged UK neighbourhoods expect less from their students (Atkinson and Kintrea 2001; Gillborn and Youdell 2000). Hastings' (2009a) case studies of 12 UK neighbourhoods suggest qualitatively that environmental service provision fails to compensate for higher levels of need in certain neighbourhoods, thereby setting in motion a mutually-reinforcing downward spiral of reactions by residents and service providers alike.

A Provisional Synthesis Regarding Evidence on Neighbourhood Effect Mechanisms

What does the foregoing evidence suggest about the importance of various neighbourhood effect mechanisms in the U.S and Western Europe, when all is said and done? With the mandatory caveat that firm conclusions are elusive here given the state of scholarship and the complexity of the topic, my evaluation provisionally suggests the following.[8]

First, in both the U.S. and Western Europe high concentrations of poverty or socially disadvantaged households (which typically are heavily Hispanic- and especially black-occupied neighbourhoods in the U.S. and immigrant-occupied neighbourhoods in Western Europe) have been consistently empirically linked to weaker cohesion and structures of informal social controls in their neighbourhoods. This situation produces, in turn, negative consequences like increased youth delinquency, criminality, and mental distress, although this mechanism has not yet been linked to other important outcomes like labour market performance. However, in both U.S. and Western European research the aforementioned concentrations of poverty or disadvantage retain their relationship with a variety of child and adult outcomes even after intra-neighbourhood levels of social control and cohesion are taken into account. Clearly, more than this mechanism is at work.

Second, the fact that neighbourhood poverty rates in the U.S. appear consistently related to a range of outcomes in a non-linear, threshold-like fashion further suggests

[8] I recognize that practitioners who deal directly with deprived neighborhoods hold divergent and conflicting opinions about which neighborhood effect mechanisms are most important (Atkinson and Kintrea 2004). The same can be said of low-income minority parents (Galster and Santiago 2006).

that the social contagion (peers) and/or the collective socialization (roles models, norms) forms of causal linkages are transpiring. There may also be some selectivity involved, as some socially weaker groups in the U.S. seem more vulnerable to these contexts than stronger ones. I do not believe that the evidence can clearly distinguish the respective contributions made by the latter two alternatives.[9] Unfortunately, with highly inconsistent evidence regarding non-linearities of neighbourhood impacts in the Western European evidence, there is no certainty about the relative importance of such processes there.

Third, in the U.S. the presence of affluent neighbours appears to provide positive externalities to their less-well off neighbours, seemingly working through social controls and collective socialization. Social networks and peer influences between the affluent and the poor, by contrast, do not appear as important in this vein. The outcomes for individuals that are most strongly related to affluent neighbours seem to be different that those most strongly related to disadvantaged neighbours. There is consistent U.S. empirical evidence to suggest thresholds here as well, though the precise threshold is unclear and likely varies by outcome being considered. The Western European evidence is much less definitive, and indeed inconsistent, in all these aforementioned regards. Finally, most U.S. and Western European evidence indicates that the influence on vulnerable individuals of advantaged neighbours is smaller in absolute value than the influence of disadvantaged neighbours, whatever the mechanism(s) at play.

Fourth, in U.S. neighbourhood contexts there is little evidence suggesting that the competition or relative deprivation mechanisms are operating in a meaningful way. The same cannot be said of Western European evidence, however, where the preponderance suggests that mixing of extremely dissimilar low- and high-income groups results in little benefit or even harms for those who are most disadvantaged.

Fifth, a large number of U.S. studies have consistently found that there is relatively little social networking between lower-income and higher-income households or children in the same neighbourhood, and this lack is compounded if there are also racial differences involved. Thus, there is little to support the version of neighbourhood effects that advantaged neighbours create valuable "weak ties" for disadvantaged ones. I could identify no Western European evidence on this point.

Sixth, local environmental differences appear substantial and likely produce important differentials in mental and physical and mental health on both sides of the Atlantic. There are huge differences in exposure to violence across U.S. neighbourhoods and this undoubtedly produces important and durable psychological consequences for children and adults that, in turn, likely have numerous but hard-to-quantify other effects. Exposure to environmental pollutants and (especially in the U.S.) to violence undoubtedly produces significant consequences for the health of children, youths and adults, though evidence on the links for many toxins besides

[9] After their review, Leventhal and Brooks-Gunn (2000) similarly concluded that the strongest support seems to be for the combined role of norms, collective efficacy (informal social controls), and peers as major neighborhood influences on adolescent behaviors.

lead is often sketchy. The longer-term consequences of these health impacts on subsequent educational outcomes, behaviours, and economic outcomes have not been adequately explored, however.

Seventh, geographic disparities related to differential accessibility to work and quality public services (especially education) are likely more severe in the U.S. than in Western Europe. At least in the U.S context, these mechanisms likely play a non-trivial role in explaining labour force and educational outcomes.

Eighth, institutional processes involving place-based stigmatization, local institutional quality and behaviour, and local market actors likely exist but quantification of their spatial variations have not been accomplished in a way that permits generalizations in either the U.S. or Western Europe. Moreover, convincing statistical models of the relationship between measured variations in these potential causal mechanisms and a wide range of behavioural or health outcomes have not been completed.

Finally, there is probably a substantial, indirect effect on children and youth than transpires through the combined effects of the social-interactive, environmental, geographic, and institutional dimensions of the neighbourhood context on their parents. This mediation of neighbourhood effects through parents is likely to affect a broad range of outcomes for their offspring, though there have been no attempts to measure comprehensively such effects.

Implications for Scholarship and Policy

Advancing Scholarship on Neighbourhood Effect Mechanisms

I return once again to a theme that introduced this paper and that echoed throughout: given the complexity of the topic there is simply far too little scholarship to make many claims about which causal links dominate for which outcomes for which people in which national contexts. I recognize that calling for "additional research" is a shop-worn conclusion for an academic paper; nevertheless, it remains unusually valid and significant in this case.

How might such additional research proceed? Given that both qualitative and quantitative approaches have different inherent strengths and limitations here, I would argue for mixed-methods strategies, ideally embedded within the same study design so the same populations, local neighbourhoods and overarching contexts can be held constant. Given the likelihood that many causal mechanisms may act cumulatively and with some durability of impact, there is a need for studies that explore residential histories and patterns of exposure to a wide variety of community conditions, not just current exposure to a narrow palette of neighbourhood measures (Rauch et al. 2001). Because there is such a wide range of potential mechanisms, quantitative studies should not satisfy themselves with easily accessed census indicators for neighbourhoods, but should strive to obtain: (1) administrative data about neighbourhood conditions (e.g., crime, low birth weight rates, child maltreatment rates); (2) data about local institutions, facilities, and schools (both their existence and quality); (3) pollution measures at a fine-grained spatial scale. In addition, I urge moving beyond distal proxies for causal processes and more efforts to collect social process data from community surveys and systematic social observations (Leventhal and Brooks-Gunn 2000; Sampson et al. 2002). Finally, there is a need for datasets that measure the amount of time spent and routines of activity in the neighbourhood and the degree to which social interaction patterns are concentrated in the neighbourhood (South 2001; Sampson 2001). Of course, these studies must also collect detailed information about family circumstances to accurately develop controls or, possibly, measures of parental mediation of neighbourhood impacts on children.

Though these data requirements represent an intimidating menu, there is one emerging study that offers unprecedented breadth in this realm. My Case Western Reserve University Mandel School colleague, Professor Anna Santiago, and I are now beginning analysis of information gathered from a natural experiment in Denver, CO. The research aims to quantify how and why a variety of outcomes (health, education, employment, behavioural and demographic) for low-income, Black and Latino children and youth residing in Denver public housing for a substantial period are statistically related to conditions in the neighbourhoods in which they were raised. Data analyzed come from surveys we conducted with 765 current and former residents of the Denver (CO) Housing Authority (DHA) whose 1,995 children met study eligibility criteria. For decades, DHA has operated public housing located units throughout the City and County of Denver. Because the initial allocation of households on the DHA waiting list to units mimics a random assignment to a wide range of neighbourhood environments, this program represents a natural experiment for overcoming parental location selection bias in estimating neighbourhood effects. We have gathered life histories for all participating children and their families, relating a wide range of outcomes to individual developmental stages. To this residential history we have merged time-coincident data from: (1) census tract indicators of socioeconomic, demographic and housing characteristics; (2) administrative data on crime, low birth weight rates, and school quality; (3) survey-based, parental-identified measures of local institutions and facilities; (4) survey-based, parental-assessed social disorder, collective efficacy, and social networks. We hope to soon add information on air quality and location of hazardous waste sites. In addition, we conducted follow-up interviews with selected children in our sample who have become young adults and their parents, to query them about their perceptions of neighbourhood effect mechanisms, parental buffering attempts at same, etc.

Implications for Public Policy

Obtaining a clearer understanding of the pathways through which neighbourhoods exert their effects is crucial for public policy formulations in at least three major domains: health, employment, and housing. Put bluntly, it is risky for policy-makers to naively observe a correlation between neighbourhood indicators and individual outcomes of interest and design programmatic strategies as if neighbourhood were a "black box." At best, inefficiencies and, at worse, negative unforeseen consequences, are all-too-likely to follow in these circumstances.

In the health domain, it is obvious from an epidemiological perspective that understanding causal pathways is of "critical importance in determining how [public health] interventions should be designed" (Sellström and Bremberg 2006, p. 553). In the employment domain, distinctive programmatic implications follow from alternative conclusions about why some able-bodied are not employed. Perhaps they: (1) lack information networks about job opportunities; (2) try to apply for work but are turned away by employers who stigmatize their neighbourhoods; (3) try to find work but cannot access jobs due to local transport inadequacies; (4) do not try to work because of negative neighbourhood peer influences; or (5) are too sick to work because of sever local pollution levels. In the housing domain, the current Western European fascination with "social mix" strategies (Galster 2007a, b) could be helpfully guided by definitive explorations about what processes are thought to follow from social mix: social-interactive, geographic, and/or institutional? (Joseph et al. 2006; Joseph 2006) If it were to prove the case that, for instance, social networks among the various neighbouring classes were the dominant mechanism of positive influence, urban design strategies designed to maximize interpersonal contacts and "community-building" activities within the mixed estates would be recommended. On the other hand, if mixing served only to remove the former external stigmatization of residents, such micro-level social processes could well be ignored by policy-makers. Finally, there are some implications that overarch particular policy domains. For example, if it were to prove the case that the vast portion of neighbourhood impacts on children occurred indirectly through mediation of parents, then it would follow that interventions designed to minimize negative neighbourhood effects should target parents, even if the ultimate goal is child development.

Acknowledgments I wish to thank Roland Atkinson, Jason Booza, Derek Hyra, Jeff Morenoff, Peter Phibbs, Kami Pothukuchi, and Lyke Thompson for stimulating conversations and helpful citations related to the subject matter of this paper. The research assistance of Sarah Marolf, Rebecca Grace Stokan, Lisa Stack, and Lonnisha Thomas is gratefully acknowledged. This paper was written with the support of the MacArthur Foundation.

References

Anderton, D. L., Anderson, A., Rossi, P. H., Oakes, J. M., Fraser, M. R., Weber, E. W., & Calabrese, E. J. (1994). Hazardous-waste facilities-environment equity issues in metropolitan areas. *Evaluation Review, 18*(2), 123–140.

Aneshensel, C. S., & Sucoff, C. A. (1996). The neighborhood context and adolescent mental health. *Journal of Health and Social Behavior, 37*(2), 293–310.

Ash, M., & Fetter, T. R. (2004). Who lives on the wrong side of the environmental tracks? Evidence from the EPA"s risk-screening environmental indicators model. *Social Science Quarterly, 85*(2), 441–462.

Atkinson, R., & Kintrea, K. (1998). *Reconnecting excluded communities: Neighbourhood impacts of owner occupation*. Edinburgh: Scottish Homes.

Atkinson, R., & Kintrea, K. (2001). Area effects: What do they mean for British housing and regeneration policy? *European Journal of Housing Policy, 2*(2), 147–166.

Atkinson, R., & Kintrea, K. (2004). Opportunities and despair, It's All in there: Practitioner experiences and explanations of area effects and life chances. *Sociology, 38*(3), 437–455.

Bayer, P., Ross, S. L., & Topa, G. (2004). *Place of work and place of residence: Informal hiring networks and labor market outcomes* (Working paper). New Haven: Economics Department, Yale University.

Beekman, T., Lyons, F., & Scott, J. (2001). *Improving the understanding of the influence of owner occupiers is mixed tenure neighborhoods, report 89*. Edinburgh: ODS Limited for Scottish Homes.

Bell, P., Greene, T., Fisher, J., & Baum, A. (1996). *Environmental psychology* (4th ed.). New York: Harcourt, Brace College Publishers.

Bernstein, J. A., Alexis, N., Barnes, C., Bernstein, I. L., Bernstein, J., Nel, A., Peden, D., Diaz-Sanchez, D., Tarlo, S. M., & Williams, P. B. (2004). Health effects of air pollution. *The Journal of Allergy and Clinical Immunology, 114*(5), 1116–1123.

Bertrand, M., Luttmer, E. F. P., & Mullainathan, S. (2000). Network effects and welfare cultures. *Quarterly Journal of Economics, 115*(3), 1019–1055.

Blasius, J., & Friedrichs, J. (2004). *Deviant norms in a poverty neighborhood*. Paper presented at the Inside Poverty Areas Conference, University of Koln.

Block, J., Scribner, R. A., & De Salvo, K. B. (2004). Fast food, race/ethnicity, and income: A geographic analysis. *American Journal of Preventive Medicine, 27*(3), 211–217.

Blokland-Potters, T. (1998). *Wat Stadsbewoners Bindt: Sociale relaties in een achterstandswijk*. Kampen: Kok Agora.

Bolster, A., Burgess, S., Johnston, R., Jones, K., Propper, C., & Sarker, R. (2004). *Neighborhoods, households and income dynamics* (CMPO Working Paper Series No. 04/106). Bristol: University of Bristol.

Booth, A., & Crouter, A. (Eds.). (2001). *Does it take a village? Community effects on children, adolescents and families*. London/Mahwah: N. J. Lawrence Erlbaum Publishers.

Boyd, M. (2008). *Jim crow Nostaligia: Reconstructing race in Bronzeville*. Minneapolis: University of Minnesota Press.

Bramley, G., Evans, M., & Noble, M. (2005). *Mainstream services and neighbourhood deprivation*. London: Neighbourhood Renewal Unit. Available at: www.neighbourhoods.gov.uk.

Briggs, X. (1997a). Moving up versus moving out: Neighborhood effects in housing mobility programs. *Housing Policy Debate, 8*(1), 195–234.

Briggs, X. (Ed.). (1997b). *Yonkers revisited: The early impacts of scattered-site public housing on families and neighborhoods*. New York: Teachers College, Columbia University.

Briggs, X. (1998). Brown kids in white suburbs: Housing mobility and the many faces of social capital. *Housing Policy Debate, 9*(1), 177–221.

Brook, R. D. (2008). Cardiovascular effects of air pollution. *Clinical Science, 115*(6), 175–187.

Brunekreef, B., & Holgate, S. (2002). Air pollution and health. *Lancet, 360*(9341), 1233–1242.

Buck, N. (2001). Identifying neighborhood effects on social exclusion. *Urban Studies, 38*(12), 2251–2275.

Bullard, R. D. (1994). *Dumping in Dixie: Race, class and environmental quality* (2nd ed.). Boulder: West View Press.

Case, A., & Katz, L. (1991). *The company you keep: The effects of family and neighborhood on disadvantaged youth* (NBER Working Paper 3705). Cambridge: National Bureau of Economic Research.

Ceballo, R., Dahl, T., Aretakis, M., & Ramirez, C. (2001). Inner-city children's exposure to community violence: How much do parents know? *Journal of Marriage and Family, 63*(4), 927–940.

Chase-Lansdale, P., Gordon, R., Brooks-Gunn, J., & Klebanov, P. (1997). Neighborhood and family influences on the intellectual and behavioral competence of pre-school and early school-age children. In J. Brooks-Gunn, G. J. Duncan, & J. L. Aber (Eds.), *Neighborhood poverty* (Context and consequences for children, Vol. 1, pp. 9–118). New York: Russell Sage.

Chay, K. Y., & Greenstone, M. (2003). The impact of air pollution on infant mortality: Evidence from geographic variation in pollution shocks induced by a recession. *Quarterly Journal of Economics, 118*(3), 1121–1167.
Chen, H., Goldberg, M. S., & Villeneuve, P. J. (2008). A systematic review of the relation between long-term exposure to ambient air pollution and chronic diseases. *Reviews on Environmental Health, 23*(4), 243–297.
Clampet-Lundquist, S. (2004). HOPE VI relocation: Moving to new neighborhoods and building new ties. *Housing Policy Debate, 15*(3), 415–447.
Clancy, L., Goodman, P., Sinclair, H., & Dockery, D. (2002). Effect of air pollution control in death rates in Dublin, Ireland. *Lancet, 360*(9341), 1210–1214.
Clougherty, J. E., & Kubzansky, L. D. (2009). A framework for examining social stress and susceptibility to air pollution in respiratory health. *Environmental Health Perspectives, 117*(9), 1351–1358.
Cole, I., Gidley, G., Ritchie, C., Simpson, D., & Wishart, B. (1997). *Creating communities of welfare housing? A study of housing association developments in Yorkshire/Humberside.* Coventry: Chartered Institute of Housing.
Condron, D., & Roscigno, V. (2003). Disparities within: Unequal spending and achievement in an urban school district. *Sociology of Education, 76*(1), 18–36.
Crane, J. (1991). The epidemic theory of ghettos and neighborhood effects on dropping out and teenage childbearing. *The American Journal of Sociology, 96*(5), 1226–1259.
Currie, J., & Neidell, M. (2005). Air pollution and infant health: What can we learn from California's recent experience? *Quarterly Journal of Economics, VCXX*(3), 1003–1030.
Darling, N., & Steinberg, L. (1997). Assessing neighborhood effects using individual-level data. In J. Brooks-Gunn, G. J. Duncan, & J. L. Aber (Eds.), *Neighborhood poverty* (Policy implications in studying neighborhoods, Vol. 2, pp. 120–131). New York: Russell Sage.
Dawkins, C., Shen, Q., & Sanchez, T. (2005). Race, space, and unemployment duration. *Journal of Urban Economics, 58*(1), 91–113.
Dean, J., & Hastings, A. (2000). *Challenging images: Housing estates, stigma and regeneration.* Bristol: The Policy Press & Joseph Rountree Foundation.
Dietz, R. (2002). The estimation of neighborhood effects in the social sciences. *Social Science Research, 31*(4), 539–575.
Diez-Roux, A. V., Nieto, F. J., Caulfield, K., Tyroler, H. A., Watson, R. L., & Szklo, M. (1999). Neighborhood differences in diet: The Atherosclerosis Risk in Communities Study (ARIC). *Journal of Epidemiological and Community Health, 53*(1), 55–63.
Dubow, E. F., Edwards, S., & Ippolito, M. F. (1997). Life stressors, neighborhood disadvantages, and resources: A focus on inner city children's adjustment. *Journal of Clinical Child Psychology, 26*(1), 130–144.
Dujardin, C., & Goffette-Nagot, F. (2007). *Neighborhood effects, public housing, and unemployment in France* (Working paper 05-05). Paris: GATE.
Duncan, C., & Jones, K. (1995). Individuals and their ecologies: Analyzing the geography of chronic illness within a multi-level modeling framework. *Journal of Health and Place, 1*(1), 27–40.
Duncan, G. J., Connell, J. P., & Klebanov, P. K. (1997). Conceptual and methodological issues in estimating causal effects of neighborhoods and family conditions on individual development. In J. Brooks-Gunn, G. J. Duncan, & J. L. Aber (Eds.), *Neighborhood poverty* (Context and consequences for children, Vol. 1, pp. 219–250). New York: Russell Sage.
Duyvendak, J., Kleinhans, R., & Veldboer, L. (2000). *Integratie door differentiatie: een onderzoek naar de effecten van gemengd bouwen.* The Hague: Ministerie van VROM.
Earls, F., McGuire, J., & Shay, S. (1994). Evaluating a community intervention to reduce the risk of child abuse: Methodological strategies in conducting neighborhood surveys. *Child Abuse & Neglect, 18*(5), 473–485.
Elder, G. H., Eccles, J. S., Ardelt, M., & Lord, S. (1995). Inner-city parents under economic pressure: Perspectives on the strategies of parenting. *Journal of Marriage and the Family, 57*(3), 771–784.
Ellen, I. G., & Turner, M. A. (1997). Does neighborhood matter? Assessing recent evidence. *Housing Policy Debate, 8*(4), 833–866.

Ellen, I. G., & Turner, M. A. (2003). Do neighborhoods matter and why? In J. Goering & J. Feins (Eds.), *Choosing a better life? Evaluating the moving to opportunity experiment* (pp. 313–338). Washington, DC: Urban Institute Press.

Ellen, I. G., Mijanovich, T., & Dillman, K. (2001). Neighborhood effects on health: Exploring the links and assessing the evidence. *Journal of Urban Affairs, 23*(3), 391–408.

Farwick, A. (2004). *Spatial isolation, social networks, and the economic integration of migrants in poverty areas*. Paper presented at the Inside Poverty Areas Conference, University of Koln.

Fernandez, R., & Harris, D. (1992). Social isolation and the underclass. In A. V. Harrell & G. E. Peterson (Eds.), *Drugs, crime, and, social isolation: Barriers to urban opportunity* (pp. 257–293). Washington: The Urban Institute.

Freeman, L. (2006). *There goes the 'hood*. Philadelphia: Temple University Press.

Friedrichs, J. (1998). Do poor neighborhoods make their residents poorer? Context effects of poverty neighborhoods on their residents. In H. Andress (Ed.), *Empirical poverty research in a comparative perspective* (pp. 77–99). Aldershot: Ashgate.

Friedrichs, J., Galster, G., & Musterd, S. (2003). Neighborhood effects on social opportunities: The European and American research and policy context. *Housing Studies, 18*(6), 797–806.

Gallagher, M. (2006). *Examining the impact of food deserts on public health in Chicago*. Chicago: Mari Gallagher Research and Consulting Group.

Gallagher, M. (2007). *Examining the impact of food deserts on public health in Detroit*. Chicago: Mari Gallagher Research and Consulting Group.

Galster, G. (2002). An economic efficiency analysis of deconcentrating poverty populations. *Journal of Housing Economics, 11*(4), 303–329.

Galster, G. (2003). Investigating behavioural impacts of poor neighbourhoods: Towards new data and analytic strategies. *Housing Studies, 18*(6), 893–914.

Galster, G. (2005). *Neighbourhood mix, social opportunities, and the policy challenges of an increasingly diverse Amsterdam*. Amsterdam: Planning & International Development Studies, Department of Geography, University of Amsterdam. http://www.fmg.uva.nl/amidst/object.cfm/objectid=7C149E7C-EC9F-4C2E-91DB7485C0839425.

Galster, G. (2007a). Neighbourhood social mix as a goal of housing policy: A theoretical analysis. *European Journal of Housing Policy, 7*(1), 19–43.

Galster, G. (2007b). Should policymakers strive for neighborhood social mix? An analysis of the western European evidence base. *Housing Studies, 22*(3), 523–546.

Galster, G. (2008). Quantifying the effect of neighbourhood on individuals: Challenges, alternative approaches and promising directions. *Journal of Applied Social Science Studies [Schmollers Jahrbuch/Zeitscrift fur Wirtschafts- und Sozialwissenschaften], 128*(1), 7–48.

Galster, G. C., & Santiago, A. M. (2006). What's the 'hood got to do with it? Parental perceptions about how neighbourhood mechanisms affect their children. *Journal of Urban Affairs, 28*(3), 201–226.

Galster, G., Andersson, R., Musterd, S., & Kauppinen, T. M. (2008). Does neighborhood income mix affect earnings of adults? New evidence from Sweden. *Journal of Urban Economics, 63*(3), 858–870.

Galster, G., Santiago, A., Cutsinger, J., & Grace-Stokan, R. (2009). *Neighborhood poverty and the development of low-income, minority children: preliminary evidence from a natural experiment in Denver.* Paper presented at APPAM meetings, Washington, DC.

Galster, G., Andersson, R., & Musterd, S. (2010). Who is affected by neighbourhood income mix? Gender, age, family, employment and income differences. *Urban Studies, 47*(14), 2915–2944.

Ganz, M. L. (2000). The relationship between external threats and smoking in central Harlem. *American Journal of Public Health, 90*(3), 367–371.

Gephart, M. (1997). Neighborhoods and communities as contexts for development. In J. Brooks-Gunn, G. J. Duncan, & J. L. Aber (Eds.), *Neighborhood poverty* (Context and consequences for Children, Vol. I, pp. 1–43). New York: Russell Sage.

Geronimus, A. T. (1992). The weathering hypothesis and the health of African-American women and infants: Evidence and speculations. *Ethnicity & Disease, 2*(3), 207–221.

Gibbons, S. (2002). *Neighbourhood effects on educational achievement: Evidence from the census and national child development study*. London: Centre for the Economics of Education, London School of Economics and Political Science.

Gillborn, D., & Youdell, D. (2000). *Rationing education: Policy, practice, reform and equity*. Buckingham: Open University Press.

Ginther, D., Haveman, R., & Wolfe, B. (2000). Neighbourhood attributes as determinants of children's outcomes. *Journal of Human Resources, 35*(4), 603–642.

Gobillon, L., Selod, H., & Zenou, Y. (2007). The mechanisms of spatial mismatch. *Urban Studies, 44*(12), 2401–2427.

Gobillon, L., Magnac, T., & Selod, H. (2010). The effect of location on finding a job in the Paris region. *Journal of Applied Econometrics*. DOI:10.1002/jae.

Goering, J., & Feins, J. (Eds.). (2003). *Choosing a better life? Evaluating the moving to opportunity experiment*. Washington, DC: Urban Institute Press.

Green, L., & Ottoson, J. M. (1999). *Community and population health* (8th ed.). Boston: McGraw-Hill.

Greenberg, M. T., Lenqua, L. J., Coie, J. D., & Pinderhughes, E. E. (1999). Predicting developmental outcomes at school entry using multiple-risk model: Four American communities. *Developmental Psychology, 35*(2), 403–417.

Guerra, N., Huesmann, L., & Spindler, A. (2003). Community violence exposure, social cognition, and aggression among urban elementary school children. *Child Development, 74*(5), 1561–1576.

Hagan, J., & Foster, H. (2001). Youth violence and the end of adolescence. *American Sociological Review, 66*(6), 874–899.

Hamilton, J. T. (1995). Testing for environmental racism: Prejudice, profits, political power? *Journal of Policy Analysis and Management, 14*(1), 107–132.

Hassing, C., Twickler, M., Brunekreef, B., Cassee, F., Doevendans, P., Kastelein, J., & Cramer, M. J. (2009). Particulate air pollution, coronary heart disease and individual risk assessment: A general overview. *European Journal of Cardiovasic Prevention and Rehabilitation, 16*(1), 10–15.

Hastings, A. (2007). Territorial justice and environmental services: UK. *Environment and Planning C, 25*(6), 896–917.

Hastings, A. (2009a). Neighbourhood environmental services and neighbourhood 'effects': Exploring the role of urban services in intensifying neighbourhood problems. *Housing Studies, 24*(4), 503–524.

Hastings, A. (2009b). Poor neighbourhoods and poor services: Evidence on the 'Rationing' of environmental service provision to deprived neighbourhoods. *Urban Studies, 46*(13), 2907–2928.

Hastings, A., & Dean, J. (2003). Challenging images: Tackling stigma through estate regeneration. *Policy and Politics, 31*(2), 171–184.

Haurin, D., Dietz, R., & Weinberg, B. (2002). *The impact of neighborhood homeownership rates: A review of the theoretical and empirical literature*. Columbus: Department of Economics Working Paper, Ohio State University.

Haveman, R., & Wolfe, B. (1994). *Succeeding generations: On the effects of investments in children*. New York: Russell Sage.

Hirschfield, A., & Bowers, K. J. (1997). The effect of social cohesion on levels of recorded crime in disadvantaged areas. *Urban Studies, 34*(8), 1275–1295.

Holguin, F. (2008). Traffic, outdoor air pollution, and asthma. *Immunology and Allergy Clinics of North America, 28*(3), 577–588.

Hyra, D. (2008). *The new urban renewal*. Chicago: University of Chicago Press.

Ioannides, Y. M., & Loury, L. D. (2004). Job information networks, neighborhood effects, and inequality. *Journal of Economic Literature, 42*(4), 1056–1093.

Jargowsky, P., & El Komi, M. (2009). *Before or after the bell? School context and neighborhood effects on student achievement* (Working Paper 28). Washington: National Center for Analysis of Longitudinal Data in Education Research, Urban Institute.

Jarrett, R. L. (1997). Bringing families back in: Neighborhoods' effects on child development. In J. Brooks-Gunn, G. J. Duncan, & J. L. Aber (Eds.), *Neighborhood poverty* (Policy implications in studying neighborhoods, Vol. 2, pp. 48–64). New York: Russell Sage.

Jencks, C., & Mayer, S. (1990). The social consequences of growing up in a poor neighborhood. In L. E. Lynn & M. F. H. McGeary (Eds.), *Inner-city poverty in the United State* (pp. 111–186). Washington, DC: National Academy Press.

Joseph, M. (2006). Is mixed-income development an antidote to urban poverty? *Housing Policy Debate, 17*(2), 209–234.

Joseph, M., Chaskin, R., & Webber, H. (2006). The theoretical basis for addressing poverty through mixed-income development. *Urban Affairs Review, 42*(3), 369–409.

Jupp, B. (1999). *Living together: Community life on mixed-tenure estates*. London: Demos.

Kain, J. F. (1992). The spatial mismatch hypothesis: Three decades later. *Housing Policy Debate, 3*(2), 371–460.

Katz, L., Kling, J., & Liebman, J. (2001). A moving to opportunity in Boston: Early results of a randomized mobility experiment. *Quarterly Journal of Economics, 116*(2), 607–654.

Kauppinen, T. M. (2004). *Neighbourhood effects in a European city: The educational careers of young people in Helsinki.* Paper presented at European Network for Housing Research meetings, Cambridge, UK.

Kawachi, I., & Berkman, L. (2003). *Neighborhoods and health.* New York: Oxford University Press.

Kelling, G., & Wilson, J. Q. (1982). Broken windows: The police and neighbourhood safety. *Atlantic Monthly, 249*(1), 29–38.

Klebanov, P. K., Brooks-Gunn, J., & Duncan, G. J. (1994). Does neighborhood and family poverty affect mothers' parenting, mental health, and social support? *Journal of Marriage and the Family, 56*(2), 441–455.

Klebanov, P. K., Brooks-Gunn, J., Chase-Lansdale, P. L., & Gordon, R. (1997). Are neighborhood effects on young children mediated by features of the home environment? In J. Brooks-Gunn, G. J. Duncan, & J. L. Aber (Eds.), *Neighborhood poverty* (Context and consequences for children, Vol. 1, pp. 119–145). New York: Russell Sage.

Kleinhans, R. (2004). Social implications of housing diversification in urban renewal: A review of recent literature. *Journal of Housing and the Built Environment, 19*(2), 367–390.

Kleit, R. (2001a). The role of neighborhood social networks in scattered-site public housing residents' search for jobs. *Housing Policy Debate, 12*(3), 541–573.

Kleit, R. (2001b). Neighborhood relations in scattered-site and clustered public housing. *Journal of Urban Affairs, 23*(3–4), 409–430.

Kleit, R. (2002). Job search networks and strategies in scattered-site public housing. *Housing Studies, 17*(1), 83–100.

Kleit, R. (2005). HOPE VI new communities: Neighborhood relationships in mixed-income housing. *Environment and Planning A, 37*(8), 1413–1441.

Kleit, R., & Carnegie, N. (2009). *Integrated or isolated? The impact of public housing redevelopment on social network homophily.* Washington: Daniel J. Evans School of Public Affairs, University of Washington [Unpublished paper].

Kohen, D. E., Brooks-Gunn, J., Leventhal, T., & Hertzman, C. (2002). Neighborhood income and physical and social disorder in Canada: Associations with young children's competencies. *Child Development, 73*(6), 1844–1860.

Lankford, H., Loeb, S., & Wyckoff, J. (2002). Teacher sorting and the plight of urban schools: A descriptive analysis. *Educational Evaluation and Policy Analysis, 24*(1), 37–62.

Leventhal, T., & Brooks-Gunn, J. (2000). The neighborhoods they live. *Psychological Bulletin, 126*(2), 309–337.

Linares, L., Heeren, T., Bronfman, E., Zuckerman, B., Augustyn, M., & Tronick, E. (2001). A Mediational model for the impact of exposure to community violence on early child behavior problems. *Child Development, 72*(2), 639–652.

Litt, J., Tran, N., & Burke, T. (2009). Examining urban brownfields through the public health 'macroscope'. In H. P. Hynes & R. Lopez (Eds.), *Urban health* (pp. 217–236). Sudbury: Jones & Bartlett Publishers.

Lopez, R., & Hynes, P. (2006). Obesity, physical activity, and the urban environment: Public health research needs. *Environmental Health, 5*(1), 25–35.

Lord, H., & Mahoney, J. (2007). Neighborhood crime and self-care: Risks for aggression and lower academic performance. *Developmental Psychology, 43*(6), 1321–1333.

Lupton, R. (2004). *Schools in disadvantaged areas: Recognising context and raising quality* (LSE STICERD Research Paper No. CASE 076). London: University of London – Institute of Education.

Manski, C. F. (1995). *Identification problems in the social sciences*. Cambridge: Harvard University Press.

Markowitz, F. E., Bellair, P. E., Liska, A. E., & Liu, J. (2001). Extending social disorganization theory: Modeling the relationships between cohesion, disorder, and fear. *Criminology, 39*(2), 293–319.

Martin, G., & Watkinson, J. (2003). *Rebalancing communities: Introducing mixed incomes into existing rented housing estates*. York: Joseph Rowntree Foundation.

Martinez, J., & Richters, P. (1993). The NIMH community violence project: II. Children's distress symptoms associated with violence exposure. *Psychiatry, 56*(1), 22–35.

McConnell, R., Berhane, K., Gilliland, F., London, S., Islam, T., & Gauderman, W. (2002). Asthma in exercising children exposed to ozone: A cohort study. *Lancet, 359*(9304), 386–391.

McConnochie, K., Russo, M., McBride, J., Szilagyi, P., Brooks, A., & Roghmann, K. (1999). Socioeconomic variation in asthma hospitalization: Excess utilization or greater need? *Pediatrics, 103*(6), 75–82.

McCulloch, A. (2001). Ward-level deprivation and individual social and economic outcomes in the British household panel survey. *Environment and Planning A, 33*(4), 667–684.

Mills, N. L., Donaldson, K., Hadoke, P. W., Boon, N. A., MacNee, W., Cassee, F. R., Sandstrom, T., Blomberg, A., & Newby, D. E. (2009). Adverse cardiovascular effects of air pollution. *National Clinical Practice Cardiovascular Medicine, 6*(1), 36–44.

Morenoff, J., Sampson, R., & Raudensbush, S. (2001). Neighborhood inequality, collective efficacy, and the spatial dynamics of homicide. *Criminology, 39*(3), 517–560.

Morland, K., Wing, S., Diez-Roux, A., & Poole, C. (2002). Neighborhood characteristics associated with the location of food stores and food service places. *American Journal of Preventive Medicine, 22*(1), 23–29.

Musterd, S. (2002). Response: Mixed housing policy: A European (Dutch) perspective. *Housing Studies, 17*(1), 139–144.

Musterd, S., & Andersson, R. (2006). Employment, social mobility and neighborhood effects. *International Journal of Urban and Regional Research, 30*(1), 120–140.

Musterd, S., Ostendorf, W., & De Vos, S. (2003). Neighbourhood effects and social mobility: A longitudinal analysis. *Housing Studies, 18*(6), 877–892.

Needleman, H., & Gastsonis, B. (1991). Meta-analysis of 24 studies of learning disabilities due to lead poisoning. *Journal of the American Medical Association, 265*(5), 673–678.

Neidell, M. (2004). Air pollution, health, and socio-economic status: The effect of outdoor air quality on childhood asthma. *Journal of Health Economics, 23*(6), 1209–1236.

Newman, K. (1999). *No shame in my game: The working poor in the inner city*. New York: Knopf & the Russell Sage Foundation.

Oberwittler, D. (2004). *Neighborhood disadvantage and adolescent adjustment: Responses of adolescents to disorder and violent subcultures*. Paper presented at the Inside Poverty Areas Conference, University of Cologne.

Oberwittler, D. (2007). The effects of neighbourhood poverty on adolescent problem behaviours: A multi-level analysis differentiated by gender and ethnicity. *Housing Studies, 22*(6), 781–804.

Ostendorf, W., Musterd, S., & de Vos, S. (2001). Social mix and the neighborhood effect: Policy ambition and empirical support. *Housing Studies, 16*(3), 371–380.

Pawson, H., Karryn, K., & MacIntosh, S. (2000). *Assessing the impact of tenure diversification: The case of Niddrie, Edinburgh*. Edinburgh: Scottish Homes.

Perlin, S., Wong, D., & Sexton, K. (2001). Residential proximity to industrial sources of air pollution: Interrelationships among race, poverty, and age. *Journal of the Environmental and Waste Management Association, 51*(3), 406–421.

Permentier, M., Bolt, G., & van Ham, M. (2007). *Comparing residents' and non-residents' assessments of neighbourhood reputations.* Paper presented at the American Association of Geographers meetings, San Francisco.

Phibbs, P. (2009). *The relationship between housing and improved health, education and employment outcomes: The view from Australia.* Paper presented at the 50th annual ACSP conference, Crystal City.

Pickett, K. E., & Pearl, M. (2001). Multilevel analyses of neighborhood, socioeconomic context and health outcomes: A critical review. *Journal of Epidemiology and Community Health, 55*(1), 111–122.

Pinkster, F. (2008). *Living in concentrated poverty*. PhD dissertation, Department of Geography, Planning, and International Development Studies, University of Amsterdam.

Pocock, S. J., Smith, M., & Baghurst, P. (1994). Environmental lead and children's intelligence: A systematic review of the epidemiological evidence. *British Medical Journal, 309*(6963), 1189–1197.

Popkin, S., Harris, L., & Cunningham, M. (2002). *Families in transition: A qualitative analysis of the MTO experience*. Washington, DC: Urban Institute Report prepared for the U.S. Department of Housing and Urban Development.

Powell, M., & Boyne, G. (2002). The spatial strategy of equality and the spatial division of welfare. *Policy & Administration, 35*(2), 181–194.

Power, A. (1997). *Estates on the edge: The social consequences of mass housing in northern Europe*. London: Macmillan.

Rauch, V. A., Andrews, H. F., & Garfinkel, R. (2001). The contribution of maternal age to racial disparities in birthweight: A multilevel perspective. *American Journal of Public Health, 91*(11), 1815–1824.

Ren, C., & Tong, S. (2008). Health effects of ambient air pollution – recent research development and contemporary methodological challenges. *Environmental Health, 7*(56).

Reyes, J. (2005). *The Impact of prenatal lead exposure on health* (Working Papers). Department of Economics, Amherst College.

Richters, P., & Martinez, J. E. (1993). The NIMH community violence project: I. Children as victims of and witnesses to violence. *Psychiatry, 56*(1), 7–21.

Ritz, B., Yu, F., Fruin, S., Chapa, G., Shaw, G., & Harris, J. (2002). Ambient air pollution and risk of birth defects in Southern California. *American Journal of Epidemiology, 155*(1), 17–25.

Rosenbaum, J. (1991). Black pioneers: Do moves to the suburbs increase economic opportunity for mothers and children? *Housing Policy Debate, 2*(4), 1179–1213.

Rosenbaum, J. (1995). Changing the geography of opportunity by expanding residential choice: Lessons from the Gautreaux program. *Housing Policy Debate, 6*(1), 231–269.

Rosenbaum, J. E., Reynolds, L., & DeLuca, S. (2002). How do places matter? The geography of opportunity, self-efficacy, and a look inside the black box of residential mobility. *Housing Studies, 17*(1), 71–82.

Rosenbaum, E., Harris, L., & Denton, N. (2003). New places, new faces: An analysis of neighborhoods and social ties among MTO movers in Chicago. In J. Goering & J. Feins (Eds.), *Choosing a better life? Evaluating the moving to opportunity experiment* (pp. 275–310). Washington, DC: Urban Institute Press.

Ross, C. E., Mirowksy, J., & Pribesh, S. (2001). Powerlessness and the amplification of threat: Neighborhood disadvantage, disorder, and mistrust. *American Sociological Review, 66*(4), 443–478.

Saha, R. (2009). A current appraisal of toxic wastes and race in the United States. In H. P. Hynes & R. Lopez (Eds.), *Urban health* (pp. 237–260). Sudbury: Jones and Bartlett Publishers.

Sampson, R. J. (1992). Family management and child development. In J. McCord (Ed.), *Advances in criminological theory* (Vol. 3, pp. 63–93). New Brunswick: Transaction Books.

Sampson, R. J. (2001). How do communities undergird or undermine human development? Relevant contexts and social mechanisms. In A. Booth & A. Crouter (Eds.), *Does it take a village? Community effects on children, adolescents and families* (pp. 3–30). London/Mahwah: Lawrence Erlbaum Publishers.

Sampson, R. J., & Groves, W. B. (1989). Community structure and crime: Testing social disorganization theory. *The American Journal of Sociology, 94*(4), 774–802.

Sampson, R., Raudenbush, S., & Earls, F. (1997). Neighborhoods and violent crime: A multilevel study of collective efficacy. *Science, 277*(5328), 918–924.

Sampson, R. J., Morenoff, J., & Earls, F. (1999). Beyond social capital: Spatial dynamics of collective efficacy for children. *American Sociological Review, 64*(5), 633–660.

Sampson, R., Morenoff, J., & Gannon-Rowley, T. (2002). Assessing 'neighborhood Effects': Social processes and new directions in research. *Annual Review of Sociology, 28*, 443–478.

Schaefer-McDaniel, N., Caughy, M. O., O'Campo, P., & Gearey, W. (2009). Examining details of neighborhood observations and the relationship to health: A literature review. *Social Science & Medicine, 70*(2), 277–292.

Schell, L., & Denham, M. (2003). Environmental pollution in urban environments and human biology. *Annual Review of Anthropology, 32*(1), 111–134.

Schill, M. (1997). Chicago's new mixed-income communities strategy: The future face of public housing? In W. Van Vliet (Ed.), *Affordable housing and urban redevelopment in the United States* (pp. 135–157). Thousand Oaks: Sage.

Sellström, E., & Bremberg, S. (2006). The significance of neighborhood context to child and adolescent health and well-being: A systematic review of multilevel studies. *Scandinavian Journal of Public Health, 34*(5), 544–554.

Shouls, S., Congdon, P., & Curtis, S. (1996). Modeling inequality in reported long term illness in the UK. *Journal of Epidemiology and Community Health, 50*(3), 366–376.

Simons, R. L., Johnson, C., Beaman, J. J., Conger, R. D., & Whitbeck, L. B. (1996). Parents and peer group as mediators of the effect of community structure on adolescent behavior. *American Journal of Community Psychology, 24*(1), 145–171.

Sinclair, J. J., Petit, G. S., Harrist, A. W., Dodge, K. A., & Bates, J. E. (1994). Age differences, and correlates of risk behavior problems. *International Journal of Behavioral Development, 17*(4), 675–696.

Small, M. L., & Newman, K. (2001). Urban poverty after the truly disadvantaged: The rediscovery of the family, the neighborhood, and culture. *Annual Review of Sociology, 27*(1), 23–45.

South, S. (2001). Issues in the analysis of neighborhoods, families, and children. In A. Booth & A. Crouter (Eds.), *Does it take a village? Community effects on children, adolescents and families*. London/Mahwah: Lawrence Erlbaum Publishers.

South, S., & Baumer, E. (2000). Deciphering community and race effects on adolescent pre-marital childbearing. *Social Forces, 78*(4), 1379–1407.

Spencer, M. B. (2001). Resiliency and fragility factors associated with the contextual experiences of low-resource urban African-American male youth and families. In A. Booth & A. Crouter (Eds.), *Does it take a village? Community effects on children, adolescents and families* (pp. 51–78). London/Mahwah: Lawrence Erlbaum Publishers.

Stansfeld, S., Haynes, M., & Brown, B. (2000). Noise and health in the urban environment. *Review of Environmental Health, 15*(1), 43–82.

Steptoe, A., & Feldman, P. J. (2001). Neighborhood problems as sources of chronic stress: Development of a measure of neighborhood problems, and associations with socioeconomic status and health. *Annals of Behavioral Medicine, 23*(3), 177–185.

Stillerman, K. P., Mattison, D. R., Giudice, L. C., & Woodruff, T. J. (2008). Environmental exposures and adverse pregnancy outcomes: A review of the science. *Reproductive Sciences, 15*(7), 631–650.

Sullivan, M. L. (1989). *Getting paid: Youth crime and work in the inner city*. Ithaca: Cornell University Press.

Taylor, M. (1998). Combating the social exclusion of housing estates. *Housing Studies, 13*(6), 819–832.
Teitler, J. O., & Weiss, C. C. (1996). *Contextual sex: The effect of school and neighborhood environments on the timing of the first intercourse.* Paper presented at the Annual Meeting of the Population Association of America, New Orleans.
Tiggs, L. M., Browne, I., & Green, G. P. (1998). Social isolation of the urban poor. *The Sociological Quarterly, 39*(1), 53–77.
Turley, R. (2003). When do neighborhoods matter? The role of race and neighborhood peers. *Social Science Research, 32*(1), 61–79.
van Beckhoven, E., & van Kempen, R. (2003). Social effects of urban restructuring: A case study in Amsterdam and Utrecht, the Netherlands. *Housing Studies, 18*(6), 853–875.
van der Klaauw, B., & van Ours, J. (2003). From welfare to work: Does the neighborhood matter? *Journal of Public Economics, 87*(5–6), 957–985.
van Ham, M., & Manley, D. (2010). The effect of neighbourhood housing tenure mix on labour market outcomes: A longitudinal investigation of neighbourhood effects. *Journal of Economic Geography, 10*(2), 257–282.
van Kempen, E. (1997). Poverty pockets and life chances. *American Behavioral Scientist, 41*(3), 430–449.
van Os, J. (2004). Does the urban environment cause psychosis? *The British Journal of Psychiatry, 184*(4), 287–288.
Vartanian, T. P. (1999a). Adolescent neighborhood effects on labor market and economic outcomes. *The Social Service Review, 73*(3), 142–167.
Vartanian, T. P. (1999b). Childhood conditions and adult welfare use. *Journal of Marriage and the Family, 61*(1), 225–237.
Veysey, B. M., & Messner, S. F. (1999). Further testing of social disorganization theory: An elaboration of Sampson and Groves's 'community structure and Crime'. *Journal of Research on Crime and Delinquency, 36*(2), 156–174.
Vrijheid, M. (2000). Health effects of residence near hazardous waste landfill sites: A review of epidemiologic literature. *Environmental health Perspectives, Supplement, 108*(1), 101–112.
Wacquant, L. (1993). Urban outcasts: Stigma and division in the black American ghetto and the French periphery. *International Journal of Urban and Regional Research, 17*(3), 366–383.
Wandersman, A., & Nation, M. (1998). Urban neighborhoods and mental health: Psychological contributions to understanding toxicity, resilience, and interventions. *American Psychologist, 53*(6), 647–656.
Weinberg, B. A., Reagan, P. B., & Yankow, J. J. (2004). Do neighborhoods affect work behavior? Evidence from the NLSY79. *Journal of Labor Economics, 22*(4), 891–924.
Wilson, W. J. (1997). *When work disappears.* New York: Knopf.
Zapata, B. C., Rebolledo, A., Atalah, E., Newman, B., & King, M. (1992). The influence of social and political violence on the risk of pregnancy complications. *American Journal of Public Health, 82*(5), 685–690.
Zenk, S. N., Schulz, A. J., Israel, B. A., James, S. A., Bao, S., & Wilson, M. I. (2005). Neighborhood racial composition, neighborhood poverty, and spatial accessibility of supermarkets in metropolitan Detroit. *American Journal of Public Health, 95*(4), 660–667.

Chapter 3
Ethnographic Evidence, Heterogeneity, and Neighbourhood Effects After Moving to Opportunity

Mario Luis Small and Jessica Feldman

Introduction

Research on neighbourhood effects is at a crossroads. Through the 1990s and early 2000s, researchers in sociology, demography, and economics were overwhelmingly concerned with estimation problems, particularly the problems associated with selection into neighbourhoods (Jencks and Mayer 1990; Goering and Feins 2003; see Small and Newman 2001; Sampson et al. 2002). By the late 1990s, the Moving To Opportunity study—a multi-million dollar randomized control trial that tracked the effects of moving to non-poor neighbourhoods—promised to address many of these concerns and give new life to the neighbourhood effects research program (Goering and Feins 2003). To the surprise of many, the results have been inconsistent, prompting disagreement over the direction the literature should take, and even over whether studying the effects of neighbourhood conditions remains a viable research agenda (see Clampet-Lundquist and Massey 2008; Ludwig et al. 2008; Sampson 2008).

In what follows, we take stock of this work and of the implications of the recent experimental studies. We argue that the first wave of neighbourhood effects research suffered from at least two problems: it assumed that the effect of neighbourhood poverty was homogeneous across subpopulations and across treatment settings, and it failed to integrate effectively ethnographic research into the quantitative empirical research program. These are separate points, but together they help make sense of the findings of MTO and pave the way for a more informed and better-targeted research agenda in neighbourhood effects.

We begin by providing a brief and broad overview of the research on neighbourhood effects up to the MTO experiments. Next, we review several recent studies that suggest that researchers should have expected heterogeneity, not homogeneity in

M.L. Small (✉) • J. Feldman
Department of Sociology, The University of Chicago, Chicago, IL, USA
e-mail: mariosmall@uchicago.edu; jessicaf@uchicago.edu

M. van Ham et al. (eds.), *Neighbourhood Effects Research: New Perspectives*,
DOI 10.1007/978-94-007-2309-2_3, © Springer Science+Business Media B.V. 2012

the effect of neighbourhood poverty—that is, that whether and how neighbourhoods mattered depended substantially on individual-, neighbourhood-, and city-level conditions. Then, we turn to the role of ethnographic evidence, arguing that the better integration between qualitative and quantitative research on poor neighbourhoods can help scholars generate clearer research expectations and interpret seemingly inconsistent findings. We suggest that an assumption of heterogeneity should permeate all aspects of the work, from how questions are asked, to how hypotheses are formulated, to how ethnographic research is conducted and interpreted. We conclude by discussing the implications of our argument for future quantitative and qualitative research on neighbourhoods.

The First Twenty Years of Neighbourhood Effects Research

The start of contemporary research on neighbourhood effects can be traced to the publication of Wilson's *The Truly Disadvantaged* (1987), which argued that American cities had experienced an economic restructuring during the 1970s and 1980s that prompted the departure of both manufacturing jobs and middle class people from the central cities, resulting in urban neighbourhoods with a greater concentration of poverty. Concentrated poverty, in turn, undermined the life chances of the poor. In the years following the book's publication, thousands of studies devoted themselves to addressing one or another of its hypotheses, and many studies specifically tested the proposition that neighbourhood poverty independently affected life chances. As Mayer and Jencks (Mayer and Jencks 1989a, b, p. 1441) wrote in an important early paper, the core hypothesis was that "poor children living in overwhelmingly poor neighbourhoods find it harder to escape poverty than poor children living in more affluent neighbourhoods." Over the years, researchers investigated the consequences of not merely neighbourhood poverty but also other neighbourhood conditions, such as disadvantage and racial homogeneity.

The ensuing neighbourhood effects literature has been canvassed many times and in great detail, obviating the need for an extensive review (Mayer and Jencks 1989a, b; Jencks and Mayer 1990; Brooks-Gunn et al. 1997; Small and Newman 2001; Sampson et al. 2002; Sampson 2008). Nevertheless, several concerns in the literature are worth noting that will prove crucial to our discussion. From its beginnings and through its first 20 years, the literature exhibited three notable concerns: a concern with selection bias, a concern with effects on average, and a concern with mechanisms.

Selection Bias

From the beginning, sceptics have argued that the effort to determine whether neighbourhood poverty affects life chances is threatened by the problem of selection bias (Jencks and Mayer 1990; see also chapters by Manley and van Ham 2011 and Hedman and van Ham 2011 in this volume). Most quantitative tests of neighbourhood effects were (and continue to be) based on observational survey data, often collected at one point in time. These data rarely contain the information required to determine with certainty why different individuals live in different neighbourhoods. For this reason, in statistical regressions, the coefficients for the effect of neighbourhood poverty may be biased due to unobserved conditions (Jencks and Mayer 1990; Tienda 1991; Duncan and Brooks-Gunn 1997; Small and Newman 2001; Goering and Feins 2003; Harding 2003a, b; Ludwig et al. 2008; DeLuca and Dayton 2009). In Clampet-Lundquist and Massey's (2008, p. 109) words, it has been difficult to determine whether "poor places make people poor, or… poor places attract poor people" (Tienda 1991). While, in the early years, researchers differed in the extent to which they worried about this issue, over time the problem became impossible to ignore.

As work in econometrics and causal inference penetrated the field, the selection bias problem came to be understood increasingly within the counterfactual model of causality (e.g. Harding 2003a, b; see Rubin 1974; Morgan and Winship 2007; Morgan 2001). In this model, experiencing neighbourhood poverty is conceived as a treatment, and each individual is assumed to have a potential outcome under the treatment state (living in a poor neighbourhood) and under the control state (living in a non-poor neighbourhood). The causal effect of the treatment for a given individual is simply the difference between her outcomes in the two treatment states. Since it is impossible to observe an individual under two different treatment states—a person cannot simultaneously live in a poor and in a non-poor neighbourhood—causal effects are estimated on average for populations (Morgan and Winship 2007, pp. 4–6). To properly estimate an average causal effect using non-experimental data one must be certain that an unobserved process did not systematically assign different kinds of individuals to different treatment states (or that those differences are ignorable). No solution is better than randomly assigning a large sample of individuals to treatment and control conditions. This approach helped generate scholarly support for randomized control trials, such as Moving To Opportunity, in the context of neighbourhood and housing research. At the same, it convinced many that in the absence of such trials, it was impossible to rule out that all neighbourhood studies that did not employ or approximate random assignment had reported spurious findings (Ludwig et al. 2008).

Effects on Average

In their attempts to discern the true effects of neighbourhood poverty, most studies implicitly assumed that the neighbourhood treatment effect was homogeneous across subpopulations and settings. For Wilson (1987) "concentration effects" were a kind of social fact believed to operate in inner cities across the nation, regardless of local political or cultural context. The early influential papers of Jencks and Mayer (1990; Mayer and Jencks 1989a, b) helped cement this orientation. Reviewing published papers and performing some analyses on their own, the authors tried to ascertain whether the findings collectively suggested that neighbourhoods do, in fact, affect life chances on average. The issue, as the authors framed it in one of their titles, was "Growing up in Poor Neighbourhoods: How Much Does it Matter?" (Mayer and Jencks 1989a, b). The question presupposed the existence of a single answer for any given outcome, regardless of location, context, or other conditions: either neighbourhoods mattered much or they did not. (There was an important exception: the authors speculated that neighbourhood SES would affect people of different SES differently.)

Following these and other early works, an entire generation of researchers concerned themselves with answering either a yes-or-no question (do neighbourhoods matter?) or a question of degree (how much do they matter?)—rather than a conditional question (under what circumstances do they matter?) (Small 2004). This orientation seemed sensible. From the perspective of the traditional, regression-based statistical models that characterized the early literature, it translated into a primary concern with estimating direct, rather than interaction, effects. Furthermore, few theories in the early literature gave researchers reason to pursue a different strategy: neither the early works of Wilson (1987), Jencks and Mayer (1990), Massey and Denton (1993), Sampson and Groves (1989), or others, nor the early ethnographic studies of Liebow (1967), Anderson (1978, 1990, 1999), Duneier (1992), or others gave researchers reason to be substantially concerned with the possibility that the effects of neighbourhood poverty depended on context, that neighbourhood poverty might substantially affect life chances in some but not other circumstances.

Mechanisms

A third overarching concern was to specify the mechanisms through which neighbourhoods affect life chances (Tienda 1991; see also chapter by Galster 2011 in this volume). Researchers have introduced scores of models. In an early review, Mayer and Jencks (Mayer and Jencks 1989a, b; Jencks and Mayer 1990) proposed three. First, having disadvantaged neighbours may affect the poor through either contagion or through the weaker ability to maintain social order. Second, having advantaged neighbours may make the poor feel relative deprivation that encourages an oppositional or deviant subculture. Third, living in a disadvantaged neighbourhood may affect the poor by limiting access to strong institutions and resources. Small and Newman (2001, p.32) reviewed the literature and argued that researchers have proposed two sets of models for how neighbourhood poverty affects life chances: socialization mechanisms, which describe how neighbourhoods socialize young residents, and instrumental mechanisms, which describe how neighbourhoods limit or otherwise affect people's ability to exercise their agency. Through socialization mechanisms, neighbourhood poverty is said to help spread negative behaviour through contagion; expose young people to fewer role models; subject them to discouraging treatment by teachers, officers, and other institutional actors; isolate them linguistically from the mainstream; and encourage them to develop an oppositional culture. Through instrumental mechanisms, neighbourhood poverty is said to limit the number of middleclass people available to meet, the amount of job information available to acquire, and the number of resources available to access. Sampson and colleagues (2002) reviewed over 40 peer-reviewed articles and identified a similar list of mechanisms, in addition to those affecting norms enforcement, collective efficacy, and routine activities.

Other researchers have argued that we should think about mechanisms differently. In a review of MTO findings, Sampson (2008) argued that mechanisms must take into account lifecycle factors—neighbourhoods are likely to matter most among young children, among those who are children and grandchildren of others who lived in poor neighbourhoods, and among those who are exposed for long periods of their life. Galster (2011) argues in his chapter in this volume that neighbourhood effects mechanisms should be thought of in terms of both how they operate and their "dosage." He groups their operation into four broad categories: social-interactive mechanisms, environmental mechanisms, geographical mechanisms, and institutional mechanisms. He then proposes that a pharmaceutical metaphor of "neighbourhood dosage" can help explain how these mechanisms produce individual responses. For example, if we take the proposed social-interactive mechanism of behavioural contagion, the dosage would refer to how often children are exposed to negative behaviours, how long the exposure occurs, and how intense the behaviours are to which children are exposed.

In spite of all this work, it is unclear that much cumulative progress was made on the question of mechanisms. First, researchers disagreed on what constitutes a mechanism. For example, while many of the mechanisms reviewed by Sampson and colleagues (2002) operate at the neighbourhood level (e.g., collective efficacy and informal social control), many of those reviewed by Small and Newman (2001) operate at the individual level (e.g., isolation and oppositional attitudes). Second, these disagreements were largely implicit, as the literature did not debate what constituted a properly specified mechanism or how they should be observed (see Hedstrom and Swedberg 1998; Hedstrom and Ylikoski 2010). Third, given the inability of many earlier studies to account for the selection problems, it was unclear which set of proposed mechanisms—and which set of variations on how mechanisms operate—to give greatest attention to or how.

The three concerns we have identified—with selection, with effects on average, and with mechanisms—did not receive equal focus. The first was an initially neglected issue that soon became an obsession; the second, more a running assumption than an intellectual preoccupation; the third, a persistent worry that never arrived at resolution. Nevertheless, the three help understand both how researchers viewed and how we ought to respond to an important study that marks a turning point in the literature, the Moving to Opportunity randomized control trials.

A Turning Point: The Moving to Opportunity Studies

Largely due to the significance of the selection problem, researchers eagerly awaited results of experimental studies, particularly of the Moving to Opportunity randomized control trials. MTO is a voucher-based housing mobility experiment that intervenes at the level of the household. Beginning in 1994, over 4,600 families living in public housing projects (in neighbourhoods with a poverty rate of at least 40%) in Baltimore, Boston, Chicago, Los Angeles, and New York City were randomly assigned to one of three experimental conditions: (a) the treatment group received vouchers and counselling assistance to move to "opportunity" neighbourhoods — in which the poverty rate was less than 10%; (b) the "Section 8 group" received a voucher but no mobility counselling or restrictions on their movement; (c) and the control group received neither vouchers nor counselling. (For comprehensive reviews of MTO design, history, interim and long-term results, see Goering and Feins 2003; Orr et al. 2003; Kling et al. 2007; Briggs et al. 2010; Sampson 2008; Ludwig et al. 2008; Clampet-Lundquist and Massey 2008). The MTO team collected data on several outcomes: economic self-sufficiency, mental health, physical health, education and risky behaviour (Orr et al. 2003). MTO promised to provide the most compelling test of the effects of neighbourhood poverty.

However, the results were inconsistent (see Orr et al. 2003; Kling et al. 2007; Ellen and Turner 2003; Briggs et al. 2010). While the number of outcomes is too large and the results are too diverse to summarize here, a few findings are worth noting. The interim studies found robust effects on adult mental health outcomes, but limited effects on physical health. On average, movers made few educational gains, and no gains overall in reading and math test scores. King et al. (2007) reported no gains to economic self-sufficiency, a composite of employment, earnings, and welfare use measures (see also Orr et al. 2003). In addition, while female youth saw gains in education, risky behaviour, and physical health outcomes, male youth experienced worse outcomes in all three measures. However, families who moved with vouchers reported greater levels of satisfaction with their living conditions, including neighbourhood attributes such as litter, graffiti, loitering and abandoned property. Voucher movers (both experimental and section 8) reported living in improved quality housing and feeling safer in their neighbourhoods (Orr et al. 2003).

Responses to the findings have ranged widely. For some, the findings confirmed what many had suspected, that early researchers greatly exaggerated the extent to which neighbourhood conditions independently affect life chances (Ludwig et al. 2008). In their mind, the selection bias problem had been as serious as believed by sceptics. (This position, in turn, obviates the need for research on mechanisms, since there is not much of an effect whose internal processes demand attention.) For others, the study violated the assumption of no interference between units, the idea that a participant's value depends only on the treatment to which the participant was assigned, not on that assigned to others. People were sampled from within housing projects in which people are assumed to interact with others with different treatments, leading to misleading results (Sobel 2006). For still others, MTO was not especially informative about neighbourhood effects; rather, it provided evidence to asses a policy intervention of voucher-based housing assistance. For example, Sampson (2008) suggested that the MTO results cannot rule out that neighbourhoods matter: only that neighbourhood poverty, for an extremely disadvantaged and limited sub-sample of the population, in a handful of cities, did not affect a particular set of outcomes during the early 2000s. Furthermore, since movers often either relocated to areas that resembled their previous neighbourhoods or failed to stay in lower-poverty areas, the MTO may understate the true treatment effect. Clampet-Lundquist and Massey (2008) argued that the experimental intervention in the MTO may not have been appropriate, as the definition of "opportunity" neighbourhood was simply one in which less than 10% of the residents were poor, and experimental and voucher movers still ended up in predominantly African American neighbourhoods.

We do not believe that the Moving To Opportunity studies provide evidence that neighbourhood conditions are unimportant. Some findings, such as the effects on mental health and feelings of safety, are robust, consistent, and easily interpretable. Other findings, such as limited effects on unemployment, do not seem surprising, given that changing neighbourhoods does not alter an adult's skill set and the intervention did not require residents to change job markets. Still other findings, such as the improvement in conditions for girls but their worsening for boys simply seem perplexing, particularly because they had not been previously hypothesized.

The principal lesson from the first round of neighbourhood studies through the early findings of MTO seems to be that whether neighbourhood poverty matters depends. The second lesson seems to be that the literature requires more information to interpret these results and more refined hypotheses to focus future research. These two lessons inform our argument that future work should be oriented toward expecting and explaining heterogeneity and that it should better integrate ethnographic fieldwork when generating hypotheses and explanations. We begin with the first of these propositions.

Moving Forward: From Homogeneous to Heterogeneous Treatment Effects

We argue that researchers should assume that neighbourhood poverty has different effects not merely (as many have shown) on different outcomes but also, and more importantly, on different kinds of individuals, in different neighbourhoods, and in different cities (Small 2004, pp. 75–89; Harding et al., forthcoming; also, Sobel 2006). In the traditional regression framework, this argument calls for paying greater attention to the interaction between neighbourhood poverty and individual-, neighbourhood-, and city-level variables. In the counterfactual framework, it calls for assuming that treatment effects are heterogeneous across subpopulations and across settings (see Morgan and Winship 2007; Hong and Raudenbush 2006; Morgan 2001; Angrist 2004). Consider several recent studies demonstrating that whether neighbourhoods matter depends on conditions at the individual, neighbourhood, and city levels.

Individual Level

Several quantitative studies have shown that neighbourhood poverty may affect different individuals differently. For example, using data from the Panel Study of Income Dynamics, Turley (2003) found that higher neighbourhood income was associated with better test scores and behaviour among white children but not among black children. Similarly, as we discussed briefly, the initial evaluation of the MTO found that moving from a poor to non-poor neighbourhood helped females more than males: teen female movers were less likely to get arrested than the control group for violent or property crimes; teen male movers were actually more likely to get arrested for property crimes (Kling et al. 2005, 2007).

Ethnographic studies have also found that neighbourhood poverty may affect different residents differently. Small (2004) studied how neighbourhood poverty affected community participation in a predominantly Puerto Rican housing complex in Boston. He found that whether residents became strongly involved or remained uninterested depended substantially on how they perceived their neighbourhood—those who perceived it as a ghetto found little justification to participate. After conducting in-depth qualitative interviews with a subsample of MTO movers, Briggs and colleagues (2010) found that respondents differed in their orientations to family—some were "kin-centered," others "kin-avoidant" or more oriented to the world of friends and acquaintances. Whether movers were more or less integrated into their new neighbourhoods depended on their familial and social obligations and on the location of those obligations. Along these general lines, Harding and colleagues (forthcoming) proposed that whether neighbourhoods affect life chances depends on the extent to which families have many or few resources. Collectively, these works call for models in which researchers better theorize and then test explicitly the presence of heterogeneity in the effects of neighbourhood conditions across subpopulations.

Neighbourhood Level

Poor neighbourhoods differ, for systematic and non-systematic reasons. For example, some poor neighbourhoods, because of the outmigration of the middle class, exhibit very low population densities; others, because of immigration by the poor, exhibit very high density (e.g., Small 2008). If the treatment is neighbourhood poverty, then the conditions under which the treatment is administered varies substantially across settings—and for reasons related to many outcomes of interest—threatening the validity of statistical inferences.

Several recent studies have made clear that neighbourhood-level differences of this kind are important. In the aforementioned study, Turley (2003) found that, for black children, higher neighbourhood incomes sometimes were related to test scores—only when they lived in those neighbourhoods with a high proportion of blacks. That is, the effect of neighbourhood income was conditional on the neighbourhood proportion black. In a study using data from the Zip Business Patterns and the U.S. Census Small and McDermott (2006) examined the relationship between neighbourhood poverty and the level of organizational density, the number of banks, credit unions, childcare centres, grocery stores, pharmacies, and other everyday establishments. They found that the negative association between neighbourhood poverty and organizational density depended on the proportion of residents in the neighbourhood who were black—the greater the proportion black, the greater the negative association. In fact, the study found that in the statistically average neighbourhood, there was no relationship between neighbourhood poverty and the presence of most of these establishments.

City Level

Finally, several studies suggest that the effect of neighbourhood poverty depends on the city. Burdick-Will and colleagues (forthcoming) reanalyzed student test score data across the five MTO demonstration sites. On average, it appeared that students in the experimental mover conditions fared no better than the control group students on tests of math and reading (evaluated 4–7 years after baseline; see Sanbonmatsu et al. 2006). However, when they disaggregated the educational test results by city, the results differed. The authors compared the treatment-on-treated effect of moving on children's verbal test scores in the full MTO sample, and then separately for Chicago and Baltimore and for New York, Boston, and Los Angeles. While the full sample showed no statistically significant effect, the differences between the first two cities and the latter three were substantial. In fact, in Chicago and Baltimore the effects were strong, with movers performing 0.3 standard deviations better than the control group. While differences in math test scores were not statistically significant between movers and non-movers for either set of cities, the difference between the sets of sites was large, of comparable magnitude, and in the same direction as the results of the verbal scores. That is, neighbourhood effects on children's performance were more likely to be found in Chicago and Baltimore. Burdick-Will and colleagues noted that in Chicago and Baltimore, residents of poor neighbourhoods were more likely to be exposed to extreme levels of violence (see also Sharkey 2009).

Our own research also demonstrates the extent to which the effects of neighbourhood poverty depend on city-level conditions. For this chapter, we prepared a test of the de-institutionalization hypotheses that produces similar results and makes clear the importance of heterogeneity across treatment settings. Wilson (1987) and others hypothesized that concentrated poverty undermines organizational density—the number of banks, clinics, bowling alleys, churches, recreation centres, and other commercial and non-profit establishments within a neighbourhood. As Wilson has argued, "poverty in ghetto neighbourhoods has sapped the vitality of local business and other institutions, and it has led to fewer… movie theatres, bowling alleys, restaurants, public parks and playgrounds, and other recreational facilities" (1995, pp. 9–10; also Wilson 1987, 1996; Wacquant 2007; Messner and Rosenfeld 2001).

However, Wilson and others derived their hypothesis based on work primarily in Chicago, without considering whether other cities would exhibit different relationships. In fact, as we discuss later, a long standing tradition has conceived of Chicago as a laboratory where phenomena occurring in the average city, or the average large city, or the average Rustbelt city, can be observed with clarity. A perspective taking heterogeneity seriously would question that assumption.

We tested Wilson's hypothesis based on data on the presence of a range of establishments for all metropolitan statistical areas of the continental United States. Based on an extensive review of the literature, we selected small establishments (fewer than 20 employees) and large ones (100 or more). Among small establishments, we identified five general types: day-to-day establishments (hardware stores, grocery stores, convenience stores, pharmacies, banks, credit unions, full- and limited-service restaurants, childcare centres, snack centres, laundries, grooming centres); small medical establishments (physicians' clinics, mental health physicians' clinics, dentists' offices, and other small offices of mental health), social service establishments (childcare centres, child and youth services, services for the elderly and disabled, and other individual and family services); recreational establishments (movie theatres, fitness and recreational sports centres, and bowling alleys); and social establishments (religious institutions, full- and limited-service restaurants, cafeterias, book stores, childcare centres, snack centres, and bars). Among large establishments we studied large hardware stores (100–999 employees), large grocery stores (100–499 employees), and large medical establishments (100 or more employees).

Consider Table 3.1, which exhibits the number of establishments per 100,000 residents in poor neighbourhoods in Chicago, and in other cities. Poor neighbourhoods are defined as zip codes in which 30% or more of the population is poor. Items in bold are composites based on the sum of the non-bold items immediately below them. The table shows that poor neighbourhoods in Chicago have fewer establishments per 100,000 than poor neighbourhoods in the average city. In fact, poor neighbourhoods in Chicago are consistently below the mean for each of the major establishment types and almost every one of the sub-types. For example, the numbers in bold indicate that the average Chicago poor neighbourhood has 120 day-to-day establishments per 100,000 residents, while the average poor neighbourhood in the average city has 220; with respect to small medical establishments, the figures are 36 for Chicago and 104 for the average city. The pattern holds for social service establishments (29 in Chicago vs. 50 in the average city), small recreational establishments (less than one in Chicago, 4 in the average city), small social establishments (99 vs. 217), large grocery stores (slightly less than 1 vs. slightly more than 1), and hospitals (2.6 vs. 3.5). In short: Chicago poor neighbourhoods are substantially less organizationally dense than the average poor neighbourhood. Perhaps surprisingly, poor neighbourhoods in Chicago also differ from those in Rustbelt cities, the subcategory of cities that many scholars suggest Chicago represents and resembles most. The Chicago figures most resemble the ten largest cities, but its poor neighbourhoods are still less organizationally dense on average.

Table 3.1 Mean number of establishments per 100,000 residents, high poverty zip codes, 2000

	Chicago	All cities	Ten largest cities	Rustbelt cities
Small day-to-day establishments	**120.21**	**219.85**	**179.67**	**209.04**
Hardware stores	3.55	3.46	3.32	2.57
Grocery stores	32.78	33.11	37.38	35.39
Convenience stores	4.70	14.29	11.82	16.16
Pharmacies	10.04	11.47	11.57	10.91
Banks	0.83	2.97	1.45	3.03
Credit unions	2.27	11.21	4.62	12.71
Childcare centers	16.00	23.07	14.67	21.64
Full-service restaurants	11.82	43.68	32.67	36.70
Limited-service restaurants	23.18	46.07	34.64	41.68
Cafeterias	0.13	2.18	1.40	1.83
Snack and non-alcoholic beverage centers	3.30	9.32	8.22	8.65
Laundries	8.31	4.94	7.08	4.81
Grooming centers	3.31	14.08	10.83	12.96
Small medical establishments	**36.00**	**104.35**	**79.61**	**87.44**
Offices of physicians (exc. mental health)	25.99	74.69	59.66	62.52
Offices of physicians, mental health	2.27	4.44	2.94	3.70
Offices of dentists	6.87	22.13	16.04	18.31
Offices of other mental health practitioner	0.86	3.09	0.97	2.91
Small social service establishments	**29.30**	**50.29**	**32.51**	**55.40**
Childcare centers	16.00	23.07	14.67	21.64
Child and youth services	3.97	5.84	4.32	7.88
Services for elderly and disabled persons	2.20	6.28	4.38	7.02
Other individual and family services	7.13	15.10	9.13	18.87
Small recreational establishments	**0.39**	**3.93**	**2.73**	**3.72**
Movie theaters	0.00	0.53	0.59	0.78
Fitness and recreational sports centers	0.22	3.18	1.92	2.50
Bowling alleys	0.17	0.23	0.22	0.44
Small social establishments	**98.96**	**216.79**	**145.12**	**228.88**
Religious orgs (e.g. churches, mosques)	34.03	60.16	33.79	77.23
Full-service restaurants	11.82	43.68	32.67	36.70
Limited-service restaurants	23.18	46.07	34.64	41.68
Cafeterias	0.13	2.18	1.40	1.83
Book stores	0.99	4.44	3.11	4.77
Childcare centers	16.00	23.07	14.67	21.64
Snack and non-alcoholic beverage centers	3.30	9.32	8.22	8.65
Alcoholic beverage drinking places	9.52	27.86	16.61	36.39
Large hardware stores	**0.00**	**0.01**	**0.00**	**0.00**
Large grocery stores	**0.74**	**1.33**	**0.98**	**1.03**
Large medical establishments	**2.57**	**3.48**	**2.54**	**3.48**
HMO medical centers	0.26	0.09	0.06	0.17
General medical and surgical hospitals	2.31	2.98	1.85	3.15
Other specialty hospitals	0.00	0.42	0.63	0.16

Source: Zip Business Patterns, 2000. U.S. Census. Zip code-level data GIS-matched to 331 metropolitan areas. Figures limited to zip codes smaller than 100 square miles with more than 100 residents, and no more than 1,000 day-to-day establishments per 100,000 residents, and those in which 30% or more of all persons are poor. There are 10 such zip codes in Chicago; 517 in all cities combined; 128 in the 10 most populous metro areas; and 124 in cities that are located in Rustbelt states (IN, IL, WI, OH, MI, MO, MN, IA, KY, WV). Small establishments have fewer than 20 employees. Large hardware stores have between 100 and 999 employees; large grocery stores have between 100 and 499 employees; large medical establishments have 100 or more employees. Categories listed in bold font are additive indices of the establishments listed beneath the category heading

Table 3.2 Difference in organisational density between poor and non-poor zip codes, 2000

	Number of establishments per 100,000 persons in poor zip codes minus non-poor zip codes			
	Chicago	All cities	Ten largest cities	Rustbelt cities
Small day-to-day establishments	–66.22	37.61	–16.94	44.23
Small medical establishments	–78.08	13.23	–43.85	16.72
Small social service establishments	4.60	21.72	3.50	28.94
Small recreational establishments	–7.53	–4.62	–6.20	–4.28
Small social establishments	–76.79	33.44	–26.37	36.47
Large hardware stores	0.00	0.00	–0.01	–0.01
Large grocery stores	–2.56	–1.27	–1.86	–1.17
Large medical establishments	1.72	2.38	1.47	2.47

Source: See Table 3.1

Table 3.2 compares organizational density between poor and non-poor neighbourhoods. It displays, in summary form, the number of establishments per 100,000 in poor zip codes minus the number in non-poor zip codes for Chicago, the country, Rustbelt cities and the ten largest cities. As shown in the table, small and large establishments differ. With respect to small establishments, poor neighbourhoods in Chicago exhibit lower organizational density than non-poor neighbourhoods. In the average U.S. city, however, the opposite is true, as is the case in Rustbelt cities. That is to say, Wilson and others accurately observed conditions in Chicago, but these observations were improperly thought to be representative of the average U.S. city and Rustbelt cities. Researchers should have hypothesized that the effect of neighbourhood poverty on organizational density depends on the city.

In sum, there is abundant evidence to suggest that heterogeneity across subpopulations and treatment settings should be a starting assumption, rather than an afterthought, in neighbourhood effects research. The MTO results can be read as confirming this fundamental thesis. Do neighbourhoods matter? The only sensible answer at this juncture is that it depends. The next phase of the literature should be devoted to understanding under what conditions neighbourhood poverty matters.

Moving Forward: Better Integrating Ethnographic Research

In addition to encouraging a concern for heterogeneity, the new phase of neighbourhood effects research calls for integrating ethnographic research more tightly into the research agenda. By "ethnographic research" we refer to both open-ended interviews and participant observation conducted by researchers interacting one-on-one with people, organizations, and public places in urban neighbourhoods. In the neighbourhood effects field, ethnographers often cite demographers or economists and vice versa. Nevertheless, we argue that building the cumulative knowledge that overcomes past limitations in neighbourhood effects research requires developing a more symbiotic relationship between quantitative and qualitative research, a relationship that, over the long run, should resemble a mixed method, rather than merely multi-method, enterprise (Tashakkori and Teddlie 2003).

Two Roles for Ethnographic Research

Ethnographic work should be integrated into the research agenda in at least two general forms: to help explain the results of prior to studies and to help generate hypotheses for future ones. We discuss each in turn.

First, ethnographic studies should be deployed to help explain the results of prior quantitative work, since only such studies provide direct access to the conditions of poor neighbourhoods and their residents' interpretations of these conditions. To the extent that researchers seek to identify the mechanisms that produced an observed effect or account for the absence of an expected relationship, they require access to how residents of poor neighbourhoods make decisions about their circumstances, a process impossible to capture fully without fieldwork. The first phase of neighbourhood effects research, culminating in the MTO studies, has produced a plethora of findings, many of them inconsistent and even contradictory, in need of explanation. Why did girls fare better than boys? Why was organizational density lower in black poor neighbourhoods but not in other poor neighbourhoods? Why did movers improve test scores in Baltimore and Chicago but not in New York, Boston, and Los Angeles? Why was neighbourhood income positively associated with white but not black children's test scores? Hundreds of findings call for explanation.

Ethnographic studies have played a rather limited role in that undertaking. Many quantitative researchers have relied less on ethnographies than on existing theory to explain their results, and many qualitative researchers have neglected to propose explanations for the results of quantitative studies. In fact, most of the ethnographic studies produced in the 1990s and early 2000s did not propose interpretations of the collective, accumulating findings of the neighbourhood effects literature (but see Pattillo-McCoy 1999; Small 2004: Harding 2010). And with the notable exception of Briggs and colleagues (2010; see also DeLuca et al. 2011 in this volume), most ethnographers did not design their research projects to help explain the particular results of previous large-scale studies, in spite of the contradictions in the quantitative findings that were arising early on. Fully making sense of the current slew of quantitative neighbourhood effects findings will prove difficult without deeply observing and interviewing people in their neighbourhoods and cities.

Second, ethnographic studies should play a central role in the process of developing hypotheses, since hypotheses based strictly on theoretical reflection, rather than at least some empirical engagement, face the risk that anthropologists have long attributed to arm-chair theories: they generate expectations that, after the fact, appear to be obviously misguided. For example, the fact that the MTO studies found no effects on unemployment seems unsurprising in retrospect, since participants were not required to change job markets. At the same time, however, no ethnographic studies had suggested that low-income job seekers would find the job hunt easier if they moved to neighbourhoods that had lower poverty rates but not necessarily more jobs. Prior fieldwork has proven repeatedly to help discipline the theorist's imagination.

To date, ethnographies have not been integral to the specific hypotheses tested in neighbourhood effects research. To be sure, several ethnographic studies have played some role. For example, Wilson (1987) explained that many of his ideas about the effects of concentrated poverty on organizational capacity derived from his personal observations of the South Side of Chicago (also Wacquant and Wilson 1989). In addition, the more systematically ethnographic studies of Edin and Lein (1997), Duneier (1992), Klinenberg (2002), Newman (1999), Pattillo-McCoy (1999), and Venkatesh (2000), have also been cited in much of the work on neighbourhood effects (see Newman and Massengill 2006; Sampson et al. 2002). Nevertheless, many of the hypotheses tested in the first wave of neighbourhood effects research, including those tested by MTO, did not seem to have much grounding in ethnographic fieldwork. For example, when proposing the mechanisms through which neighbourhoods should affect life chances, Jencks and Mayer (1990) referred to few ethnographies (in part because these lacked the comparative assessments being reviewed in their study). Similarly, in their paper generating hypotheses for why neighbourhoods might matter, Ellen and Turner (2003) proposed mechanisms such as lower quality of local institutions, weaker norms and collective efficacy, and ineffective social networks, but cited virtually no ethnographies to inform these conjectures. Along these lines, when hypothesizing the possible pathways through which neighbourhood poverty may affect youth outcomes in their New York City MTO evaluation, Leventhal and Brooks-Gunn (2003) relied on well-established theoretical models but not on ethnographic field research. And in their studies of the effects of neighbourhood disadvantage on collective efficacy and other outcomes, Sampson and colleagues derive their hypotheses less from current ethnographic studies than from prior theories and large-n studies of the neighbourhood-level precursors of crime (Shaw and McKay 1942; see Sampson et al. 1997).

We must also note, however, that ethnographic studies had only rarely produced concrete hypotheses for quantitative studies to evaluate. On the contrary, while many ethnographers dating back to the 1960s and 1970s painted vivid pictures of conditions in poor urban neighbourhoods (Liebow 1967; Hannerz 1969; Stack 1974; Anderson 1978), few of them structured their work with the express purpose of generating the explicit, testable hypotheses that guide large-n research. At this juncture, the field beckons for such hypotheses from fieldworkers.

We see little hope of progress in the neighbourhood effects literature unless these two approaches to qualitative work—accounting for the results of prior quantitative research and generating hypotheses for future quantitative studies—become more central than they have been to the larger enterprise. The contradictory findings of MTO and prior studies call for a coherent interpretation, or set of interpretations, which require a strong grounding in carefully targeted ethnographic research. Furthermore, the cacophony of ideas about how neighbourhood matters—including

Table 3.3 Role of ethnographic research under different assumptions about nature of neighbourhood effects

	Homogeneity assumption	Heterogeneity assumption
Explain existing results (post quantitative work)	Explain observed effects	Explain why effects are observed in some settings but not others, and for some populations but not others
Generate new hypotheses (pre quantitative work)	Hypothesise how neighbourhood poverty will affect life chances	Hypothesise for which populations and under which settings neighbourhood poverty will affect life chances

a large and ever-growing set of notions about how to conceive of the underlying mechanisms of neighbourhood effects—has greater chances of arriving at some harmony if it is disciplined by a strong engagement with the field. Of the (now) scores of extant hypotheses about how neighbourhoods matter, those supported by targeted fieldwork should be prioritized strongly by researchers.

Ethnographic Data and Heterogeneity

Nevertheless, while the findings of the first major wave of neighbourhood effects research call for integrating ethnography more effectively, this integration is unlikely to prove useful unless researchers also change their practical and theoretical assumptions toward a model in which, until proven otherwise, treatment effects are assumed to be heterogeneous and relationships are assumed to be conditional. This reorientation affects what questions are asked, what answers are expected, how cities and neighbourhoods and individuals are selected for study, how hypotheses are produced, what kinds of hypotheses are generated, and, in general, how neighbourhoods are expected to affect behaviour.

Consider Table 3.3. The left column represents the orientation toward neighbourhood effects that, from both quantitative and qualitative perspectives, has characterized most of the literature until recently. Under that model, if ethnographies were more tightly integrated to quantitative research, they would take the following form. When developing explanations for observed neighbourhood effects, ethnographers would begin with the discovery of direct effects (such as greater fear of safety), find an assumed representative city (such as Chicago), select an assumed representative poor neighbourhood (such as Woodlawn on the South Side) or population (such as black poor mothers), and try to find the mechanisms linking neighbourhood poverty to the outcome (the reasons those who feel unsafe seem to do so).

From the perspective on the left column, only this procedure would ensure an explanation likely to be applicable regardless of context. When generating hypotheses for future quantitative studies, researchers would also find an assumed representative city, select an assumed representative poor neighbourhood, housing project, or population; and generate hypotheses about how neighbourhood poverty affects a given outcome—for example, neighbourhood poverty would be hypothesized to increase social isolation (outcome) by increasing distrust (mechanism). From the perspective on the left column, this procedure would help ensure the generalizability of the proposition.

The core assumptions behind that model, particularly the idea that researchers should begin by identifying a representative city, are deeply entrenched in the literature. First, that model is a manifestation of one of the tenets of the Chicago School of Sociology: cities take form as a result of inherent ecological processes, and the city of Chicago is an ideal "laboratory" for observing them. Because of this assumption, and the long-standing tradition of urban research at the University of Chicago, an overwhelming number of studies in the neighbourhood effects literature are based in the city of Chicago. Most of the ethnographic research cited in the neighbourhood effects literature was conducted, like Wilson's work, in Chicago (Wacquant and Wilson 1989; Duneier 1992; Pattillo-McCoy 1999; Venkatesh 2000; Klinenberg 2002). In fact, some of the most highly used datasets to examine neighbourhood effects are sited in Chicago. The important Project on Human Development in Chicago Neighborhoods (PHDCN) has been used to support many of the propositions that neighbourhoods matter (Sampson et al. 1997). The earlier Urban Poverty and Family Life Survey, which was the basis of Wilson (1996) was also fielded in Chicago. In fact, many of the multi-city studies, such as the Multi-City Study of Urban Inequality, MTO, and the Three-City Study, included Chicago in their samples. No American city has been studied more carefully by urban social scientists, in part because of the idea that it exhibits universal phenomena. Second, the idea that researchers should identify a representative neighbourhood is a natural extension of the logic of demographic quantitative research—select a sample that is representative—applied to ethnography (e.g. King et al. 1994; for a critique, see Small 2009). In the ethnographic tradition, researchers have rarely selected neighbourhoods presumed to be representative in a statistical sense (but see McDermott 2006); most commonly, they have studied neighbourhoods that seemed to typify concentrated poverty, such as large public housing projects or predominantly black poor neighbourhoods (e.g., Venkatesh 2000; Wacquant 2007; see Small 2004, 2007, 2008). Either way, the model has been to look for cases that appear to be typical poor neighbourhoods.

Despite its popularity, continuing to operate under that model for producing explanations and hypotheses will help neither the integration of ethnographies nor the progress of literature, since it remains mired in the assumptions of homogeneous treatment effects, unconditional relationships, and inherent neighbourhood poverty forces that led researchers to overestimate what the MTO would show and has left them at a loss to explain a slew of disparate findings.

By contrast, consider the right column of Table 3.3, which illustrates the heterogeneity assumption that, we argue, is called for by the most recent wave of research on neighbourhood effects and should inform ethnographic study. Rather than centre solely or even primarily on direct effects, researchers would probe the extent to which the effects of neighbourhood poverty depend on city-, neighbourhood-, or individual-level conditions. From the traditional regression perspective, it may seem curious to pursue what appear to be interaction effects in favour of direct effects, but the first 20 years of work clearly suggest that whether neighbourhood poverty matters depends on the circumstances. From the counterfactual perspective, our proposition probably appears more straightforward; it calls for assuming heterogeneity across subpopulations and across treatment settings unless proven otherwise. This heterogeneity—e.g., that poor neighbourhoods affect boys and girls differently, that some neighbourhoods are depopulated while others overcrowded, that those in New York differ from those in Chicago—which once appeared secondary, can no longer be considered unimportant; it should form the foundation of ethnographers' efforts to provide explanations and identify the mechanisms underlying observed associations. Ethnographers should begin, for example, with the fact that the MTO treatment group improved verbal scores in Baltimore and Chicago but not in New York, Boston, or Los Angeles (Burdick-Will et al. forthcoming). As Small (2004, p. 176) argued based on research in a Puerto Rican housing complex in Boston, researches should "use heterogeneity in responses to neighbourhood poverty as the starting point rather than [something] to ignore…." Observed differences between populations, neighbourhoods, and cities should constitute the point of departure for those developing explanations.

Furthermore, as we indicate in the bottom right panel of Table 3.3, ethnographers generating new hypotheses should abandon altogether the effort to produce hypotheses about how neighbourhoods (universally) affect life chances, and instead hypothesize for which kinds of individuals and in which kinds of neighbourhoods or cities neighbourhood poverty should matter (Small 2004). The question of how to generate hypotheses based on fieldwork gains poignancy when selecting sites for analysis. An ethnographer operating under the old model would seek neighbourhoods and cities that appeared representative. The fact that ethnographies, by necessity, are usually limited in scope to one or two neighbourhoods in one or two cities resulted in the persistent worry that the cases might not be representative or typical.

An ethnographer operating under the heterogeneity model differs in several respects. First, the ethnographer who assumes that effects, to greater or lesser extent, depend on conditions at the city, neighbourhood, and individual levels must seek comparisons at one or more of those levels. Comparative designs become favoured. Second, since no city, neighbourhood, or individual is assumed to capture the essence of neighbourhood effects, the ethnographer would abandon the notion that any site or kind of actor is representative. Rather than studying Chicago on the assumption that it bears evidence to universal phenomena, the researcher would study under-explored cities such as Memphis, Pittsburgh, Phoenix, Eugene, New Haven, or Denver, under the assumption that conditions may differ substantially from oft-repeated stories and that these differences may condition the effects of neighbourhood poverty. Rather than studying, or only studying, predominantly black housing projects in areas losing low-skilled manufacturing jobs, the researcher would study, or also study, comparatively under-explored sites such as Chinese-American neighbourhoods with high proportions of poor immigrants, predominantly white poor neighbourhoods with high levels of drug abuse, or aging multi-ethnic neighbourhoods with high proportions of residents on fixed retirement incomes. These differences form the bases for new hypotheses about how neighbourhoods matter. Third, the ethnographer adopting this model would generate a different kind of hypothesis. Since the objective is to hypothesize the causes behind differences, the hypotheses would necessarily focus on the mechanisms through which either different kinds of individuals respond to the stimulus of neighbourhood poverty (in the case of heterogeneity across subpopulations) or different kinds of neighbourhood or city contexts alter the stimulus (in the case of heterogeneity across settings). Rather than providing expectations about inherent descriptive traits, they would provide expectations about the mechanisms generating differences.

Conclusion

We have argued that the conflicting and inconsistent results of the first wave of studies on neighbourhood effects call for placing heterogeneity at the centre of the research agenda and for better mobilizing ethnographic research to explain these results and generate new expectations. We have also argued that better integrating ethnographies requires that this process, too, be structured toward heterogeneity, toward explaining why different kinds of actors respond differently to neighbourhood poverty, why poor neighbourhoods of different kinds exhibit different patterns, and why poor neighbourhoods in different cities vary so substantially and appear to affect individuals differently. In this endeavour, the old notion that particular kinds of neighbourhoods or cities—notably Chicago—represent ideal laboratories to observe universal social processes is increasingly unhelpful.

The early phase of neighbourhood effects research posed a clear set of questions—does neighbourhood poverty affect life chances (net of selection bias), and, if so, how?—that provided a clear target for a diverse set of sociologists, geographers, developmental psychologists, demographers, and economists to pursue. The first of these questions helped generate strong academic support for an ambitious randomized control trial whose results were eagerly awaited. It seems increasingly clear, however, that much of this work would be better served if informed by clearer theories and stronger fieldwork, and that the greater empirical payoffs would have been found in the search for conditional relationships. As researchers pursue new kinds of questions, they would do well to transcend the limitations of the past.

Acknowledgements The authors would like to thank the University of Chicago for generous research support and Julia Burdick-Will and the editors of the volume for valuable comments on an earlier version of this paper. All errors remain our own.

References

Anderson, E. (1978). *A place on the corner*. Chicago: University of Chicago Press.
Anderson, E. (1990). *Streetwise: Race, class and change in an urban community*. Chicago: University of Chicago Press.

Anderson, E. (1999). *Code of the street: Decency, violence and the moral life of the inner city*. New York: W.W. Norton.

Angrist, J. D. (2004). Treatment effect heterogeneity in theory and practice. *The Economic Journal, 114*(494), C52–C83.

Briggs, X., Popkin, S., & Goering, J. (2010). *Moving to opportunity: The story of an American experiment to fight Ghetto poverty*. New York: Oxford University Press.

Brooks-Gunn, J., Duncan, G., & Aber, J. L. (1997). *Neighborhood poverty: Context and consequences for children*. New York: Russell Sage Foundation.

Burdick-Will, J., Ludwig, J., Raudenbush, S. W., Sampson, R. J., Sanbonmatsu, L., & Sharkey, P. (forthcoming). *Converging evidence for neighborhood effects on children's test scores: An experimental, quasi-experimental, and observational comparison*. Brookings Institution.

Clampet-Lundquist, S., & Massey, D. S. (2008). Neighborhood effects on economic self-sufficiency: A reconsideration of the moving to opportunity experiment. *The American Journal of Sociology, 114*(1), 107–143.

DeLuca, S., & Dayton, E. (2009). Switching social contexts: The effects of housing mobility and school choice programs on youth outcomes. *Annual Review of Sociology, 35*, 457–91.

DeLuca, S., Duncan, G. J., Keels, M., & Mendenhall, R. (2011). The notable and the null: Using mixed methods to understand the diverse impacts of residential mobility programs 2011. In M. van Ham, D. Manley, N. Bailey, L. Simpson, & D. Maclennan (Eds.), *Neighbourhood effects research: New perspectives*. Dordrecht: Springer.

Duncan, G., & Brooks-Gunn, J. (1997). *The consequences of growing up poor*. New York: Russell Sage Foundation.

Duneier, M. (1992). *Slim's table*. Chicago: University of Chicago Press.

Edin, K., & Lein, L. (1997). *Making ends meet: How single mothers survive welfare and low-wage work*. New York: Russell Sage Foundation.

Ellen, I. G., & Turner, M. (2003). Do neighborhoods matter and why? In J. Goering & R. Feins (Eds.), *Choosing a better life? Evaluating the moving to opportunity experiment* (pp. 313–338). Washington, DC: Urban Institute Press.

Galster, G. (2011). The mechanism(s) of effects: Theory, evidence and policy implications. In M. van Ham, D. Manley, N. Bailey, L. Simpson, & D. Maclennan (Eds.), *Neighbourhood effects research: New perspectives*. Dordrecht: Springer.

Goering, J., & Feins, J. D. (2003). *Choosing a better life? Evaluating the moving to opportunity social experiment*. Washington DC: Urban Institute Press.

Hannerz, U. (1969). *Soulside: Inquiries into Ghetto culture and community*. New York: Columbia University Press.

Harding, D. J. (2003a). Counterfactual models of neighborhood effects: The effect of neighborhood poverty on high school dropout and teenage pregnancy. *The American Journal of Sociology, 109*(3), 676–719.

Harding, D. J. (2003b). *Living the drama: Community, conflict, and culture among inner-city boys*. Chicago: University of Chicago Press.

Harding, D. J., Gennetian, L., Winship, C., Sanbonmatsu, L., & Kling, J. (forthcoming). *Unpacking neighborhood influences on education outcomes: Setting the stage for future research*. In G. Duncan & R. Murnane (Eds.), Social inequality and educational disadvantage. New York: Russell Sage Foundation.

Hedman, L., & van Ham, M. (2011). Understanding neighbourhood effects: Selection bias and residential mobility. In M. van Ham, D. Manley, N. Bailey, L. Simpson, & D. Maclennan (Eds.), *Neighbourhood effects research: New perspectives*. Dordrecht: Springer.

Hedstrom, P., & Swedberg, R. (1998). *Social mechanisms: An analytical approach to social theory*. New York: Cambridge University Press.

Hedstrom, P., & Ylikoski, P. (2010). Causal mechanisms in the social sciences. *Annual Review of Sociology, 36*(1), 49–67.

Hong, G., & Raudenbush, S. W. (2006). Evaluating Kindergarten Retention Policy: A case study of casual inference for multi-level observational data. *Journal of the American Statistical Association, 101*(474), 901–910.

Jencks, C., & Mayer, S. (1990). The social consequences of growing up in a poor neighborhood. In L. Lynn & M. McGreary (Eds.), *Inner city poverty in the United States* (pp. 111–186). Washington DC: National Academy Press.

King, G., Keohane, R. O., & Verba, S. (1994). *Designing social inquiry*. Princeton: Princeton University Press.

Klinenberg, E. (2002). *Heat wave*. Chicago: University of Chicago Press.

Kling, J., Ludwig, J., & Katz, L. F. (2005). Neighborhood effects on crime for female and male youth: Evidence from a randomized housing mobility experiment. *Quarterly Journal of Economics, 120*(1), 87–130.

Kling, J., Liebman, J., & Katz, L. (2007). Experimental analysis of neighborhood effects. *Econometrica, 75*(1), 83–119.

Leventhal, T., & Brooks-Gunn, J. (2003). New York City site findings: The early impacts of moving to opportunity on children and youth. In J. Goering & J. D. Feins (Eds.), *Choosing a better life?* (pp. 213–243). Washington, DC: Urban Institute Press.

Liebow, E. (1967). *Tally's corner: A study of negro streetcorner men*. Boston: Little, Brown.

Ludwig, J., Liebman, J. B., Kling, J. R., Duncan, G. J., Katz, L. F., Kessler, R. C., & Sanbonmatsu, L. (2008). What can we learn about neighborhood effects from the moving to opportunity experiment? *The American Journal of Sociology, 114*(1), 189–231.

Manley, D., & van Ham, M. (2011). Neighbourhood effects, housing tenure and individual employment outcomes. In M. van Ham, D. Manley, N. Bailey, L. Simpson, & D. Maclennan (Eds.), *Neighbourhood effects research: New perspectives*. Dordrecht: Springer.

Massey, D., & Denton, N. (1993). *American apartheid*. Cambridge: Harvard University Press.

Mayer, S. E., & Jencks, C. (1989a). Growing up in poor neighborhoods: How much does it matter? *Science, 243*(4934), 1441–1445.

Mayer, S. E., & Jencks, C. (1989b). Growing up in poor neighborhoods: How much does it matter? *Science, 243*(4897), 1441–1445.

McDermott, M. (2006). *Working class white*. Berkeley: University of California Press.

Messner, S. F., & Rosenfeld, R. (2001). *Crime and the American dream* (3rd ed.). Belmont: Wadsworth.

Morgan, S. (2001). Counterfactuals, causal effect heterogeneity, and the catholic school effect on learning. *Sociology of Education, 74*(4), 341–374.

Morgan, S., & Winship, C. (2007). *Counterfactuals and causal inference*. New York: Cambridge University Press.

Newman, K. (1999). *No shame in my game*. New York: Knopf.

Newman, K., & Massengill, R. P. (2006). The texture of hardship: Qualitative sociology of poverty, 1995–2005. *Annual Review of Sociology, 32*, 423–446.

Orr, L., Feins, J. D., Jacob, R., Beecroft, E., Sanbonmatsu, L., Katz, L. F., Liebman, J. B., & Kling, J. R. (2003). *Moving to opportunity interim impacts evaluation*. Washington DC: U.S. Department of Housing and Urban Development.

Pattillo-McCoy, M. (1999). *Black picket fences*. Chicago: University of Chicago Press.

Rubin, D. (1974). Estimating causal effects of treatments in randomized and nonrandomized studies. *Journal of Educational Psychology, 66*(5), 688–701.

Sampson, R. (2008). Moving to inequality: Neighborhood effects and experiments meet social structure. *The American Journal of Sociology, 114*(1), 189–231.

Sampson, R. J., & Byron Groves, W. B. (1989). Community structure and crime: Testing social-disorganization theory. *The American Journal of Sociology, 94*(4), 774–802.

Sampson, R., Morenoff, J., & Gannon-Rowley, T. (2002). Assessing 'Neighborhood Effects': Social processes and new directions in research. *Annual Review of Sociology, 28*, 443–4778.

Sampson, R. J., Raudenbush, S., & Earls, F. (1997). Neighborhoods and violent crime: A multilevel study of collective efficacy. *Science, 277*(5328), 918–24.

Sanbonmatsu, L., Kling, J. R., Duncan, G. J., & Brooks-Gunn, J. (2006). Neighborhoods and academic achievement: Results from the MTO experiment. *Journal of Human Resources, 41*(4), 649–691.

Sharkey, P. (2009). *The acute effect of local homicides on children's cognitive performance.* Working Paper, University, New York.
Shaw, C., & McKay, H. ([1942] 1969). *Juvenile delinquency and urban areas.* Chicago: University of Chicago Press.
Small, M. L. (2004). *Villa Victoria: The transformation of social capital in a Boston Barrio.* Chicago: University of Chicago Press.
Small, M. L. (2007). Is there such a thing as 'The Ghetto': The perils of assuming that the south side of Chicago represents poor black neighborhoods. *City, 11*(3), 413–21.
Small, M. L. (2008). Four reasons to abandon the idea of 'The Ghetto'. *City and Community, 7*(4), 389–398.
Small, M. L. (2009). 'How Many Cases Do I Need?' On science and the logic of case selection in field based research. *Ethnography, 10*(1), 5–38.
Small, M. L., & McDermott, M. (2006). The presence of organizational resources in poor urban neighborhoods: An analysis of average and contextual effects. *Social Forces, 84*(3), 1697–1724.
Small, M. L., & Newman, K. (2001). Urban poverty after the truly disadvantaged: The rediscovery of the family, the neighborhood and culture. *Annual Review of Sociology, 27*(1), 23–45.
Sobel, M. E. (2006). What do randomized studies of housing mobility demonstrate?: Causal inference in the face of interference. *Journal of the American Statistical Association, 101*(476), 1398–1407.
Stack, C. (1974). *All our kin.* New York: Basic Books.
Tashakkori, A., & Teddlie, C. (2003). *Handbook of mixed methods in social and behavioural research.* Thousand Oaks: Sage.
Tienda, M. (1991). Poor people and poor places: Deciphering effects on poverty outcomes. In J. Huber (Ed.), *Macro-micro linkages in sociology* (pp. 244–263). Newbury Park: Sage.
Turley, R. N. L. (2003). When do neighborhoods matter? The role of race and neighborhood peers. *Social Science Research, 32*(1), 61–79.
Venkatesh, S. (2000). *American project: The rise and fall of a modern Ghetto.* Cambridge: Harvard University Press.
Wacquant, L. (2007). *Urban outcasts: A comparative sociology of advanced marginality.* Cambridge: Polity Press.
Wacquant, L., & Wilson, W. J. (1989). The cost of racial and class exclusion in the inner city. *The Annals of the American Academy of Political and Social Science, 501*(1), 8–25.
Wilson, W. J. (1987). *The truly disadvantaged.* Chicago: University of Chicago Press.
Wilson, W. J. (1995). The political economy and urban racial tensions. *The American Economist, 39*(1), 3–14.
Wilson, W. J. (1996). *When work disappears: The world of the new urban poor.* New York: Knopf.

Chapter 4
Understanding Neighbourhood Effects: Selection Bias and Residential Mobility

Lina Hedman and Maarten van Ham

Introduction

The number of studies investigating neighbourhood effects has increased rapidly over the last two decades. Although many of these studies claim to have found evidence for neighbourhood effects, most 'evidence' is likely the result of reversed causality. The main challenge in modelling neighbourhood effects is the (econometric) identification of causal effects. The most severe problem is selection bias as a result of selective sorting into neighbourhoods (Jencks and Mayer 1990; Tienda 1991; Duncan et al. 1997; Galster 2008; Hedman 2011). Selection bias occurs when the selection mechanism into neighbourhoods is not independent from the outcome studied. For example, unemployed people are more likely to move into deprived neighbourhoods than employed people. If this selection mechanism is not adequately controlled for in modelling the effect of living in a deprived neighbourhood on unemployment, a correlation between unemployment and neighbourhood deprivation might be mistaken for a neighbourhood effect.

The selection bias problem is well-known and many recent studies on neighbourhood effects attempt to correct for it in one way or another, often using econometric modelling techniques. This chapter argues that selection bias in neighbourhood effects research is more than a statistical error and that understanding selection into and out of neighbourhoods is at the heart of understanding neighbourhood effects. Neighbourhood selection is the result of residential mobility choices made by households within a restricted choice set (van Ham and Feijten 2008; Feijten and

L. Hedman (✉)
Institute for Housing and Urban Research, Uppsala University, Gävle, Sweden
e-mail: lina.hedman@ibf.uu.se

M. van Ham
OTB Research Institute for the Built Environment, Delft University of Technology, Delft, The Netherlands
e-mail: m.vanham@tudelft.nl

M. van Ham et al. (eds.), *Neighbourhood Effects Research: New Perspectives*, DOI 10.1007/978-94-007-2309-2_4, © Springer Science+Business Media B.V. 2012 van Ham 2009; van Ham and Clark 2009). These residential choices sort households with different characteristics into different (types of) neighbourhoods, producing and affecting patterns of residential segregation. The sorting process can either reproduce or alter neighbourhood characteristics over time. Especially neighbourhoods with high levels of population turnover may experience high levels of change within relatively short time frames.

If we are to understand mechanisms behind neighbourhood effects, more knowledge is needed about residential mobility and the selective sorting into and out of neighbourhoods (see also Hedman 2011). Residential mobility should be an integral part of the conceptual framework which we use to understand neighbourhood effects. This chapter will empirically and conceptually illustrate the connections between the problem of selection bias and selective mobility patterns and processes. We will argue that a better understanding of mobility and neighbourhood sorting is key in understanding neighbourhood effects and that processes of selective mobility should be incorporated into models of neighbourhood effects. The remainder of this chapter is organised as follows. First, we will illustrate the importance of selective mobility using data for three neighbourhoods in Stockholm, Sweden. Second, we will discuss theories of residential mobility and the results of existing empirical studies. Third, we present an approach to incorporate residential mobility in models of neighbourhood effects. The final section offers a conclusion and discussion.

An Empirical Illustration of Selective Mobility Patterns

There are several ways in which patterns of residential mobility can be described as 'selective'. For instance, the propensity to move for different groups within neighbourhoods is not the same: younger adults tend to be more mobile than older adults. According to Andersson (2000), half of the moves the average Swede makes during his or her life time (about ten) take place between the ages of 18 and 34. Other features associated with mobility are household composition (single households move relatively often) and housing tenure (renters move more often than homeowners). In the context of understanding neighbourhood effects, we are not so much interested in the differences between groups in mobility rates, but more in mobility patterns and their causes: who moves to (or stays in) certain types of neighbourhoods and why?

The sorting of people with different characteristics into and out of different types of neighbourhoods is one of the main causes of residential segregation. Selective mobility can cause the neighbourhood population composition to change dramatically within a relatively short period of time. This has been illustrated by studies of white flight in the US (see Ellen 2000). They argue that ethnic segregation is partly caused by the "flight" of white people from neighbourhoods where the share of black inhabitants has reached a critical point. The overrepresentation of whites among out-movers results in the neighbourhood concentration of "black" residents increasing, in turn continuing and even accelerating the out-mobility of white inhabitants. It is more common however, that selective mobility patterns reproduce existing neighbourhood characteristics. For example, neighbourhoods with high concentrations of immigrants often attract new immigrants. This mechanism can partly be explained through residential preferences (they want to live in these neighbourhoods because of, for example, ethnic specific services), and partly through restrictions within the local housing markets (the ethnic concentration neighbourhoods are relatively affordable and thus available for low-income groups). The relative weight of these two explanations will differ between (ethnic and socio-economic) groups and between housing markets.

This section of the chapter provides some empirical examples of selective mobility patterns and their relations to neighbourhood change or reproduction. We present moving patterns and neighbourhood trends for the 1990–2008 period for three neighbourhoods in Stockholm, Sweden. These neighbourhoods are by no means representative for the city as a whole but they are chosen because of their different characteristics and mobility patterns. The three neighbourhoods are *Bjursätra*, located in the south of Stockholm and part of the relatively deprived larger area of Rågsved; *Ängbylunden*, which is a villa neighbourhood known for its green spaces, located in the west of Stockholm and part of the larger area Bromma; and *Rinkeby*, one of Sweden's best-known deprived and immigrant-dense neighbourhoods located in the North West of Stockholm. The data used for this study is derived from the GeoSweden database, which is a longitudinal micro-database of the entire Swedish population drawn from a number of different administrative registers. This data is unique, as it consists of linked annual demographic, geographic and socio-economic data for each individual living in Sweden, for the whole 1990–2008 period. With the data it is possible to follow people over the full 18 year period and to study their residential histories, including their neighbourhood histories.

Figures 4.1 and 4.2 show the share of foreign born people and the share of employed people (aged 20–64) in the three neighbourhoods for the 18 year period, and the averages across all neighbourhoods in Stockholm municipality. The data clearly illustrates how some neighbourhoods can change dramatically over time. The data shows that there are substantial differences in the share of foreign born inhabitants among the three neighbourhoods, from 7% in Ängbylunden in 2008 to 61.7% in Rinkeby. Over the 18 year period, the share of foreign born inhabitants has increased in the municipality as a whole and in two of the three neighbourhoods; Rinkeby and Bjursätra. The increase in Rinkeby is fairly modest, 5.7 percentage points, while Bjursätra has witnessed a dramatic increase of foreign born people: over the 18 years, the share of foreign born in the neighbourhood has more than doubled, from 20.1% in 1990 to a level where almost half the population was born in another country. In Ängbylunden, the share of foreign born has actually declined over the period.

Figure 4.2 shows that employment levels have increased in Stockholm municipality since 1994 (due to a change in the measurement of employment in 1993, data for that year and earlier was left out the figure). Employment levels were relatively low in the early 1990s because of the economic recession in Sweden. However,

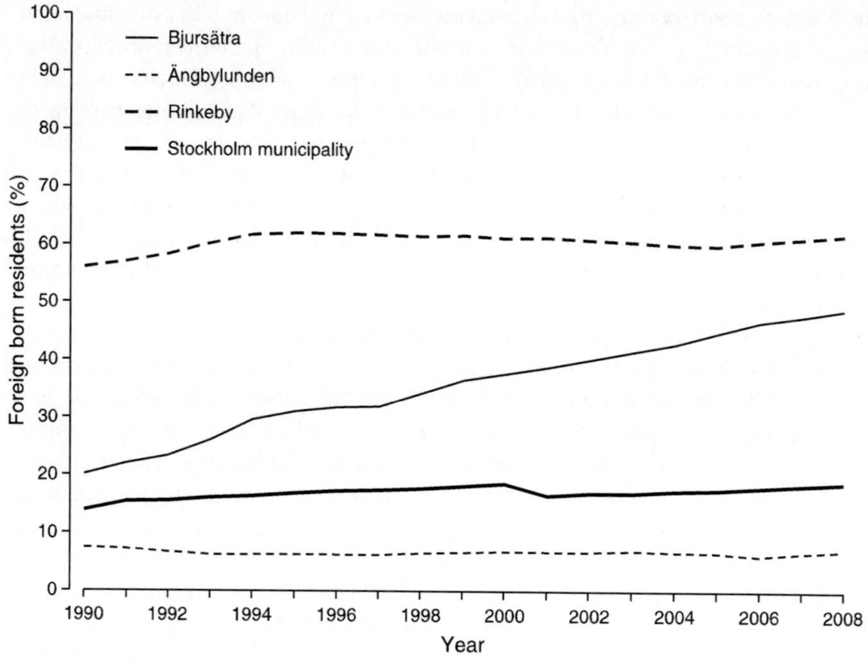

Fig. 4.1 Share of foreign born residents, 1990–2008

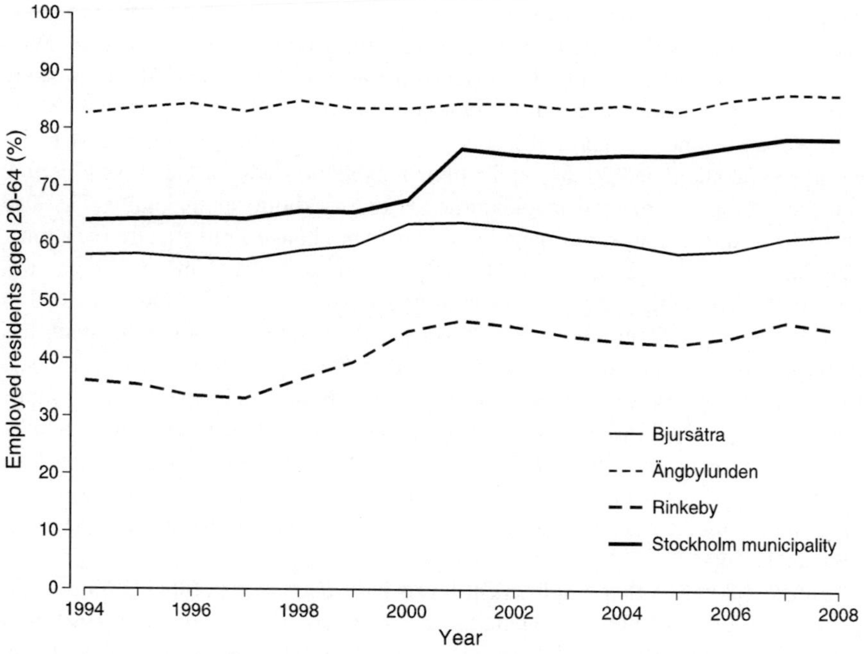

Fig. 4.2 Share of employed residents (aged 20–64), 1994–2008

Table 4.1 Number of inhabitants and turnover rates, 1990 and 2007/2008

	Bjursätra	Ängbylunden	Rinkeby
Number of inhabitants			
1990	9,022	4,505	13,222
2008	10,576	5,227	14,996
Turnover rate[a]			
1990	0.13	0.08	0.13
2007	0.13	0.08	0.11

[a]The turnover rate is based on the number of out-movers from the neighbourhood. 2007 is the last year for which a turnover rate can be calculated.
Source: Author's calculations using data from GeoSweden

while the municipal average signals a substantial increase in employment levels, the trends in Bjursätra and Ängbylunden are quite stable, albeit at different levels. While Ängbylunden had an employment rate of 86.3% in 2008, the equivalent for Bjursätra was 62.0%. Rinkeby is the only one of the three neighbourhoods that experienced an increase in employment levels similar to the average, but the neighbourhood remains at a level well below the municipal average. In 2008, the employment level in Rinkeby was 45.3% (Stockholm average 78.7%). The ranking of the three neighbourhoods remains the same over the 18 years: Ängbylunden is the neighbourhood with the highest employment levels and Rinkeby is the neighbourhood with the lowest employment levels. In other words, employment patterns in these three neighbourhoods are reproduced during the period. This is important in the context of understanding neighbourhood effects: these neighbourhoods most likely receive people with employment characteristics similar to those who are already living there.

There are several mechanisms which can explain the patterns in the share of ethnic minorities and employment levels in these three neighbourhoods. For example, the share of ethnic minorities in a neighbourhood could change because of demographic events (see Finney and Simpson 2009) but also due to selective mobility of the majority and minority population. Similarly, changes in the employment rates in a neighbourhood could be explained by labour mobility (in and out of employment) of neighbourhood residents, but also by selective mobility of employed and unemployed residents into and out of neighbourhoods. If employment rates in a neighbourhood stay relatively low over a longer period of time, which is the case in Rinkeby, this might be caused by in-mobility of people with similar employment characteristics as the neighbourhood population.

Table 4.1 shows population turnover rates for the three neighbourhoods, based on the number of people leaving the neighbourhood. A move is defined as a change of neighbourhood between 2 years, so within-neighbourhood mobility is not included in the figures. Table 4.1 shows us that population turnover rates are fairly similar (between 8 and 13%) in the three areas, despite their very different characteristics, and turnover rates are very stable over time. Ängbylunden, the villa neighbourhood, has the lowest level of population turnover with 8% a year. All three

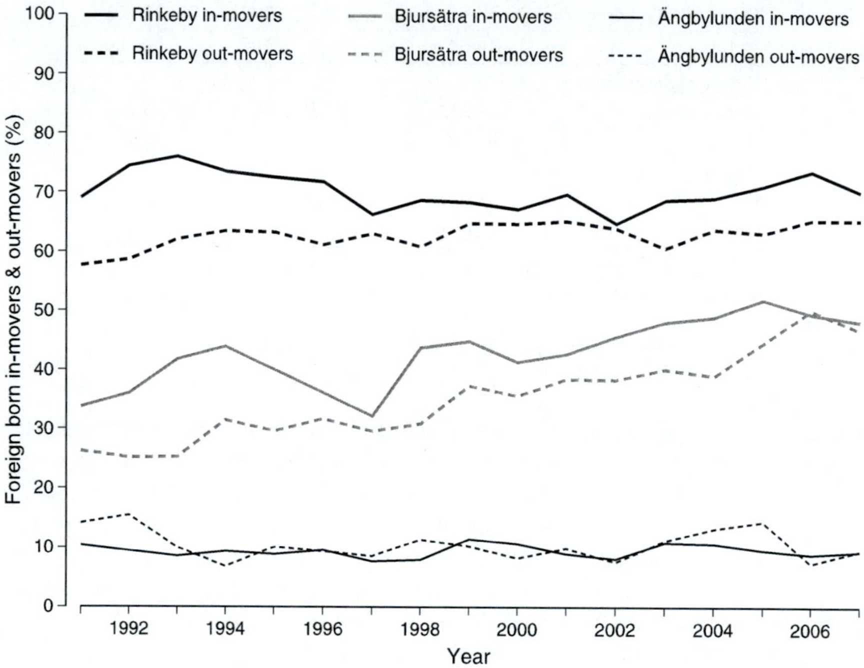

Fig. 4.3 Share of foreign born among in-movers and out-movers, 1991–2007 (*Source*: Author's calculations using data from GeoSweden)

neighbourhoods experienced an increase in population over the 18 year period, which suggests that the out-movers were replaced by in-movers. Differences in characteristics between out-movers and in-movers, and natural change of the neighbourhood population are responsible for changes in the population composition of the neighbourhoods.

To better understand the changes in ethnic composition and employment levels (Figs. 4.1 and 4.2) we show the shares of foreign born and employed people among in- and out-movers to the three neighbourhoods during the same time period (Figs. 4.3 and 4.4). Figure 4.3 clearly shows how the share of foreign born among in-movers in both Bjursätra and Rinkeby consistently exceeds the share of foreign born among out-movers. This produces the increase of immigrants in the two areas as observed in Fig. 4.1. The stable ethnic composition of Ängbylunden is explained by similar shares of foreign born among in- and out-movers. We can also see that the most immigrant-dense area (Rinkeby) attract the highest share of immigrant in-movers while few move to the Swedish majority neighbourhood of Ängbylunden. The share of immigrants moving to Bjursätra is constantly increasing during the period indicating a positive relationship between the share of immigrant in-movers and the share of foreign born in the neighbourhood. Whereas Rinkeby and Bjursätra both had a positive "immigrant gap", both neighbourhoods experienced a negative employment gap; the share of employed people is higher among those leaving the

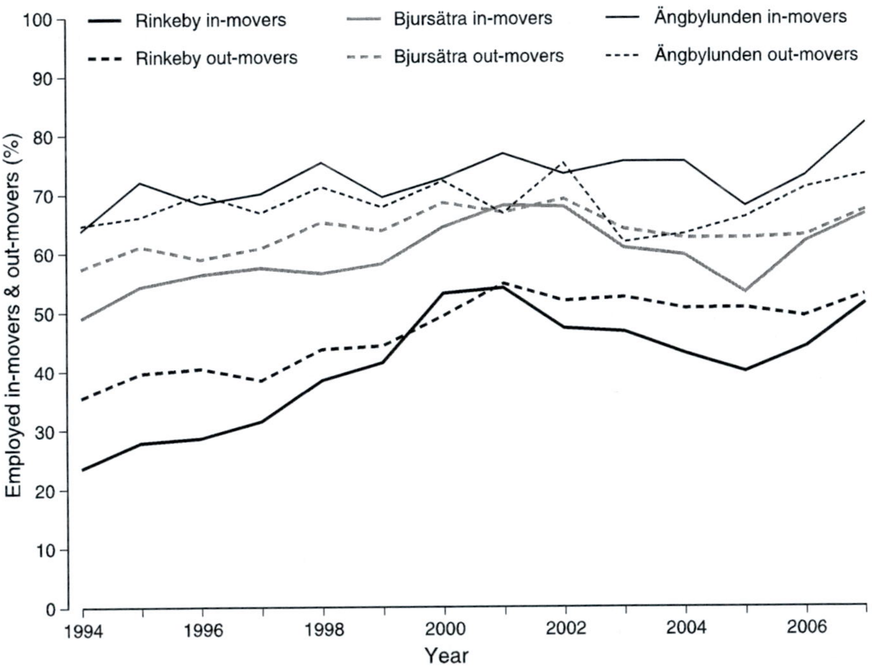

Fig. 4.4 Share employed (age 20–64) among in-movers and out-movers, 1994–2007 (*Source*: Author's calculations using data from GeoSweden)

respective neighbourhoods than among those moving to the neighbourhoods (Fig. 4.4). Thus, the low levels of employment in especially Rinkeby are reproduced through selective moving patterns. This is an important observation as it shows that selective mobility is likely to explain (at least part of) the correlation between individual level unemployment and neighbourhood unemployment levels. There is also a very small employment gap between in-movers and out-movers in Ängbylunden, but going in the opposite direction. This is not surprising given that moving to Ängbylunden requires more financial resources than moving to any of the other two areas.

Figures 4.3 and 4.4 clearly illustrate how the characteristics of in- and out-movers can affect the neighbourhood population composition in terms of ethnic composition and employment levels. Selective mobility can result in either reproduction of existing characteristics, or in a change in characteristics over time. To further illustrate how individuals with different characteristics sort into different neighbourhoods, the characteristics of in-movers to each of the three neighbourhoods in 2008 are described in Table 4.2. The table shows no large differences in terms of age and gender composition but in-movers to Ängbylunden have a substantially stronger socio-economic position than in-movers to the other two neighbourhoods, and especially Rinkeby: they have higher levels of education, higher levels of employment, higher incomes, and are less dependent on social benefits. Thus, in accordance with

Table 4.2 Characteristics of in-movers, 2008

	Bjursätra	Ängbylunden	Rinkeby
Age			
Mean age	30.4	27.6	28.1
Share children (0–18)	17.5	20.9	21.0
Share elderly (65+)	4.3	2.8	1.6
Sex			
Share females	48.9	45.7	49.1
Family composition (20+)			
Share singles	63.6	50.8	57.1
Share with children	24.4	35.0	27.2
Education (20+)			
Share compulsory school (–9 years)	19.3	4.4	36.3
Share high school (10–12 years)	42.1	30.3	32.1
Share higher education (13+year)	33.9	64.8	21.6
Employment level (20–64)			
Share employed	66.9	80.1	45.3
Income (20–64)			
Mean annual disposable income (*100 kr)	1,533	2,741	1,141
Mean annual work income (*100 kr)	1,471	2,527	839
Share social benefits (all ages)	19.2	1.5	36.0
Country of birth			
Share Swedish born	49.5	89.1	23.0
Share non-western	32.1	5.6	63.7

Source: Author's calculations using data from GeoSweden

many previous studies, the results show that individuals with stronger socio-economic positions move to more affluent neighbourhoods, while those with less resources move to more deprived areas. These mobility patterns reproduce the neighbourhoods' relative status and patterns of urban socio-economic segregation over time. A similar conclusion can be drawn for ethnic patterns, where nine out of ten in-movers to the Swedish majority neighbourhood Ängbylunden are Swedish born, while the respective figures for the more immigrant-dense areas of Bjursätra and Rinkeby are 49.5% and 23.0%.

The above data analyses clearly show how mobility patterns are selective and how this selectivity reproduces or changes neighbourhood characteristics. Selective in-mobility of people into neighbourhoods is an issue that needs to be addressed in studies of neighbourhood effects. If in a neighbourhood with relatively low employment levels those who get a job leave the neighbourhood, and are replaced by others without a job, it is not the neighbourhood which causes unemployment. It is the neighbourhood housing stock attracts unemployed people who cannot afford to live elsewhere. If the share of employed people among in-movers is much lower than among out-movers, this will seriously bias models of neighbourhood effects if not adequately controlled for.

Selective mobility and the resulting residential patterns would cause serious problems in neighbourhood effects studies if segregation was extreme, for example, when all low income residents live in the most deprived neighbourhoods. In such a case there would be no variation and therefore no control groups. Such extreme segregation would make it impossible to assess the effect of living in certain types of neighbourhoods. This is however rarely the case so at least in theory, selective mobility should not be a problem if we can adequately measure the selection mechanisms. Both the ethnicity and employment status, as used in the above examples, of in- and out-movers are easily measureable. However, selective mobility patterns become a problem if they are caused by unmeasured characteristics (characteristics not available in the data used) that are correlated with the outcomes of a neighbourhood effect study. Examples of such characteristics are soft skills, initiative, risk taking behaviour, which can all be expected to be also related to the probability to find and keep a job. To better understand how people decide whether to move or not and where to move to we must look into theories and studies of residential mobility.

Selective Mobility and the Selection Problem

Understanding why people end up in certain types of neighbourhoods is complicated by the fact that housing is a composite good (see van Ham 2012a). A dwelling can be described by its various characteristics such as tenure, size, style, quality, and (relative) location. One cannot buy a single aspect of a dwelling separately as dwellings are bundles of characteristics, including the neighbourhood (van Ham 2012a). The type of housing available in certain neighbourhoods, in terms of tenure (rented or owner-occupied) and price is important in understanding how households sort over neighbourhoods. Some neighbourhoods will never be considered by some types of households because they are either beyond their financial means, or because they do not offer the right types of dwellings. It is important to be critical of the concept of *choice* here as most households choose their dwelling, and associated neighbourhood, within a very constrained choice set (van Ham 2012a, b). In fact, some households have no choice at all, especially when they depend on housing offered by social landlords, which is often concentrated in a limited number of neighbourhoods within a city. Notwithstanding the above, it can be argued that self-selection also plays a role in the choice of neighbourhoods. Households also choose their neighbourhood based on the reputation of the place (Permentier et al. 2009), and based on other personal preferences. So both structural factors (the housing market) and individual preferences sort households into neighbourhoods.

Some of the factors which influence neighbourhood choice are easily observable using standard surveys. Others are more difficult to measure and are often not observed in the data (see earlier in this chapter). In neighbourhood effect studies a problem arises when these unobserved characteristics also influence the individual outcome under study, which will lead to biased estimates. Several authors list selection bias as one of the most urgent issues to solve in the field of neighbourhood

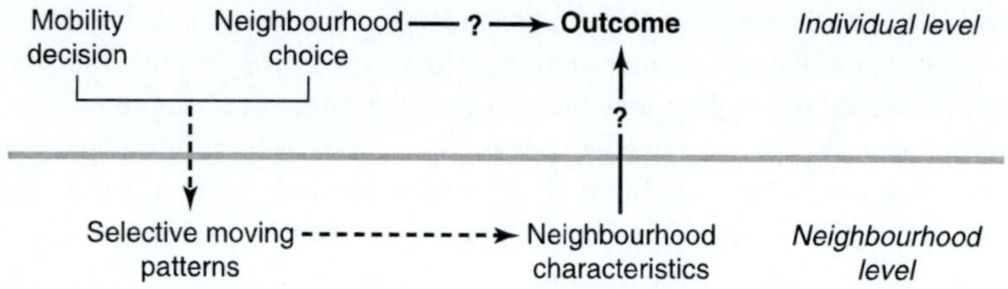

Fig. 4.5 An illustration of the relationship between selective mobility and the selection problem

effects research (Jencks and Mayer 1990; Tienda 1991; Duncan et al. 1997; Galster 2008; van Ham and Manley 2010; Hedman 2011). However, not everyone agrees on the direction of selection bias. Jencks and Mayer (1990) and Tienda (1991) argue that selection bias results in overestimated neighbourhood effects while Brooks-Gun et al. (1997) suggest that the opposite could also occur. This would, for example, be the case if those being the most negatively affected by a neighbourhood factor also are the first ones to leave. The extent of selection bias is also unclear. Some authors (e.g. Dawkins et al. 2005; Galster et al. 2007, 2008) find statistically significant evidence of neighbourhood effects even after controlling for selection. Others argue that their results indicate that selection explains all the correlation between neighbourhood characteristics and individual outcomes: in other words, what scholars believe to be a neighbourhood effect is nothing but a selection effect. For example, Oreopoulos (2003) found a positive correlation between living in wealthier areas or Toronto and income, employment and welfare participation, but only for those living in private housing. He found no such evidence for those living in social housing. He explained this by arguing that the housing allocation of those in social housing is more or less random (housing officers allocate housing based on need and waiting lists), while the choice of neighbourhood for those in private housing is strongly related to their socio-economic status. Oreopoulos (2003) concluded that he did not find evidence for causal neighbourhood effects. Van Ham and Manley (2010) found similar results using longitudinal data for Scotland. They only found correlations between neighbourhood characteristics and individual labour market outcomes for homeowners, but not for social renters. The results by Oreopoulos and van Ham and Manley do not show that neighbourhood effects do not exist, but they clearly show that selection effects are a significant problem when estimating neighbourhood effects.

The problem of selection bias is illustrated by Fig. 4.5. A correlation between neighbourhood characteristics and an individual level outcome can either be the result of selective mobility, or of neighbourhood characteristics (see the chapter by Galster 2011 in this volume for an extensive list of mechanisms through which the neighbourhood can affect individual level outcomes). To be able to draw the conclusion that the neighbourhood has a real *causal* effect on individual level outcomes, the selection effect has to be accounted for. Figure 4.5 also illustrates how mobility decisions of individuals and groups result in selective mobility patterns (the dashed arrows). Such mobility patterns do in turn affect the composition and characteristics of neighbourhoods and potentially also the entire urban neighbourhood hierarchy, as illustrated by our own empirical analysis earlier in this chapter.

To understand why selection bias may occur, and to understand the possible extent and direction of bias, we must look into the issue of neighbourhood sorting; how and why do households sort themselves into different types of neighbourhoods. Although there is a very large literature on residential mobility choices in terms of the type and tenure of dwellings chosen, the literature studying neighbourhood choices is relatively small. One reason is that the most common framework for studying residential mobility treats housing choice as the result of interaction between household needs, demands and preferences and the characteristics of *dwellings*. Neighbourhood choice is in such a framework a by-product of the distribution of available and attractive dwellings. A relatively recent set of studies are however arguing that more attention must be directed to the role of the neighbourhood in the residential choice process (Lee et al. 1994; Kearns and Parkes 2003; Clark et al. 2006; Clark and Ledwith 2006; van Ham and Feijten 2008; Feijten and van Ham 2009; van Ham and Clark 2009). Two sets of (interrelated) literatures are of special relevance for understanding neighbourhood choice; the literature focusing on how residential mobility choices are made, and the literature focusing on understanding residential segregation and how it is maintained by selective mobility patterns.

Factors Affecting Neighbourhood Choice

In their now classical study on residential mobility, Brown and Moore (1970) divide the mobility decision into two separate choices: the decision to leave and the choice of destination. Both of these decisions are undertaken within a set of needs, preferences, and constraints. The choice of destination is often biased to certain areas. Brown and Moore use the concept of awareness space to refer to those places the household had knowledge of before the search process began. The search space, they argue, is often found within this awareness space; "[t]he household will search only those areas contained within its awareness space that satisfy the environmental and locational criteria of its aspirations, i.e., its search space" (Brown and Moore 1970, p. 9). In other words, some areas are excluded from the search space. This exclusion, as well as the ultimate choice, is based on needs, preferences and constraints. Households preferring an owner-occupied detached dwelling will only search areas where such housing is available. Households which cannot afford to buy their own house will only search in neighbourhoods where rental dwellings are available. Furthermore, Brown's and Moore's argument claims that household will primarily search in neighbourhoods that are familiar to them, so information about alternatives plays an important role in the choice process (see also van Ham 2012a, b).

Besides information, financial resources are another important factor determining the outcome of the residential choice process. The more one earns, the larger the choice set of dwellings and neighbourhoods. Banks and other mortgage lenders have a large influence on the resources available to households to buy property on the market. And, as discussed earlier, the structure of the local housing market has an important effect on the residential choice process. People can only move to neighbourhoods where there is vacant housing available. The constraints faced by some households are so severe that is misleading to talk about housing *choice* (van Ham 2012a, b). For example, households in desperate and urgent need of housing are often forced to accept the first available option, often in less desirable neighbourhoods (van Ham and Manley 2009). Most cities are spatially (and socially) segregated along socio-economic lines, and the purchasing power of households is an important factor in neighbourhood sorting (Hedman et al. 2011). The most attractive and expensive areas are inaccessible for a large share of the population. Several studies have shown how people leave the most distressed areas as their incomes increase, leaving vacancies that are filled by those whose options are very limited (e.g. Skifter Andersen 2003; Andersson and Bråmå 2004; Sampson and Sharkey 2008).

Although some households face severe constraints in the housing market, most households have some degree of choice. A household's choice set may be constrained to cheap dwellings in less attractive areas but if there are vacancies in a variety of such neighbourhoods, households can still choose the alternative which best matches their specific (locational) needs and preferences, even if none of the alternatives is considered very good. Many factors have been suggested to affect the choice of destination (and also the decision to move). Brown's and Moore's (1970) review list five categories: *accessibility* (to city centre, communications, service, green areas etc.), *physical characteristics of the neighbourhood* (physical condition of street and sidewalk, layout, beauty), *services and facilities* (quality and accessibility), *social environment* (socio-economic, ethnic, and demographic composition, friends and friendliness), and *individual site and dwelling characteristics* (costs, housing size etc.). Obviously, households differ in terms of which aspects they find most important and how they value these aspects, and their attitudes and preferences change over time. In general, families with children tend to value child friendliness and access to good quality schools while young singles often move toward the city centres.

In a recent survey[1] (spring 2009) among a sample of "movers" and "stayers"[2] in four neighbourhoods in Uppsala, Sweden, respondents in the age group 28–55 were

[1]The survey is part of the project "Den etniska segmenteringens mekanismer - exemplet bostadsmarknaden" [Ethnic housing segmentation and discrimination – a study of institutional practices and preferences], sponsored by the Swedish Council for Working Life and Social Research, and conducted by Roger Andersson, Irene Molina, and Lina Hedman at the Institute for Housing and Urban Reseach, Uppsala University, and by Åsa Bråmå at the Centre for Municipality Studies, Linköping University.

[2]All respondents had stayed 2 years (2005–2007) in one of the four selected neighbourhoods (Luthagen, Svartbäcken/Tunabackar, Stenhagen, Gottsunda). About 50% (where possible, otherwise the entire moving population) had moved to a different neighbourhood in 2008 while the other 50% remained in the same neighbourhood. The total number of respondents was 1,257, the response rate was 48.2%.

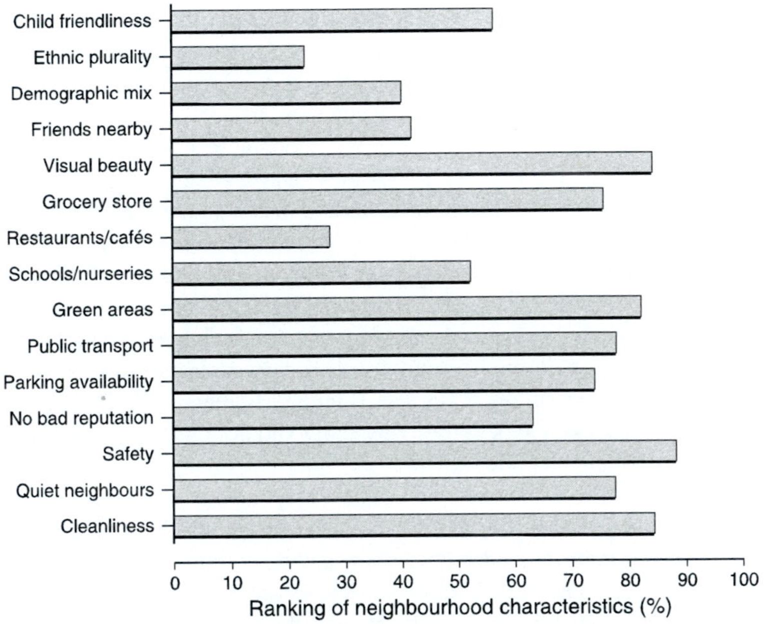

Fig. 4.6 Share of respondents ranking different neighbourhood factors to be "important" or "very important" (4 or 5 on scale 1–5) (*Source*: Own survey, all respondents, ages 28–55)

asked how important certain neighbourhood features were to them when thinking of potential destinations. Those that seemed most important to the respondents were neighbourhood safety, cleanliness, aesthetic beauty, access to green spaces, quiet neighbours, and a more general provision of services such as grocery stores and public transport (see Fig. 4.6). Previous studies support these results as they have found neighbourhood satisfaction and attachment to be correlated with physical conditions and appearance, quality and amount of services, safety, and stigmatization (Burrows and Rhodes 2000; Parkes et al. 2002; Kearns and Parkes 2003; Permentier et al. 2009). Some minor differences were found among the different age categories, where the youngest age groups (aged 28–35) placed more importance on good access to schools and their working places where the older respondents (aged 45–55) found access to green areas and neighbourhood cleanness to be more important. Families with children found neighbourhood reputation, neighbourhood safety, child friendliness, access to schools, sports facilities and parking lots more important than singles, who in turn favoured access to restaurants and cafés and public communications more than did the families. However, there was no difference between the age categories in terms of which of the factors were the most important.

No major differences regarding the most important neighbourhood characteristics were found between the "mover" and "stayer" categories in the survey. Other studies have shown that neighbourhood satisfaction and neighbourhood attachment are correlated with moves and moving intentions where those who are the most satisfied are the least likely to (want to) leave (Galster 1987; Lee et al. 1994; Clark and Ledwith 2006; Guest et al. 2006; van Ham and Feijten 2008; Feijten and van Ham 2009; Permentier et al. 2009). The previous discussion indicates that those living in safe, clean, quiet neighbourhoods with good access to services and a good reputation are the least likely to have a desire to leave. Distressed neighbourhoods are the least likely to have all these attributes and they also tend to have the highest turnover rates, although it is unclear whether this is due to neighbourhood characteristics or population composition (see e.g. Bailey and Livingston 2007; van Ham and Clark 2009). Moving plans are also affected by neighbourhood change, where households experiencing change, or believing that their neighbourhood will decline (in their opinion) will be more likely to express a wish to leave (Galster 1987; Lee et al. 1994; Kearns and Parkes 2003; van Ham and Feijten 2008; Feijten and van Ham 2009; van Ham and Clark 2009). The survey did not find any large differences in neighbourhood characteristics regarded as important among inhabitants of different neighbourhoods, with the exception of how respondents valued neighbourhood population diversity. The inhabitants of the most immigrant-dense neighbourhood expressed stronger preferences for both ethnic and demographic diversity compared to the residents of other neighbourhoods. These preferences were also stronger among "stayers" compared to those who had left the area.

The role of ethnic preferences (among both majority and minority groups) in relation to ethnic residential segregation has been a topic that has received much attention since the publication of Schelling's (1969, 1971) seminal papers in which he demonstrated that small differences in preferences between two groups could cause high levels of segregation due to adjusted, ethnically selective mobility patterns. Schelling's hypothesis was confirmed by Clark (1991) who even argued that the differences in preferences of ethnic composition between whites and blacks were much larger than those hypothesized by Schelling and that especially white people expressed strong preferences for ethnic homogeneity in the neighbourhoods (see also Clark 1992). Ethnic preferences (of the majority population) are an important explanation for the ethnically selective mobility patterns creating ethnic residential segregation as demonstrated by the white flight/white avoidance theories. White flight theory suggest that the white/native population leaves neighbourhoods when the share of minorities becomes too high; white avoidance theories hypothesise that when whites choose a neighbourhood they avoid areas with high shares of minority inhabitants (Ellen 2000). The theories have found empirical support in both the U.S. (e.g. Crowder 2000; Quillian 2002) and in European countries. For example, van Ham and Feijten (2008), Feijten and van Ham (2009), and van Ham and Clark (2009) have found for the Netherlands that native Dutch people express stronger wishes to leave when the percentage of ethnic minorities in their neighbourhood increases, and Bråmå (2006) shows that native Swedes tend to avoid immigrant-dense neighbourhoods.

The Schelling argument emphasizes within-group preferences with regard to living among similar others, whereas the white flight/avoidance theories also touch upon attitudes towards other groups. Several scholars have tried to explain the aversion of especially whites (or natives) towards living in ethnically mixed neighbourhoods. Some studies suggest that race[3] is an independent factor, and thus that racism (frequently of the white population towards minority groups) is an important variable explaining ethnic segregation (e.g. Zubrinsky and Bobo 1996; Emerson et al. 2001). Other studies argue that it is not race per se that makes whites less willing to move into black neighbourhoods but that race is a proxy for other attributes, such as housing prices, crime levels or general neighbourhood standards (e.g. Clark 1992; Harris 1999; Crowder 2000). Ethnic minorities are more likely to be poor, and are therefore also more likely to live in poverty areas. So according to the race-proxy argument, whites do not avoid ethnic minority areas, but they avoid poor neighbourhoods which are often also ethnic minority areas. The spatial assimilation model argues that ethnic segregation is the outcome of differences in socio-economic resources together with a lack of cultural assimilation of minorities. According to the theory, the residential mobility patterns of ethnic minorities will become increasingly similar to those of the majority population as they become more integrated and their socio-economic position becomes stronger. A complementary theory is the place stratification model, which emphasises the importance of other forms of constraints than economical and cultural ones. It argues that, for example, discrimination and housing availability or housing allocation rules restrict the opportunities for ethnic minorities to have housing careers similar to natives and that these differences will remain even after immigrants have become more integrated (see Bolt and van Kempen 2003 for an overview of the place stratification and spatial assimilation models). Studies have found evidence of housing market discrimination for several countries (for the U.S. see Turner et al. 2002).

An alternative theory argues that ethnic segregation is the result of the voluntary clustering of ethnic minorities. This literature emphasises how minorities can gain benefits from living together, such as keeping their own language and culture, having access to ethnic specific services such as specialised stores or places of worship, and having access to a local economic system. It has also been suggested that ethnic clustering is used as a means of defence against a hostile host society, but also that ethnic clustering might help to integrate into the host society (see for example Portes and Manning 1986). Bowes et al. (1997) have found that the Pakistani population in Scotland accepted living in a deprived neighbourhood if that enables them to live with more co-ethnics. Theories of voluntary clustering rely on relatively homogeneous ethnic concentration neighbourhoods. In Sweden (and many other parts of Europe) such neighbourhoods do not exist; some immigrant-dense neighbourhoods in Sweden contain between 50 and 100 different nationalities, making theories of voluntary clustering less applicable. Molina's (1997) study found that ethnic minorities living in an immigrant-dense area in Uppsala, Sweden, had the same housing and neighbourhood ambitions as the native population but were less able to realize them, thus pointing at different forms of constraints.

[3]The U.S. literature consistently refers to "race" while the European literature refers to "ethnicity" and "immigrant status".

The above literature review suggest that the selective in- and out-mobility patterns found for Bjursätra, Ängbylunden and Rinkeby are most likely the result of a complex mixture of constraints of various forms and preferences for different types of dwellings and neighbourhoods. Existing studies provide knowledge of how different dwelling characteristics are related to mobility but we know less about how factors at the neighbourhood level affect neighbourhood choice and sorting. We know that some groups are more likely to move to, for example, ethnic concentration neighbourhoods than others but there is less knowledge on how households value ethnic composition in relation to, for example, socio-economic status of neighbours, neighbourhood safety, or reputation. There is also a need of research looking at how households choose between neighbourhoods: why does a household choose one "safe" neighbourhood over another, and how do households with very restricted choice sets rank neighbourhoods and dwellings available to them?

To further explore the neighbourhood dimension in mobility processes is not only an important task for the residential mobility literature but also something that would be of great importance for studies on neighbourhood effects. Such studies would provide us with a better knowledge of neighbourhood sorting processes and thus the problem of potential selection bias. It would make it possible to better incorporate sorting into models and measurements of neighbourhood effects.

Methods for Dealing with Neighbourhood Sorting and Selection Bias

In this section of the chapter, we argue that studies on neighbourhood choice and neighbourhood sorting would benefit from using alternative modelling strategies. We also discuss how these strategies can be incorporated into models of neighbourhood effects to control for the bias caused by selective mobility to neighbourhoods.

Most quantitative studies of neighbourhood sorting use logistic regression models, binary or multinomial, to model the probability that households move to a certain type of neighbourhood. These models treat neighbourhood choice as a function of the characteristics of individuals or households: for example, the models estimate how different individual characteristics, such as income, employment status or ethnicity, affect the likelihood of moving to neighbourhoods in category A relative to those in category B. Binary logistic regression models are by definition restricted to using only two outcome categories (for example whether or not a neighbourhood is deprived or ethnically concentrated). Multinomial logistic regression models can handle more alternatives but these become impractical to work with if the number is too large. The need to reduce the number of alternatives means that neighbourhoods need to be categorized based on, for example, poverty levels, levels of ethnic concentrations, housing tenure, or combinations of these. The models are of limited use to understand neighbourhood choice because they can only take a limited number of neighbourhood characteristics into account and neighbourhoods typically need to be grouped using a simple categorisation. In other words, studies using binary or multinomial logistic regression models teach us a lot about simple processes of neighbourhood sorting (where do people go) but little about how different neighbourhood factors affect the sorting process.

A type of model that can include multiple neighbourhood characteristics simultaneously is the conditional logit model,[4] introduced in the context of social sciences and residential mobility by McFadden in 1974. The conditional logit treats choice as a function of the characteristics of the *alternatives* within a choice set. In a residential mobility setting, this means that each individual can choose from a number of neighbourhood alternatives based on a large range of characteristics of these neighbourhoods. In a recent paper Hedman et al. (2011) modelled neighbourhood choice where each moving household was assumed to choose between 10 alternative neighbourhoods. Their model estimated the probability that households choose their actual destination (the other nine are random alternative neighbourhoods) based on neighbourhood characteristics such as the share of foreign born, the share of public renters and the median neighbourhood income. A weakness of the conditional logit model is that it cannot include individual characteristics directly because these do not vary within choice sets of neighbourhoods (only the characteristics of the neighbourhoods within a choice set vary, not the characteristics of the households making the choice). Individual characteristics can only be included when interacted with neighbourhood characteristics. The conditional logit model is well known in, for example, the transport choice literature, but is rarely used to estimate neighbourhood choice. To our knowledge, only three (very recent) studies have used it to estimate neighbourhood choice: Ioannides and Zabel's (2008) study on neighbourhood effects on housing structure demand; Quillian and Bruch's (2010) study modelling neighbourhood choice in relation to race and class; and Hedman et al. (2011) estimating neighbourhood sorting in Sweden.

We have so far in this chapter argued that neighbourhood sorting processes are a major source of selection bias in studies of neighbourhood effects and that more knowledge is needed on these processes to fully understand the selection problem (see also Sampson and Sharkey 2008). Most neighbourhood effect studies still do not adequately take neighbourhood sorting into account but an increasing number of studies attempt to address the issue. Oreopoulos (2003) and van Ham and Manley (2010) both argued that selection bias in models of neighbourhood effects is only a problem for those groups who have some degree of choice over where they live. Both studies found no neighbourhood effects for those in (to some extend randomly allocated) social housing, but did find 'effects' for those in private housing. They concluded that these 'effects' are most likely the result of reversed causation. By comparing those in private housing and those in social housing, Oreopolous and van Ham and Manley simulated an experimental setting. The best known examples of quasi-experiments are Chicago's Gatreaux assisted housing program and the

[4]For formal descriptions of the conditional logit model and how it differs from the multinomial logistic model, see McFadden (1974), Hoffman and Duncan (1988).

American Moving to Opportunity (MTO) program. These programs allocated households to dwellings and neighbourhoods more or less randomly, thereby attempting to overcome selection bias. Although these studies are seen as the gold-standard in neighbourhood effects research, it has been argued that they still suffer from selection bias as allocation to neighbourhoods was not completely random.

For most studies it is not possible to execute large random trials and these studies often depend on register or survey data. A number of approaches can be used to limit the effects of selection bias when using such data. One approach is the use of sibling data to control for unmeasured parental characteristics by measuring differences in outcomes if the siblings have experienced different neighbourhood environments during their life-time (e.g. Aaronson 1998). Instrumental variable approaches are also used to control for selection bias, which requires instruments to replace the variable that is correlated with the outcome of interest (Galster et al. 2007). Difference models are also popular as they eliminate all time-invariant unobservable individual characteristics (see Galster et al. 2008). All these techniques theoretically remove bias, at least to some extent. However, unfortunately, they do not address the issue of selective mobility.

A classic technique to take selection mechanisms into account is the Heckman two-step model. It involves a first step in which the probability of sorting into a group is estimated, and a second step in which the outcome of interest is estimated for the specific group, thereby correcting for potential bias caused by the selection mechanism. In the setting of a neighbourhood effect study, step one could be to model the probability of living in a certain neighbourhood and step two could be a model of neighbourhood effects. Although the two-step model has some disadvantages and has been criticized in the econometric literature (e.g. Puhani 2000), it has the enormous benefit of including real measures of selective mobility and neighbourhood sorting. To our knowledge, only one study has attempted a two-step technique to account for bias in modelling neighbourhood effects: Ioannides and Zabel's 2008 paper in which they estimate how the neighbourhood context affects housing demand. The paper is highly technical, but uses a promising advanced method to estimate neighbourhood sorting and to control for bias. The approach not only tackles the urgent problem of selection bias in the neighbourhood effect literature but also gives insight into neighbourhood sorting. It is thus a technique that benefits two literatures separately while also incorporating them methodologically into a holistic framework that increases our general understanding of neighbourhoods and residential segregation.

Concluding Remarks

This chapter has argued that in order to further our understanding of neighbourhood effects we should incorporate neighbourhood sorting into our models. Many approaches for dealing with selection bias treat neighbourhood sorting as a statistical nuisance. These approaches might do the job, but reveal nothing about the processes behind the potential bias. Neighbourhood sorting is of interest in its own right and surprisingly few studies focus on why certain households 'choose' certain neighbourhoods. An understanding of neighbourhood sorting is also key in understanding residential segregation and the production and reproduction of neighbourhoods of different characteristics and status. In other words, an understanding of neighbourhood sorting is important to understand the dynamic contexts (neighbourhoods) that neighbourhood effect theory assumes affect inhabitants.

Neighbourhood effect studies are thus in the situation where the processes behind one of its key methodological problems (selection bias) are also critical to fully understand the neighbourhood context itself. It is thus remarkable that residential mobility and neighbourhood sorting have been almost completely ignored in neighbourhood effects research. This chapter has demonstrated how selective mobility patterns affect the neighbourhood context and have the potential to cause selection bias when modelling neighbourhood effects. In line with previous studies, we recognize the need for more studies focusing on neighbourhood sorting, especially how neighbourhood factors influence mobility decisions. Furthermore, most studies looking at neighbourhood choice focus on a range of general neighbourhood characteristics, such as employment rates and the percentage of ethnic minorities. More research is needed on the factors that cause households to choose, for example, one deprived neighbourhood over another. This point is important as some theories assume that processes and transmission mechanisms behind neighbourhood effects are neighbourhood-specific.

References

Aaronson, D. (1998). Using sibling data to estimate the impact of neighborhoods on children's educational outcomes. *Journal of Human Resources, 33*(4), 915–946.

Andersson, R. (2000). Varför flyttar man? *Hemort Sverige* (pp. 50–65). Norrköping: Board of Integration.

Andersson, R., & Bråmå, Å. (2004). Selective migration in Swedish distressed neighbourhoods: Can area-based urban policies counteract segregation processes? *Housing Studies, 19*(4), 517–539.

Bailey, N., & Livingston, M. (2007). *Population turnover and area deprivation*. Bristol: Joseph Rowntree Foundation.

Bolt, G., & van Kempen, R. (2003). Escaping poverty neighbourhoods in the Netherlands. *Housing, Theory and Society, 20*(4), 209–222.

Bowes, A., Dar, N., & Sim, D. (1997). Tenure preference and housing strategy: An exploration of Pakistani experiences. *Housing Studies, 12*(1), 63–84.

Bråmå, Å. (2006). "White flight"? The production and reproduction of immigrant concentration areas in Swedish cities, 1990–2000. *Urban Studies, 43*(7), 1127–1146.

Brooks-Gunn, J., Duncan, G., & Aber, J. (Eds.). (1997). *Neighborhood poverty: Vol. 1 Context and consequences for children*. New York: Russell Sage.

Brown, L. A., & Moore, E. G. (1970). The intra-urban migration process: A perspective. *Geografiska Annaler B, 52*(1), 1–13.

Burrows, R., & Rhodes, D. (2000). The geography of misery: Area disadvantage and patterns of neighbourhood dissatisfaction in England. In J. Bradshaw & R. Sainsbury (Eds.), *Researching poverty* (pp. 191–213). Aldershot: Ashgate.

Clark, W. A. V. (1991). Residential preferences and neighborhood racial segregation: A test of the Schelling Segregation Model. *Demography, 28*(1), 1–19.

Clark, W. A. V. (1992). Residential preferences and residential choices in a multiethnic context. *Demography, 29*(3), 451–466.

Clark, W. A. V., Deurloo, M., & Dieleman, F. (2006). Residential mobility and neighbourhood outcomes. *Housing Studies, 21*(3), 232–342.

Clark, W. A. V., & Ledwith, V. (2006). Mobility, housing stress and neighbourhood contexts: Evidence from Los Angeles. *Environment and Planning A, 38*(6), 1077–1093.

Crowder, K. (2000). The racial context of white mobility: An individual-level assessment of the White Flight Hypothesis. *Social Science Research, 29*(2), 223–257.

Dawkins, C. J., Shen, Q., & Sanchez, T. W. (2005). Race, space, and unemployment duration. *Journal of Urban Economics, 58*(1), 91–113.

Duncan, G., Connell, J., & Klebanov, P. (1997). Conceptual and methodological issues in estimating causal effects of neighborhoods and family conditions on individual development. In J. Brooks-Gunn, G. Duncan, & J. Aber (Eds.), *Neighborhood poverty* (Context and Consequences for Children, Vol. 1, pp. 219–250). New York: Russell Sage.

Ellen, I. (2000). *Sharing America's neighborhoods*. Cambridge: Harvard University Press.

Emerson, M. O., Yancey, G., & Chai, K. J. (2001). Does race matter in residential segregation? Exploring the preferences of White Americans. *American Sociological Review, 66*(6), 922–935.

Feijten, P., & van Ham, M. (2009). Neighbourhood change ... reason to leave? *Urban Studies, 46*(10), 2103–2122.

Finney, N., & Simpson, L. (2009). Population dynamics: The roles of natural change and migration in producing the ethnic mosaic. *Journal of Ethnic and Migration Studies, 35*(9), 1479–1496.

Galster, G. (1987). *Homeowners and neighborhood reinvestment.* Durham: Duke University Press.

Galster, G. (2008). Quantifying the effect of neighbourhood on individuals. Challenges, alternative approaches and promising directions. *Journal of Applied Social Sciences. Schmollers Jahrbuch, 128*(1), 7–48.

Galster, G. (2011). The mechanism(s) of neighbourhood effects: Theory, evidence, and policy implications. In M. van Ham, D. Manley, N. Bailey, L. Simpson, & D. Maclennan (Eds.), *Neighbourhood effects research: New perspectives*. Dordrecht: Springer.

Galster, G., Andersson, R., Musterd, S., & Kauppinen, T. M. (2008). Does neighborhood income mix affect earnings of adults? New evidence from Sweden. *Journal of Urban Economics, 63*(3), 858–870.

Galster, G., Marcotte, D. E., Mandell, M., Wolman, H., & Augustine, N. (2007). The influence of neighborhood poverty during childhood on fertility, education and earnings outcomes. *Housing Studies, 22*(5), 723–751.

Guest, A. M., Cover, J. K., Matsueda, R. L., & Kubrin, C. E. (2006). Neighbourhood context and neighbouring ties. *City and Community, 5*(4), 363–385.

Harris, D. R. (1999). "Property values drop when blacks move in, because...": Racial and socioeconomic determinants of neighborhood desirability. *American Sociological Review, 64*(3), 461–479.

Hedman, L. (2011). The impact of residential mobility on measurements of neighbourhood effects. *Housing Studies, 26*(4), 501–519.

Hedman, L., van Ham, M., & Manley, D. (2011). Neighbourhood choice and neighbourhood reproduction. *Environment and Planning A, 43*(6), 1381–1399.

Hoffman, S. D., & Duncan, G. J. (1988). Multinomial and conditional logit discrete-choice models in demography. *Demography, 25*(3), 415–427.

Ioannides, Y. M., & Zabel, J. E. (2008). Interactions, neighbourhood selection and housing demand. *Journal of Urban Economics, 63*(1), 229–252.

Jencks, C., & Mayer, S. (1990). The social consequences of growing up in a poor neighbourhood. In L. Lynn & M. McGeary (Eds.), *Inner-city poverty in the United States* (pp. 187–222). Washington, DC: National Academy Press.

Kearns, A., & Parkes, A. (2003). Living in and leaving poor neighbourhoods in England. *Housing Studies, 18*(6), 827–851.

Lee, B. A., Oropresa, R. S., & Kanan, J. W. (1994). Neighborhood context and residential mobility. *Demography, 31*(2), 249–270.
McFadden, D. (1974). Conditional logit analysis of qualitative choice behaviour. In P. Zarembka (Ed.), *Frontiers in econometrics*. New York: Academic.
Molina, I. (1997). *Stadens rasifiering. Etnisk boendesegregation i folkhemmet* (Racialization of the City. Ethnic Residential Segregation in the Swedish Folkhem), Geografiska regionstudier Nr 32, Uppsala University.
Oreopoulos, P. (2003). The long-run consequences of living in a poor neighbourhood. *Quarterly Journal of Economics, 118*(4), 1533–1575.
Parkes, A., Kearns, A., & Atkinson, R. (2002). What makes people dissatisfied with their neighbourhoods? *Urban Studies, 39*(13), 2413–2438.
Permentier, M., van Ham, M., & Bolt, G. (2009). Neighbourhood reputation and the intention to leave the neighbourhood. *Environment and Planning A, 41*(9), 2162–2180.
Portes, A., & Manning, R. (1986). The immigrant enclave: Theory and empirical examples. In S. Olzak & J. Nagel (Eds.), *Competitive ethnic relations*. New York: Academic.
Puhani, P. (2000). The Heckman Correction for sample selection and its critique. *Journal of Economic Surveys, 14*(1), 53–68.
Quillian, L. (2002). Why is black–white residential segregation so persistent?: Evidence on three theories from migration data. *Social Science Research, 31*(2), 197–229.
Quillian, L., & Bruch, E. (2010). *Race and class in neighborhood mobility: A conditional logit model of neighbourhood migration*. Paper presented at the meeting of the Population Association of America, Dallas, Texas, the United States.
Sampson, R., & Sharkey, P. (2008). Neighborhood selection and the social reproduction of concentrated racial inequality. *Demography, 45*(1), 1–29.
Schelling, T. (1969). Models of segregation. *The American Economic Review, 59*(2), 488–493.
Schelling, T. (1971). Dynamic models of segregation. *Journal of Mathematical Society, 1*(1), 143–186.
Skifter Andersen, H. (2003). *Urban sores: On the interaction between segregation, urban decay and deprived neighbourhoods*. Aldershots: Ashgate.
Tienda, M. (1991). Poor people and poor places: Deciphering neighborhood effects on poverty outcomes. In J. Huber (Ed.), *Macro-micro linkages in sociology* (pp. 204–212). Newbury Park: Sage Publications.
Turner, M. A., Ross, S. L., Galster, G. C., & Yinger, J. (2002). *Discrimination in metropolitan housing markets: National results from Phase I HDS 2000*. Washington, DC: The Urban Institute.
van Ham, M. (2012a). Housing behaviour. In D. Clapham, W.A.V. Clark & K. Gibb (Eds.), *Handbook of housing studies*. Chapter 3. London: Sage.
van Ham, M. (2012b). The economics of housing choice. In S. J. Smith, M. Elsinga, L. Fox-O'Mahony, S. E. Ong & S. Wachter (Eds.), *The international encyclopedia of housing and home*. Entry 94. Oxford: Elsevier.
van Ham, M., & Clark, W. A. V. (2009). Neighbourhood mobility in context: Household moves and changing neighbourhoods in the Netherlands. *Environment and Planning A, 41*(6), 1442–1459.
van Ham, M., & Feijten, P. (2008). Who wants to leave the neighbourhood? The effect of being different from the neighbourhood population on wishes to move. *Environment and Planning A, 40*(5), 1151–1170.
van Ham, M., & Manley, D. (2009). Social housing allocation, choice and neighbourhood ethnic mix in England. *Journal of Housing and the Built Environment, 24*(4), 407–422.
van Ham, M., & Manley, D. (2010). The effect of neighbourhood housing tenure mix on labour market outcomes: A longitudinal investigation of neighbourhood effects. *Journal of Economic Geography, 10*(2), 257–282.
Zubrinsky, C. L., & Bobo, L. (1996). Prismatic metropolis: Race and residential segregation in the City of the Angels. *Social Science Research, 25*(4), 335–374.

Chapter 5
Social Mix, Reputation and Stigma: Exploring Residents' Perspectives of Neighbourhood Effects

Kathy Arthurson

Introduction

This chapter focuses on reputation as a neighbourhood characteristic which may have important effects on individual residents' opportunities and experiences and social inclusion. In other words it explores the debates about neighbourhood effects, specifically through viewing the role of stigma or reputation of an area as a key independent mechanism for affecting the life chances of residents. Australian debates, in part, cite lack of 'social mix', whereby social housing neighbourhoods consist of large concentrations of homogenous housing and tenants experience high levels of disadvantage, as a significant part of the problem in contributing to the negative reputations. The media has played an active role in supporting and embellishing pathological depictions of social housing estates as sites of disorder and crime, drawing on explanations that cite individual agency and behaviour as the problems (Arthurson 2004). Residing in neighbourhoods with poor reputations is thought to have numerous impacts on residents' prospects and opportunities. The range of pertinent factors mentioned in the literature include, access to employment and educational opportunities and the shaping of residents' social networks and reactionary behaviours (Atkinson and Kintrea 2001). In turn, the perceived reputation of the neighbourhood is an important predictor of residents' intentions to leave the neighbourhood. Findings such as this question the sustainability of urban renewal policy directions that artificially create mixed tenure neighbourhoods without focusing on enhancing neighbourhood reputation (Permentier et al. 2009). Those with choice may move out of the neighbourhood leaving only the most disadvantaged residents behind, in effect working against improvement to the neighbourhood reputation (Permentier et al. 2009; van Ham and Manley 2010).

K. Arthurson (✉)
Faculty of Health Sciences, Southgate Institute for Health, Society and Equity,
Flinders University, Bedford Park, SA, Australia
e-mail: kathy.arthurson@flinders.edu.au

M. van Ham et al. (eds.), *Neighbourhood Effects Research: New Perspectives*,
DOI 10.1007/978-94-007-2309-2_5, © Springer Science+Business Media B.V. 2012

The remaining residents may then feel trapped in the neighbourhood adding to the problematic reputation (Kearns and Parkinson 2001). A related literature is also developing that explores the effects of experiencing stigma on residents' health and wellbeing (Scrambler 2009; Warr 2005; Palmer et al. 2004, 2005).

A stigmatised neighbourhood is seen to affect residents' access to employment as some employers, for instance, discriminate against potential employees residing in neighbourhoods with poor reputations on the basis of 'postcode' (Bradbury and Chalmers 2003; Palmer et al. 2005; Ziersch and Arthurson 2005). Businesses may be reluctant to locate in or near these stigmatised neighbourhoods reducing the availability of quality retail outlets and local employers (Atkinson and Kintrea 2001). Other related adverse implications raised in the debates are about the sub-standard local services and amenities, including schools that may have difficulty attracting quality teachers or a diversity of pupils (Galster 2007). A counter argument is that within some impoverished neighbourhoods specialised services are often available based on the high concentrations of residents in need that otherwise may not be available if this need falls below a certain service level 'threshold' (Atkinson and Kintrea 2001).

Related arguments are that the experience of living in an ill reputed neighbourhood may cause residents to adopt self-defeating behaviours. For instance, educational horizons and personal ambitions may be curtailed by fatalistic values linked to place of residence and the effects of experiencing spatially concentrated disadvantage or what some argue constitutes 'a culture of poverty' (Murray 1994). Murray argues that a 'culture of poverty' is sustained through the workings of the welfare state, in this instance through the development of concentrations of homogenous social housing that facilitates sameness and tenants' dependency and feckless behaviour rather than building individual agency, aspirations and capacity for change.

In a more structural conceptualisation of the issue of neighbourhood reputation that considers some of the broader societal determinants of poverty and inequality, MacIntyre and Ellaway (2000, p. 343) identify the reputation of an area as a separate dimension. Reputation is viewed as one of two "collective social functionings and practices" that are "socially patterned" but nonetheless impact on the availability of material or infrastructure resources. Within this framework they conceptualise these latter features as 'opportunity structures'. By 'opportunity structures' they refer to the features of the physical and social environment, factors that are envisaged as outside of individual control, which may be health enhancing or health damaging'. From this viewpoint the way that residents, policy makers and the business sector perceive the reputation of the neighbourhood has potential impacts on opportunity structures and behaviours of residents. The reputation affects the self-esteem and morale of the residents, the available infrastructure and who is likely to move into or out of the neighbourhood. There are other pertinent examples of how 'collective social functionings and practices' impact in a practical sense. Hastings (2009) found, for instance, that staff may vary the quality of the services provided, depending on their perception of the reputation and subsequent merit of the neighbourhood, suggesting that stigma has detrimental consequences for peoples' lives. Likewise, experiencing fear of crime (without necessarily being a victim of it) and the negative perceptions of lack of safety that are often associated with stigmatized neighbourhoods, are linked to lowered health and wellbeing outcomes for residents (Ziersch and Baum 2004; Warr 2005).

Interconnected with these debates is the proposition that the perception of control that residents have over the processes of experiencing stigmatisation is an important factor impacting on health and well being (Marmot and Wilkinson 2001). In neighbourhoods with poor reputations harsh judgements are made about residents, including depictions about the receipt of welfare by an 'undeserving poor' (Palmer et al. 2005; Warr 2005). Residents often have little control over these processes and the resultant feelings of shame, blame, devaluation and depictions of deviating from the 'normal' are interrelated with the health related effects of stigma, including decreased morale and self-esteem and increased anxiety levels (Scrambler 2009). The feelings associated with the occurrence of stigma are likened by some to the experience of racial prejudice and may have analogous detrimental effects on health (Krieger et al. 2005; Kelaher et al. 2008). In totality, in current debates about neighbourhood effects, residents' experiences of living in neighbourhoods of concentrated social housing with poor reputations that are viewed as 'dysfunctional' places are thought to doubly reinforce many of the difficulties of already socially excluded individuals and their ability to reach their full potential and become socially included.

The association of social housing neighbourhoods with stigma and poor reputations has important damaging ramifications not only for social housing residents but for other eligible individuals and families experiencing housing affordability problems. Recent social surveys undertaken in Australia by Burke et al. (2005) report that as many as 46% of Australian households, living in private rental accommodation and in receipt of Commonwealth Rental Assistance[1], claim they would never consider applying for public housing because of its poor reputation.

Thus, an overall but often understated aim of contemporary social housing estate regeneration projects, in Australia, is to improve the reputation of the neighbourhoods. This aim is thought to be achievable, in part, through changing the mix of housing dwellings in terms of quality, size and tenure type along with physical upgrading of social housing. Regeneration involves demolition, subdividing existing large backyards into a number of smaller allotments to construct two or more houses or a group of units, where there may have once been a single dwelling, thereby increasing the density of housing in the neighbourhood. Related initiatives seek to change the socioeconomic mix of the estates. Key approaches include building new housing for private sale to attract home owners into the neighbourhood and permanently relocating many social housing residents to other neighbourhoods, in effect attempting to rebalance neighbourhood social mix. Some studies suggest that increasing the balance of home owners in areas of concentrated social housing through regeneration activities is associated with enhanced reputations of the overall neighbourhoods (See for instance, Beekman et al. 2001;

[1] Commonwealth Rental Assistance is an income support payment for low income households to assist in meeting their housing costs when renting in the private market.

Atkinson and Kintrea 2001; Martin and Watkinson 2005). Likewise studies of long established mixed tenure neighbourhoods, which were originally planned that way, rather than being created through estate regeneration activities suggest that social housing residents in these types of 'mixed' neighbourhoods do not identify themselves as stigmatised. This seems in part due to residents' awareness that people residing outside of the neighbourhood and the wider public do not directly associate the neighbourhoods with social housing (Ruming et al. 2004). However, on some estates with long histories of poor reputations, despite changes implemented to social mix and other regeneration activities the problems of stigma appear more intractable to change (Robertson et al. 2008; Hastings and Dean 2003).

This chapter seeks to contribute to knowledge and understandings of the dynamics of neighbourhood stigma and reputation from residents' perspectives and whether the situations are improved post regeneration, especially with changes made to social mix. Despite the growing debates about neighbourhood effects and the question of whether living in disadvantaged areas contributes to or compounds social exclusion for already socioeconomically disadvantaged residents, little is known about the way residents perceive the reputations of their neighbourhoods (Permentier et al. 2009). As Link and Phelan (2001, p. 365) argue, much of the research is uniformed by the lived experiences of people who are deemed as stigmatised. In particular, little is known about the extent to which the experiences and dynamics of neighbourhood reputation and stigma differs between housing tenure groups (Permentier et al. 2009). Throughout history the middle classes have tended to speak for the disadvantaged as if they know what is best for them with some exceptions such as the work of Mark Peel (2003), which has enabled people to tell their own stories (see also Darcy and Gwyther 2011 in this volume). With these discrepancies in mind, while this chapter draws first on survey data it also utilises qualitative findings from in-depth interviews conducted with social housing, tenants, home owners and private renters to inform the spectrum of residents' perceptions of changes to social mix and the impacts on neighbourhood reputation. The exploration focuses on reputation as a neighbourhood characteristic which may have effects on individual residents' opportunities and outcomes. In other words it explores some of the debates about neighbourhood effects, specifically presenting the role of stigma or reputation as an independent factor.

The Case Study Neighbourhoods

The data collection for the research was conducted in three neighbourhoods, Mitchell Park, Hillcrest and Northfield all located within the metropolitan region of Adelaide. Prior to the regeneration projects commencing, all three neighbourhoods were characterised by high levels of socio-economic disadvantage and concentrations of social housing. The housing in the neighbourhoods was highly identifiable as social housing, concentrated and in run down condition. As shown in Table 5.1, post-regeneration the concentrations of social housing were reduced by as much as 50%. The neighbourhoods were extensively revitalised over the past 15–20 years

Table 5.1 Changes in concentrations of social housing in the three regenerated neighbourhoods

	Social housing concentration			
	Before (%)	After (%)	Before (n)	After (n)
Mitchell Park	75	35	1,000	350
Hillcrest	60	10.2	350	118
Northfield	27	19.9	226	238

Source: City of Port Adelaide Enfield (2010a), City of Port Adelaide Enfield (2010b), South Australian Housing Trust (2005), Phillips (1994)

with changes made to the social mix of the areas through demolition and sales of public housing, urban infill and building of new housing for private sale to attract homebuyers into the neighbourhoods. At Northfield, although as in the other two case study areas the concentration of social housing within the neighbourhood was reduced (27% to 19.9% respectively), it differed in that the overall number of social housing dwellings increased slightly (from 226 to 238), (Table 5.1). This was due to the specific project focus on urban consolidation and utilisation of vacant land that was released for new housing construction.

A questionnaire survey was posted to a random sample of 800 households across the three case study neighbourhoods and 325 surveys were completed and returned. Respondents consisted of 117 males (37%) and 199 females (63%) and there was no significant association between tenure and gender (Chi-squared=4.080, df=3, n=299, p=.253, Cramer's V=.117). After accounting for the non-deliverables (i.e. insufficient address; empty house, non-residential, n=78) the overall response rate was 45%.

Participants for the in-depth interview phase of the study were recruited through an expression of interest form that was included with the survey questionnaire. Sixty-five people returned the forms indicating their interest in participating in an interview. Forty interviews were conducted. Of these, 16 were classified as home owners living in homes that they either 'owned outright' or were 'owned with a mortgage', 14 lived in social housing and 10 were renting in the private sector. The interviews were recorded and transcribed providing insights into residents' understandings of social mix and the relationship to neighbourhood reputation. The transcripts were collated by drawing together thematic issues in order to identify patterns, similarities and differences (Rice and Ezzy 1999).

Survey Findings

Stigma and the Neighbourhood

In reflecting on the gaps in the neighbourhood effects literature it was considered important in the current study to explore the internal dynamics of social mix and stigma for home owners (owned outright and owned with mortgage) and public (social housing) and private renters. This addresses the specific point identified that little is known about whether there are any differences in residents' perspectives about neighbourhood reputation across the individual housing tenure groups, Studies also indicate that residents' perceptions of the neighbourhoods are strongly influenced by how they think others from outside the area will view it (Permentier et al. 2008, 2009). It seems that residents may internalise similar representations of their neighbourhoods as those of outsiders and give similar rankings (Blockland 2008; Curtis and Jackson, 1977 in Permentier et al. 2008). Thus, if residents feel that from the viewpoint of outsiders the neighbourhood has retained a negative reputation, despite efforts at urban renewal and changes to social mix, then we have to question whether much has improved (Permentier et al. 2009). Permentier et al. (2008) argues that for these reasons it is not only important to explore how residents understand their neighbourhood (internal perceptions) but also how they think others will assess their neighbourhood (external perceptions). In the current study it was therefore considered important to assess internal and also external perceptions of how residents thought outsiders would view their neighbourhood. Outsiders are more likely than residents to assess the reputation on a limited number of characteristics. Physical features including general aesthetics, building density, housing condition and the availability of green spaces are all important factors that influence their perceptions of the reputation of an area (De Decker and Pannecoucke 2004; Wassenberg 2004). These sorts of objective neighbourhood characteristics have been found to be more important in explaining perceived reputation than neighbourhood satisfaction but subjective assessments of neighbourhood attributes are more important in explaining neighbourhood satisfaction than perceived reputation (Permentier et al. 2010).

To this end residents were not asked the direct question of how people would rate the 'reputation' of the neighbourhood. Instead questions were posed to residents about four aspects of the neighbourhood related to stigma and then they were asked what outsiders think. Respondents were asked to rank their neighbourhood (on a scale of 1 unfavourable to 7 favourable) on four measures that encompassed the features of the housing and quality and safety of the neighbourhood. The four aspects were concerned with *House Condition*, *Attractiveness*, *Safety* and *Density*. The first two characteristics were particularly functional and physical and utilised because they generally have an obvious and immediate visual impact on both residents and non-residents perceptions of stigma. The perceptions of safety as detailed are interconnected to the health related aspects of experiencing stigma. Studies suggest that the density or assessments of how crowded the neighbourhoods are, is also implicated, as increased density is associated with lower quality of more stigmatised neighbourhoods (Permentier et al. 2008).

The ranking had two aspects as respondents were first asked to rate their neighbourhood on how they viewed it (Self-rating), and second how they thought people who lived outside of the area would rate it (External Rating). Both the internal and external rankings were included to address the point identified in the literature that the internal representations held by residents and what they consider outsiders think

Table 5.2 Self-rating of house condition by tenure groups

	Q21 House condition – self rating					
	Poor condition (1–3)		Neutral (4)		Good condition (5–7)	
	Count	%	Count	%	Count	%
Private rental	6	11.3	8	15.1	39	73.6
Owned outright	4	12.9	3	9.7	24	77.4
Owned with mortgage	10	8.5	18	15.4	89	76.1
Total	9	9.0	24	24.0	67	67.0
	29	9.6	53	17.6	219	72.8

Table 5.3 External rating of house condition by tenure groups

	Q22 House condition – outside rating					
	Poor condition (1–3)		Neutral (4)		Good condition (5–7)	
	Count	%	Count	%	Count	%
Private rental	7	14.3	8	16.3	34	69.4
Owned outright	4	13.8	8	27.6	17	58.6
Owned with mortgage	14	12.3	24	21.1	76	66.7
Total	21	21.2	17	17.2	61	61.6
	46	15.8	57	19.6	188	64.6

of the neighbourhood are highly interrelated. The difference between the two says something about the reputation of the place as perceived by residents. Consequently, stigma may seem much reduced when residents are asked for their internal perceptions but if they still feel that the neighbourhood is highly stigmatised from the viewpoint of outsiders then they may experience some of the negative health effect or other negative effects of stigma. For instance, whilst estate residents consider the presence of home owners as beneficial for the reputation of the neighbourhood, owners often express the view that outsiders still see the estates as negative (Atkinson and Kintrea 2001; Hastings 2004). In view of these sorts of findings it was important to ascertain not only whether residents thought the neighbourhoods were no longer stigmatised but whether they thought outsiders still attached stigma to the neighbourhoods despite the substantial changes undertaken to the tenure mix.

The Condition of Housing

Across the different housing tenure groups the majority of respondents gave favourable internal and external ratings when asked what they thought was the condition of the housing in their neighbourhood (Tables 5.2 and 5.3). No significant associations

Table 5.4 Self and external rating of house condition by tenure

	Overall	Social housing	Private rental	Owned outright	Owned with mortgage
	Mean (SD)	Mean (SD)	Mean (SD)	Mean (SD)	Mean (SD)
Self-rating	5.32 (1.39)	5.33 (1.69)	5.07 (1.44)	5.56 (1.31)	5.13 (1.23)
External rating	4.89 (1.51)	5.04 (1.73)	4.79 (1.47)	5.04 (1.45)	4.74 (1.34)
t value (df)	7.536 (304)	1.886 (48)	1.612 (28)	5.596 (112)	3.927 (98)
p value (η^2)	<.001 (.16)	.065 (.07)	.118 (.08)	<.001 (.22)	<.001 (.14)

Note: Critical alpha with Bonferroni adjustment is $\alpha = .01$

were found between tenure and internal rating of house condition and external rating of house condition.

When rating the condition of the housing from their own perspective (Table 5.4), owned outright gave the most favourable mean rating (5.56), whereas private renters allocated the lowest mean rating (5.07). Likewise, in evaluating how people outside of the area would rate the condition of the housing (Table 5.4), the most favourable rating was also provided by owned outright (5.04), along with social housing tenants (5.04). The least favourable ratings were from owned with mortgage (4.74) and private renters (4.79). However, none of these findings differed significantly. Respondents generally rated the condition of the housing (5.32) significantly more favourably than they thought people living outside of their area (4.89) would. However, self-rating was significantly higher than external rating only for owned outright and owned with mortgage.

Attractiveness of the Neighbourhood

Within each tenure group the majority of respondents also gave favourable internal and external ratings when asked about the attractiveness of their neighbourhood (Tables 5.5 and 5.6). The exception was private renters, where fewer respondents than in other tenures thought that outsiders saw their area as attractive, and more respondents thought that outsiders would rate their neighbourhood as unattractive (Table 5.6). There was no significant association between tenure and internal and external ratings of neighbourhood attractiveness.

When providing self and external ratings on neighbourhood attractiveness (Table 5.7), respondents in social housing gave the most favourable mean ratings (5.23, 5.00, respectively), while respondents in private rentals gave the lowest mean ratings (4.59, 4.00, respectively). No significant differences were found in the mean internal self-rating scores across the four tenure groups but the mean external rating scores were significantly different although the actual differences in mean scores between the tenure groups were quite small. The mean for private rental (4.00) differed significantly from that of own home (4.68) and from social housing (5.00).

Table 5.5 Self-rating of attractive neighbourhood by tenure groups

	Q21 House condition – self rating					
	Poor condition (1–3)		Neutral (4)		Good condition (5–7)	
	Count	%	Count	%	Count	%
Private rental	6	11.8	9	17.6	36	70.6
Owned outright	7	22.6	5	16.1	19	61.3
Owned with mortgage	10	9.0	22	19.8	79	71.2
Total	15	15.0	17	17.0	68	68.0
	38	13.0	53	18.1	202	68.9

Table 5.6 External rating of attractive neighbourhood by tenure groups

	Q22 House condition – outside rating					
	Poor condition (1–3)		Neutral (4)		Good condition (5–7)	
	Count	%	Count	%	Count	%
Private rental	6	12.5	9	18.8	33	68.8
Owned outright	11	37.9	7	24.1	11	37.9
Owned with mortgage	21	18.4	27	23.7	66	57.9
Total	23	23.2	17	17.2	59	59.6
	61	21.0	60	20.7	169	58.3

Table 5.7 Self and external rating of attractive neighbourhood by tenure

	Overall	Social housing	Private rental	Owned outright	Owned with mortgage
	Mean (SD)	Mean (SD)	Mean (SD)	Mean (SD)	Mean (SD)
Self-rating	5.06 (1.48)	5.23 (1.63)	4.59 (1.74)	5.22 (1.40)	4.93 (1.29)
External rating	4.72 (1.55)	5.00 (1.62)	4.00 (1.54)	4.83 (1.47)	4.68 (1.47)
t value (df)	5.835 (300)	2.040 (47)	2.999 (28)	4.665 (109)	2.128 (98)
p value (η^2)	<.001 (.10)	.047 (.08)	.006 (.24)	<.001 (.17)	.036 (.04)

Note: Critical alpha with Bonferroni adjustment is $\alpha = .01$

As in the ratings for house condition respondents on average rated the attractiveness of the neighbourhood (5.06) more favourably than they thought people living outside of their area (4.72) would. Self-ratings were significantly higher than external ratings only for owned outright.

Table 5.8 Self-rating of safe neighbourhood by tenure groups

	Q21 Safe neighbourhood – self-rating					
	Dangerous (1–3)		Neutral (4)		Safe (5–7)	
	Count	%	Count	%	Count	%
Social housing	11	20.8	11	20.8	31	58.5
Private rental	7	22.6	8	25.8	16	51.6
Owned outright	10	8.7	24	20.9	81	70.4
Owned with mortgage	10	10.0	25	25.0	65	65.0
Total	38	12.7	68	22.7	193	64.5

Table 5.9 External rating of safe neighbourhood by tenure groups

	Q22 Safe neighbourhood – outside rating					
	Dangerous (1–3)		Neutral (4)		Safe (5–7)	
	Count	%	Count	%	Count	%
Social housing	10	19.6	13	25.5	28	54.9
Private rental	10	34.5	7	24.1	12	41.4
Owned outright	21	18.8	30	26.8	61	54.5
Owned with mortgage	30	30.3	28	28.3	41	41.4
Total	71	24.4	78	26.8	142	48.8

Table 5.10 Rating of safe neighbourhood by tenure

	Overall	Social housing	Private rental	Owned outright	Owned with mortgage
	Mean (SD)	Mean (SD)	Mean (SD)	Mean (SD)	Mean (SD)
Self-rating	4.99 (1.47)	4.80 (1.71)	4.48 (1.46)	5.17 (1.48)	4.97 (1.28)
External rating	4.54 (1.62)	4.73 (1.73)	4.21 (1.84)	4.69 (1.53)	4.29 (1.53)
t value (df)	6.617 (305)	.574 (50)	.969 (28)	4.312 (110)	5.682 (98)
p value (η^2)	<.001 (.13)	.569 (.01)	.341 (.03)	<.001 (.14)	<.001 (.25)

Note: Critical alpha with Bonferroni adjustment is $\alpha = .01$

Safety of the Neighbourhood

The majority of respondents within each tenure group gave favourable internal and external ratings when asked to rank their neighbourhood on safety (Tables 5.8 and 5.9). There was no significant association between tenure and the internal ratings or external ratings.

When providing the internal and external ratings of their neighbourhoods on safety (Table 5.10), respondents that owned their home outright gave the most favourable mean ratings (5.17; 4.69 respectively), and respondents in private rentals gave the lowest mean ratings (4.48; 4.21 respectively). None of these differences were significant.

Overall, however, respondents rated the safety of their neighbourhood (4.99) significantly more favourably than they thought that people living outside of their area (4.54) would. Self-rating was significantly higher than external rating for owned outright and owned with mortgage respondents but not for social housing or private renters (Table 5.10).

Table 5.11 Self-rating of crowded neighbourhood by tenure groups

	Q21 Crowded neighbourhood – self-rating					
	Crowded (1–3)		Neutral (4)		Uncrowded (5–7)	
	Count	%	Count	%	Count	%
Social housing	5	10.0	14	28.0	31	62.0
Private rental	4	13.3	6	20.0	20	66.7
Owned outright	14	12.4	24	21.2	75	66.4
Owned with mortgage	14	14.0	18	18.0	68	68.0
Total	37	12.6	62	21.2	194	66.2

Table 5.12 External rating of crowded neighbourhood by tenure groups

	Q22 Crowded neighbourhood – outside rating					
	Crowded (1–3)		Neutral (4)		Uncrowded (5–7)	
	Count	%	Count	%	Count	%
Social housing	7	14.3	10	20.4	32	65.3
Private rental	7	24.1	4	13.8	18	62.1
Owned outright	20	17.5	31	27.2	63	55.3
Owned with mortgage	18	18.2	28	28.3	53	53.5
Total	52	17.9	73	25.1	166	57.0

Table 5.13 Rating of crowded neighbourhood by tenure

	Overall	Social housing	Private rental	Owned outright	Owned with mortgage
	Mean (SD)	Mean (SD)	Mean (SD)	Mean (SD)	Mean (SD)
Self-rating	5.01 (1.49)	5.06 (1.60)	4.93 (1.62)	5.02 (1.49)	4.93 (1.45)
External rating	4.74 (1.50)	4.98 (1.55)	4.59 (1.70)	4.69 (1.52)	4.64 (1.42)
t value (df)	4.637 (301)	.551 (47)	2.069 (28)	3.047 (110)	3.216 (98)
p value (η^2)	<.001 (.07)	.584 (.01)	.048 (.13)	.003 (.08)	.002 (.10)

Note: Critical alpha with Bonferroni adjustment is $\alpha = .01$

Tenure and Neighbourhood Density

Most respondents gave favourable internal and external ratings when asked about the density of the neighbourhood (Tables 5.11 and 5.12). There was no significant association between tenure and internal rating or external rating of neighbourhood density.

When rating their neighbourhoods on density (Table 5.13) respondents in social housing gave the most favourable mean internal and external ratings (5.06, 4.98 respectively), and private renters gave the lowest ratings (4.93, 4.59 respectively). No significant differences were found in the mean internal and external self-rating scores across the four tenure groups.

Respondents generally rated the density of their neighbourhood (5.01) significantly more favourably than they thought people living outside of their area (4.74) would view it. Self-rating was significantly higher than external rating only for owned outright and owned with mortgage.

Summary and Discussion of Survey Results

In summary, the majority of respondents gave favourable ratings when asked to rate their neighbourhood internally (from their own viewpoint) and externally (from how they thought people outside of the neighbourhood viewed it) on four measures related to stigma – *House Condition*, *Attractiveness*, *Safety* and *Density*. The exception was private renters where fewer respondents than expected thought that outsiders saw their area as attractive, and more than in the other housing tenure groups rated their neighbourhood as unattractive to outsiders (external rating). The findings pose the question of why the private renters gave the least favourable rankings on attractiveness of their neighbourhood.

Overall, when analysing the differences between internal and external ratings within the four housing tenure groups, on all four measures respondents' internal (self) ratings from their own perspectives, were more favourable than their external judgements of how they felt that people from outside the area would view the neighbourhoods. However, for those in social housing tenure, the internal and external ratings were never significantly different while for owned outright the differences were significant across all four measures. For owned with mortgage the findings were significant for house condition, safety and density but not for attractiveness while for private renters the internal and external ratings were only significantly different for attractiveness. These differences may reflect the situation whereby home owners (owned outright and owned with mortgage) unlike renters chose to purchase in particular neighbourhoods and are more likely to have a commitment to making the most of living there (Brown et al. 2003) or consider it acceptable for other reasons such as affordability of the housing or a location close to the city. Hence, they may view the neighbourhood favourably for a variety of reasons but still perceive that it is probably less attractive to outsiders. Conversely, social housing tenants have limited choice about where they live so may adapt to the local neighbourhood. Post regeneration most social housing tenants receive improved housing that is generally no longer easily identifiable as social housing and this may reflect the similarity in their internal and external ratings across the four characteristics. In order to see if further light could be shed on these findings the more nuanced accounts of social mix and stigma were drawn on from the in-depth qualitative interviews with residents.

Interview Findings

In supporting the findings of the survey, interviewees overall expressed the view that post regeneration the neighbourhoods were more attractive and the condition of the housing was much improved. Responses in relation to 'attractiveness of the neighbourhood' and 'condition of the housing' are discussed in tandem as the two aspects were inextricably linked. The findings on safety and density were more independent but still linked to the other aspects and each is discussed in turn.

Attractiveness of the Neighbourhood and Condition of the Housing

Interviewees commonly cited the upgrading of the housing, landscaping and other physical improvements as enhancing the attractiveness of the neighbourhoods. The names of the areas were changed, for instance, one precinct of the old Mitchell Park social housing neighbourhood was re-branded as 'The Vines'. Nevertheless, they also argued that despite these changes the stigma of the neighbourhoods had not completely dissipated, especially from the perspective of people living outside of the neighbourhoods:

> I know a lot of people would say you wouldn't want to go and live there [Mitchell Park] but er, I think it is just wonderful the development that has happened.......... The houses that are obviously privately owned and the trust houses that I would imagine that have come into private ownership they all seem to be blending in so well together and taking pride. I think it's wonderful. I say 'I live at Mitchell Park' and people sort of raise an eyebrow and then suddenly they remember 'ah that's right there's been a huge development going on there hasn't there?' and you say 'yes it's so good, it's like living at Mawson Lakes[2] with all the fancy houses! (Mitchell Park, participant 2, public housing).

In other instances it was suggested that although the density of social housing was reduced its presence lowered the attractiveness of the neighbourhood:

> Cause I can tell you going down my street which ones are the housing trust, which ones are the rentals by the rubbish they are leaving out in the street. Dumping it alongside the road, that sort of attitude, and what it actually does is actually instead of pulling up those who are in the lower socio-economic group it actually dumbs down, it drags down the neighbourhood (Mitchell Park, participant 7, owned outright).
>
> There are one or two streets that I wouldn't want to live in. That's mainly probably because they are Housing Commission homes and you might find that obviously the people that live in those homes are maybe of a poorer quality of life or something like that but that I suppose is being judgemental. It might be a very nice street to live in but I wouldn't live in it. When you look at the home and the way it has been let go, you wouldn't want to live next to somewhere like that I think (Mitchell Park, participant 118, private rental resident).
>
> The problem with it [public housing] is that there are still pockets. Like that street opposite me, it is a pocket of housing trust people and you can tell it. I think they need to be more in between and the houses need to look the same like my house and my neighbour's house. But you know they don't, they look run down, shabby. And so, you get a little group of those types of houses where these people live and they don't look after them because they don't have to. Whereas if they had houses that look the same as everyone else's there might be more incentive to look after them if the people around them were actually homeowners (Mitchell Park, participant 9, owned outright).
>
> They [South Australian Housing Trust] are all mixed in everywhere but they're not as noticeable now as they used to be. They've blended them in so that you really don't know

---

[2]Mawson Lakes is an extensive new housing development North of Adelaide that this respondent, and many other South Australians, consider a desirable place to live.

which ones are the housing trust homes and which are the bought ones. They've done it that way on purpose I think, so they don't stand out. But I know how they stand out. You look for the black numbers on the walls. So you really can't tell until you get to see the pattern as you drive around and you look and you really pay attention then you get to know which ones are which, but overall they blend in really well (Hillcrest, participant 98, owned outright).

At Hillcrest and Northfield the improved attractiveness of the neighbourhoods was linked specifically to the influx of home owners along with the association of the regeneration projects with the marketing and promotion of the adjacent and desirable new private sector housing development of Oakden:

[Oakden] was a very upmarket sort of sales promotion thing and that. They then started Hillcrest advertising when they did the redevelopment right next door to Oakden. They attached it to that. You saw it becoming more pleasurable, more likeable, more upmarket as things progressed (Hillcrest, participant 35, public housing).

If I say I live at Hillcrest they kind of look down their nose, but as soon as I tell them it's on the border of Oakden they go ahh… because it's trendy and new and modern and more expensive. Whereas they think Hillcrest is still old and crusty…. I know most of the houses in my street and in my block are new but people seem to think it's still the old Housing Trust homes and the dilapidated old homes that were here before. Perhaps because they haven't been here for a long time (Hillcrest, participant 7, owned with mortgage)

An unexpected finding was the belief, expressed from home owners (both owned outright and owned with mortgage) and social housing residents that the increased mix of private renters in the neighbourhood was detracting from the attractiveness of the neighbourhoods. In particular, reference was made to investors purchasing newer houses for sale and the older non refurbished social housing without a commitment to upgrading it, but merely to rent on the private rental market:

Probably we have more trouble with the private rental ones, of the old transportable ones – one down the street here. We've had problems with various people who have been in there (Hillcrest, participant 35, public housing).

We have one next door [private rental] and they don't look after it, he couldn't care less (Northfield, participant, 161, owned outright).

This may shed some understandings on why there were differences in the survey findings between the internal and external ratings for owned outright and owned with mortgage on attractiveness of the neighbourhood and the condition of housing. Atkinson and Kintrea (2001) found that post regeneration owners still thought outsiders viewed the neighbourhood as more negative. On the other hand for social housing residents any stigma related to the condition of housing is likely to be substantially reduced as efforts are made to blend in the regenerated or new social housing with private housing. A tenant at Mitchell Park, for instance, reported that one of her neighbours in their group of units did not want others in the neighbourhood to know that the units were public housing and that this was now possible because of the extensive refurbishments:

And he said 'ah I'd never tell anyone this is housing trust', I said 'really why?'. 'Ah no he said'. But there's nothing, no one would know, you know, they'd just think, ah a nice group of units. All the garden out the front was established by the trust and it's all nice and neat and tidy. We've each done our own things in our back yard and I thought that's really sad …It's a beautiful unit. How lucky are we, how lucky are we! (Mitchell Park, participant 2, public housing).

This supports the findings by Reutter et al. (2009, p. 298) that people living in poverty have a profound awareness of stigma and a sense that in some ways they are culpable for their predicament. Their coping strategies include concealing their discreditable status and managing the sense of dislocation between how they think they are perceived (virtual) and how they feel (actual). The changes to social housing meant that it was no longer recognisable as social housing at least for these tenants. In turn, this may help to explain why social housing residents' internal and external ratings were never significantly different in the survey as they now felt more able to conceal the fact they were living in social housing. Thus, it is not surprising that residents on the whole experience greater contentment with their homes and for social housing residents the stigma attached to the condition of their housing in the newly regenerated neighbourhood is lessened (Baker and Arthurson 2007).

Safety

While the survey found that respondents overall gave favourable ratings for the safety of the neighbourhoods, in interviews concerns were raised about safety in some specific areas where social housing was still concentrated post urban renewal. Home owners at Mitchell Park, for instance, highlighted two particular streets of social housing concentration describing it as the problematic part of the neighbourhood where they would not want to live. These streets were singled out as being 'danger zones', and 'bad places' where you would not walk down at night and were also commonly known as 'the South Australian Housing Trust part of the neighbourhood':

> [I] don't like to stereotype or whatever but there are some bad areas, streets I don't like to walk down at night [name of street] being one of them…I have heard of people, there's a lady riding her bike has had things thrown at her as she rides her bike she works at night as she works as a cleaner up at Flinders (Mitchell Park, participant 3, owned with mortgage).
>
> One day it was like being in New York. I looked out my window, and I could see these cars and these police officers in vests with guns, and swarming around the outside of the house. Then there was this big attack, and they grabbed the girl and dragged her, and she bit someone, and they had the ambulance. And it was like the streets of New York here! (Mitchell Park, participant 9, owned outright).

Density

Interviewees mostly reported the density of the neighbourhood as favourable as in the survey findings. One older couple talked about how they moved to Hillcrest specifically in order to purchase a house on a bigger than average size block of land. The original housing at Hillcrest (and the other case study neighbourhoods) was constructed on single land parcels of 725 square metres or larger and this was a common characteristic of the neighbourhood before urban renewal. As this couple home owners and had purchased prior to the renewal project commencing they were little affected by the subdivision of social housing blocks into smaller allotments to increase the density of the neighbourhood. They described their previous experience of living in a higher density neighbourhood as too noisy, too close to the neighbours and without enough land for gardening and other activities. These were common themes for other respondents that did not approve of the higher density of the neighbourhoods. Another family renting social housing had only decided to move to Hillcrest after they were offered a house on a similar size large block of land. Some of the other home owners that also still had access to larger land holdings expressed the view that they did not favour the higher density parts of the neighbourhoods:

> I would not want to live in the areas where the houses touch each other (Hillcrest, participant 40, owned outright)
>
> The negative aspect of the neighbourhood is having twice as many people and putting two houses on one block (Hillcrest, participant 55, owned with mortgage)
>
> There is more traffic than before which makes it a lot nosier (Northfield, participant 282, owned with mortgage)
>
> I do not like the urban consolidation ….there is not enough space the houses are too close together (Northfield, participant 149, owned with mortgage)

Conclusions

In general, the findings support those of other studies on neighbourhood effects, at least from the internal viewpoint of residents and across the four measures considered. Introducing home owners onto social housing estates as part of urban renewal initiatives to some extent appears to improve the overall reputation of the neighbourhoods (e.g. Beekman et al. 2001). In turn, this should help to moderate the impact of reputation in structuring opportunities and experiences for residents. A limitation of the current study was that it was not possible to conduct a before or after measure of the reputation of the neighbourhoods. Nevertheless, in the in-depth interviews many of the original social housing residents that had relocated temporarily and then moved back into the neighbourhoods talked about how the stigma previously attached to the neighbourhoods before the urban renewal projects was much improved. Likewise, home owners reported that they would not have considered living in the areas before the urban renewal projects were implemented due to the poor reputations.

The findings also add to our understandings of residents' perspectives about neighbourhood reputation across individual housing tenure groups; for owned outright, owned with mortgage, and public and private renters. The majority of respondents across housing tenure groups did not perceive their neighbourhoods as having poor reputations, although as many of them articulated all of these neighbourhoods were stigmatised prior to regeneration taking place. However, as Atkinson and Kintrea (2001) found in a study of three Scottish estates, where owner occupation was introduced in the 1990s, whilst the stigma was reduced it had not completely disappeared. The more nuanced accounts in the interviews suggested that owned outright and owned with mortgage residents still associated specific clusters of social housing in particular streets and parts of the neighbourhood as stigmatised and with safety issues. The current findings were similar to those of Beekman et al. (2001) study of ten case study estates with diversified tenure, in that owners were associated with an enhanced area reputation but in turn owners were more likely to identify problems, such as inappropriate social behaviour, as due to public housing tenants. This is not surprising given that individuals entering public housing are increasingly high need and complex tenants. In view of this situation the stigma attached to social housing is likely to increase rather than dissipate. On the whole social housing tenants did not seem overly concerned about stigma as the condition of their new improved housing meant to some extent they could not be as readily identified as social housing tenants as they had been in the past.

In general respondents felt that outsiders would rate their neighbourhoods more negatively than they themselves did. The differences were significant across all four measures of reputation for owned outright and owned with mortgage with the exception for the later of attractiveness. Although on the whole the ratings were positive it does indicate as suggested by other findings (Permentier et al. 2010) that successful urban renewal policies need to focus attention on making sure that the neighbourhoods are also seen as attractive to those living outside of them.

An unexpected finding was the issue raised about the increasing numbers of tenants in private rental as contributing to lowered attractiveness of the neighbourhood. This finding may to some extent explain the differences in internal and external ratings of private rental tenants on attractiveness of the neighbourhood. Respondents noted that often the houses in private rental were not well maintained as the function was merely to obtain a rental income for an absentee landlord. The findings suggest that from the viewpoint of many of the residents interviewed, the private rental tenure is increasingly becoming associated with stigma in regenerated neighbourhoods. This finding raises questions as the balance of housing assistance in Australia is moving to favour provision of subsidies for private rental assistance, and affordable rental housing funded through private landlords as opposed to social housing supplied and administered through government.

References

Arthurson, K. (2004). From stigma to demolition: Australian debates about housing and social exclusion. *Journal of Housing and the Built Environment, 19*(3), 255–270.

Atkinson, R., & Kintrea, K. (2001). Disentangling area effects: Evidence from deprived and non-deprived neighbourhoods. *Urban Studies, 38*(12), 2277–2298.

Baker, E., & Arthurson, K. (2007). Housing, place or social networks: What's more important for relocating tenants? *Australian Planner, 44*(4), 25–38.

Beekman, T., Lyons, F., & Scott, J. (2001). *Improving the understanding of the influence of owner occupiers in mixed tenure neighbourhoods*. Edinburgh: Report to Scottish Homes.

Blockland, T. (2008). You got to remember you live in public housing: Place-making in an American housing project. *Housing, Theory and Society, 25*(1), 31–46.

Bradbury, B., & Chalmers, J. (2003). *Housing, location and employment final report*. Melbourne: AHURI.

Brown, B., Perkins, D., & Brown, B. (2003). Place attachment in a revitalizing neighbourhood: Individual and block levels of analysis. *Journal of Environmental Psychology, 23*(3), 259–271.

Burke, T., Neske, C., & Ralston, L. (2005). *Which households eligible for public housing do not apply and why? AHURI Research and Policy Bulletin*. Melbourne: AHURI.

City of Port Adelaide Enfield. (2010a). Profile of Hillcrest, based on the 2001 and 2006 Australian Bureau of Statistics Census of Population and Housing.

City of Port Adelaide Enfield. (2010b). Profile of Northfield, based on the 2001 and 2006 Australian Bureau of Statistics Census of Population and Housing.

Darcy, M., & Gwyther, G. (2011). Recasting research on 'neighbour hood effects': A collaborative, participatory, trans-national approach. In M. van Ham, D. Manley, N. Bailey, L. Simpson, & D. Maclennan (Eds.), *Neighbourhood effects research: New perspectives*. Dordrecht: Springer.

De Decker, P., & Pannecoucke, I. (2004). The creation of the incapable social tenant in Flanders, Belgium. An appraisal. *Journal of Housing and the Built environment, 19*(3), 293–309.

Galster, G. (2007). Neighbourhood social mix as a goal of housing policy: A theoretical analysis. *European Journal of Housing Policy, 7*(1), 19–43.

Hastings, A. (2004). The stigmatisation of social housing estates: Beyond pathological explanations. *Journal of Housing and the Built Environment, 19*(3), 233–254.

Hastings, A. (2009). Poor neighbourhoods and poor services: Evidence on the 'rationing' of environmental service provision to deprived neighbourhoods. *Urban Studies, 46*(13), 2907–2927.

Hastings, A., & Dean, J. (2003). Challenging images: Tackling stigma through estate regeneration. *Policy & Politics, 31*(2), 171–184.

Kearns, A., & Parkinson, M. (2001). The significance of neighbourhood. *Urban Studies, 38*(12), 2103–2110.

Kelaher, M., Paul, S., Lambert, H., Ahmad, W., & Davey Smith, G. (2008). The impact of different measures of socioeconomic position on the relationship between ethnicity and health. *Annals of Epidemiology, 18*(5), 351–356.

Krieger, N., Smith, K., Naishadham, D., Hartman, C., & Barbeau, E. (2005). Experiences of discrimination: validity and reliability of a self-report measure for population health research on racism and health. *Social Science and Medicine, 61*(7), 1576–1596.

Link, B., & Phelan, J. (2001). Conceptualising stigma. *Annual Review of Sociology, 27*(1), 363–385.

MacIntyre, S., & Ellaway, A. (2000). Ecological approaches: The rediscovery of the role of the physical and social environment. In L. Berkman & I. Kawachi (Eds.), *Social epidemiology* (pp. 332–348). Oxford: Oxford University Press.

Marmot, M., & Wilkinson, R. G. (2001). Psychosocial and material pathways in the relation between income and health: A response to Lynch et al. *British Medical Journal, 322*(7296), 1233–1236.

Martin, G., & Watkinson, J. (2005). *Rebalancing communities by mixing tenures on social housing estates*. York: Joseph Rowntree Foundation.

Murray, C. (1994). *Underclass: The crisis deepens*. London: Institute of Economic Affairs.

Palmer, C., Ziersch, A., Arthurson, K., & Baum, F. (2004). Danger lurks around every corner fear of crime and its impact on opportunities for social interaction. *Urban Policy and Research, 22*(4), 411–426.

Palmer, C., Ziersch, A., Arthurson, K., & Baum, F. (2005). Challenging the stigma of public housing: preliminary findings from a qualitative study in South Australia. *Urban Policy and Research, 23*(4), 393–411.

Peel, M. (2003). *Lowest rung: Voices of Australian poverty*. New York: Cambridge University Press.

Permentier, M., van Ham, M., & Bolt, G. (2008). Same neighbourhood....different views? A confrontation of internal and external neighbourhood reputations. *Housing Studies, 23*(6), 833–855.

Permentier, M., van Ham, M., & Bolt, G. (2009). Neighbourhood reputation and the intention to leave the neighbourhood. *Environment and Planning A, 41*(9), 2162–2180.

Permentier, M., Bolt, G., & van Ham, M. (2010). Determinants of neighbourhood satisfaction and perception of neighbourhood reputation. *Urban Studies, 48*(5), 977–996.

Phillips, J. (1994). *Redevelopment strategy, policy and planning branch*. Adelaide: Department of Housing and Urban Development.

Reutter, L., Veenstra, G., Love, R., Raphael, D., & Makwarimba, E. (2009). Who do they think they are, anyway? Perceptions of and responses to poverty stigma. *Qualitative Health Research, 19*(3), 297–311.

Rice, P., & Ezzy, D. (1999). *Qualitative research methods: A health focus*. Melbourne: Oxford University Press.

Robertson, D., Smyth, J., & McIntosh, I. (2008). *Neighbourhood identity: Effects of time, location and social class*. York: Joseph Rowntree Foundation.

Ruming, K., Mee, K., & McGuirk, P. (2004). Questioning the rhetoric of social mix: Courteous community of hidden hostility. *Australian Geographical Studies, 42*(2), 234–248.

Scrambler, G. (2009). Health-related stigma. *Sociology of Health and Illness, 31*(3), 441–455.

South Australian Housing Trust. (2005). *Mitchell Park Urban Renewal Project, Analysis and determination of the social, physical and financial changes and outcomes in Mitchell Park*, Adelaide: Study Brief.

van Ham, M., & Manley, D. (2010). The effect of neighbourhood housing tenure mix in labour market outcomes: A longitudinal investigation of neighbourhood effects. *Journal of Economic Geography, 10*(2), 257–282.

Warr, D. (2005). Social networks in a 'discredited' neighbourhood. *Journal of Sociology, 41*(3), 285–305.

Wassenberg, F. (2004). Renewing stigmatised estates in the Netherlands: A framework for image renewal strategies. *Journal of Housing and the Built environment, 19*(3), 271–292.

Ziersch, A., & Arthurson, K. (2005). Social networks in public and community housing: The impact on employment outcomes. *Urban Policy and Research, 23*(4), 429–445.

Ziersch, A., & Baum, F. E. (2004). Involvement in civil society groups: Is it good for your health? *Journal of Epidemiology and Community Health, 58*(6), 493–500.

Chapter 6
Theorising and Measuring Place in Neighbourhood Effects Research: The Example of Teenage Parenthood in England

Ruth Lupton and Dylan Kneale

Introduction

This chapter is designed to make a dual contribution to this volume on neighbourhood effects. In one respect, it can be read as an empirical investigation of neighbourhood effects on the likelihood of becoming a teenage parent in England. Policies to reduce teenage parenthood have been a prominent element of government policy on social exclusion since the late 1990s, and are apparently informed in part by the purported existence of neighbourhood effects, although evidence of these effects is remarkably weak. Here we test for such effects by matching neighbourhood data for the first time to a longitudinal study of people born in 1970. Using this rich data source enables us to take account of the antecedents and circumstances of people who become parents in their teens, as well as their neighbourhood characteristics at the age of 16.

The chapter also aims, however, to highlight some of the conceptual problems in much existing neighbourhood effects research around the role of place and the importance of geography, and to illuminate in a transparent way some of the difficulties in putting these right. We hope that the chapter will prompt further conceptual and methodological advances in the measurement of neighbourhood effects in general, as well as adding to the evidence base on teenage parenthood.

R. Lupton (✉)
Centre for Analysis of Social Exclusion (CASE), London School of Economics and Political Science, London, UK
e-mail: r.lupton@lse.ac.uk

D. Kneale
International Longevity Centre, London, UK

Centre for Longitudinal Studies, Institute of Education, London, UK
e-mail: dylankneale@ilcuk.org.uk

M. van Ham et al. (eds.), *Neighbourhood Effects Research: New Perspectives*,
DOI 10.1007/978-94-007-2309-2_5,

Some Problems with Neighbourhood Effects Research

In a previous paper (Lupton 2003), which reviewed the state of 'neighbourhood effects' research and its usefulness for policy purposes, Lupton concluded that the field was hampered by two significant problems. One was its disciplinary divide. Qualitative researchers have for many years been interested in understanding the ways in which neighbourhoods work as social and physical spaces; the idea that neighbourhoods have 'effects' on individuals is implied, although the term is rarely used. Many studies only focus on one neighbourhood so are not in a position to compare whether some neighbourhoods have greater 'effects' than others – the purpose is to describe the processes at work. More recently, a substantial body of quantitative research has emerged, testing the extent to which neighbourhood characteristics influence measurable individual outcomes i.e. whether (all other things being equal) it makes a difference being in one kind of neighbourhood versus another. The paper argued that whilst in theory the former (qualitative) kind of research has the potential to generate sophisticated understandings of neighbourhood/individual interactions which could then be quantitatively tested, in practice quantitative research tends to use very simple measures of neighbourhood, usually driven by the data available, and often fairly meaningless in terms of the hypotheses being tested. A further difficulty is that both qualitative and quantitative researchers whose primary interest is in neighbourhood effects may explore multiple 'outcomes' within the same studies – for example school drop-out, teenage parenthood and unemployment. This generalist approach means that the importance of neighbourhood for any given 'outcome' is tested without an in-depth knowledge of those particular specialist fields which could inform an understanding of the mechanisms at work and the ways in which they operate spatially. Specialists in these issues, by the same token, are not necessarily interested in neighbourhood dimensions.

The second problem was that inappropriate spatial scales are often used for the mechanism ostensibly being tested, usually because researchers are limited by the boundary data available to them, but also for the reasons given above – researchers may be operating without any theoretical underpinning for testing particular variables at particular spatial levels. A by-product is that different studies use different geographies, making it very hard to discern consistent patterns emerging as the volume of research findings expands. The paper concluded that these difficulties contribute to inconclusive and contradictory results which do not yet provide a convincing evidence base. It argued for a much closer and also a more critical collaboration between quantitative and qualitative researchers so that qualitative understandings of place are better reflected in quantitative models.

Before and since this paper was written, others have come to some of the same conclusions. Galster's (2011) chapter in this volume argues that the field needs to advance from tests for evidence of neighbourhood effects to specific testing of particular mechanisms, and it is notable that the seminar series on neighbourhood effects which has spawned this volume has explicitly aimed to bring together qualitative and quantitative contributions. So progress is being made. However, we should not underestimate the conceptual and methodological difficulties in overcoming these problems. Our hope is that in laying some of these difficulties bare within this chapter, we can encourage others to tackle them explicitly.

The approach we take is as follows. Taking the issue of teenage parenthood, which is a specialist area of research for one of us, we approach an enquiry into neighbourhood effects from a review of the existing (qualitative and quantitative) literature on the social and economic processes which affect the timing of parenthood. We ask to what extent these processes could be spatialised, and at which spatial levels. Using the British Cohort Study (BCS70) we then test for evidence that spatial differences matter, first using the standard geographies available and then using bespoke geographies designed to represent better the spatial scales over which we believe the relevant mechanisms operate. We describe in a transparent way the difficulties we face in matching theory and data, and assess the implications this has for the meaning of our findings.

Influences on Teenage Parenthood

Since the 1960s, Britain has seen declining birth rates, increased childlessness, and more out-of-wedlock births, along with a growing divergence in fertility patterns between women in different social classes (Joshi 2002). Middle class women have increasingly delayed fertility while patterns of young parenthood have persisted more strongly among working class women.

This divergence is typically explained in terms of the opportunity cost of childbearing. Women who face the highest levels of wage penalties and missed chances for career progression through taking time out of the labour market as mothers (opportunity costs) will be those who delay this process the most (Becker 1991; Joshi 1998, 2002). Several studies have consistently outlined the link between higher educational levels and delayed parenthood (for example Kiernan and Hobcraft 1997; Lappegard and Ronsen 2005; Rendall et al. 2005; Rendall and Smallwood 2003; Smith and Ratcliffe 2009).

An alternative theory is that early parenthood reflects non-normative values around fertility. In recent years, this position has been framed within a discourse of 'social exclusion' (Burchardt et al. 2002) which posits that exclusion from mainstream social, productive and consumptive activities in society may lead to the adoption of values or behaviours (such as early motherhood) that reject or render impossible approved or normative routes towards 'inclusion'. However, other authors prefer an analysis that suggests that value differences around fertility between social classes are shaped not only by current labour market opportunities but also by historic patterns, gender roles, and attitudes towards family and community. The social exclusion perspective tends to assume that early parenthood is ultimately unwanted, privileging the values of the included class as normative and desirable, although some existing literature finds early motherhood to be specifically desired (Afable-Munsuz et al. 2005; Kiernan 1997), and early mothers to be strongly tied to the activities of motherhood (Holmlund 2005).

Table 6.1 Influences on teenage motherhood: potential place effects and scales

Influence	How place might have an effect	Relevant scale
Opportunity cost	The state of the local labour market	Labour market
	School quality	School attended by respondent
Social class, values and social exclusion	Transmission of social class values in which early motherhood is seen as a valued transition into adulthood Perceptions of opportunity costs	Probably neighbourhood, although likely that social class norms and values also operate over larger areas: localities, and perhaps even cities and sub-regions
Characteristics of social networks	Peer group characteristics Levels of community efficacy and social control Intergenerational support and high social capital	Probably neighbourhood, although characteristics of wider and neighbouring areas possibly important
Other neighbourhood factors	Levels of residential turnover, levels of neighbourhood change, quality of local youth, health and advisory services	Probably neighbourhood. For some measures (e.g. labour market change, larger scales may be necessary)

A further theoretical perspective on fertility emphasises that ideas are diffused across social networks, and that both socioeconomic changes and changes in attitudes are necessary before new ideas on fertility are adopted (Casterline 2001; Lesthaeghe and Neels 2002; Seltzer et al. 2005). Applying this to the case of early parenthood is analogous to saying that women are choosing to become early parents because this is sanctioned within their social network. Clearly part of the sanctioning will come from shared values around motherhood, contraception and abortion among people of the same social class who occupy the same networks. Class differences in the structure of social networks may also be influential. For example, the closer geographical proximity of generations within traditional working class communities may enable stronger transmission of attitudes. However, the nature of social networks can vary by locality, independent of class, and is partly conditioned by neighbourhood design, community facilities, housing market conditions, level of population turnover and so on. These factors may also be important in their own right. Institutional theories emphasize the role of schools, businesses, political organizations and social service agencies, and their moderating effects. South and Crowder (1999) outline the case for the behaviour of external (institutional) adults within the neighbourhood, for example teachers, the police and so on, in predicting early parenthood. This has also been discovered in UK empirical studies of early parenthood, where the educational expectations of teachers were found to be significant predictors of teenage parenthood (Kneale 2010).

Evidently all of these influences on teenage parenthood could be expected to vary across space, although not all at the same spatial scale. Social networks are most likely to operate at the neighbourhood level, labour market effects and socio-cultural effects at a variety of different geographies (Table 6.1)

In a companion paper to this chapter (Lupton and Kneale 2010) we review what the existing literature tells us about place effects at these different spatial levels. Two UK studies that examine the effect of generalised neighbourhood poverty do not find significant neighbourhood deprivation effects on teenage parenthood after taking account of individual predictive factors (McCulloch 2001; Sloggett and Joshi 1998). Sloggett and Joshi used an electoral ward-level deprivation measure, McCulloch a composite of local authority districts (a much larger scale). McCulloch did suggest that a place effect remained that was not specifically related to neighbourhood poverty. A number of US studies *do* find associations between generalised poverty and teenage parenthood even within small areas (see for example: Crane 1991; South and Crowder 1999). Galster et al. (2007), however, do not, once instrumental variables were used to minimise selection effects and endogeneity. Importantly this study only examined births up to and including age 17, a particular subset of those considered in the wider literature.

Some studies have looked at place effects at a larger geographic level in tests for labour market effects. Using a limited range of family background controls (family of origin income and structure), Ermisch and Pevalin (2003), found that the 1-year lagged unemployment rate in the travel-to-work area (TTWA) was positively related to the hazard of becoming a teenage mother. Similarly, Del Bono (2004) also used the employment rate at the county level in her study, which was found to be significant in increasing the risk of non-marital fertility.

Quantitative tests of the influence of 'cultural' or social class values are relatively rare in the UK literature, perhaps because direct measures of attitudes at sub-national level are hard to find. Some studies have explicitly explored the impact on early parenthood of living in places with different social networks and social capital. Driscoll et al. (2005) found that both community opportunity and efficacy influenced the transition to teenage birth, with higher levels reducing the likelihood of parenthood occurring. Haveman and colleagues' study (1997) found the proportion of young adolescents belonging to religious organisations in a census tract area lowered the likelihood of teen childbearing, controlling for individual characteristics. This was alongside other neighbourhood characteristics, including state expenditure on family planning policies and so on. Outside the immediate teenage parenthood field, there is also a large number of studies looking at neighbourhood influences on adolescent sexual behaviour. For example, Cleveland and Gilson (2004) included the proportion of single parent families as an indicator of social control and found this to be associated with the number of sexual partners for both males and females, although less so for females where the effect was mitigated by individual level family structure. Browning and colleagues found a similar dynamic between family-centred processes and neighbourhood as was the case for males in Cleveland and Gilson's study, although in their study this applied to both men and women and collective efficacy was only effective where individual family level processes were weaker (Browning et al. 2005). More specific mechanisms are outlined in Bell's UK study of teenage fertility. Here, an algorithm of a 'carnivalistic' attitude of young people towards sex as a reaction to the temporary nature of the surrounding local population led to higher engagement in risk behaviours for early pregnancy. Seasonal employment patterns in these areas led many young people to shift their goals away from becoming a good employee to becoming a good parent, and this resulted in a wider neighbourhood acceptance of early parenthood and a low acceptance of abortion (Bell et al. 2004). Bell's findings tie in with the social disorganisation/community efficacy theory proposed by several studies of inauspicious fertility events.

We can thus see that existing work comprises some generalist studies (testing teen parenthood alongside other outcomes) and some that attempt to test specific hypotheses, with labour market and network explanations most fully tested amongst these. Often the choice of geography is not explicitly justified, being dependent on the data available. It is unusual for more than one geography to be tested in the same study, or at least to be reported. It is possible that 'non-results' at other geographies are not reported, although these may tell an important story in themselves.

Testing for Place Effects on Teenage Parenthood in the UK Using the BCS70

In this chapter, we attempt to test specific influences on teenage parenthood. We use data from the British Cohort Study (BCS70), which started following around 17,000 people from their birth in 1970 to the present day, collecting detailed information about all major domains of life, including health, intelligence and cognitive function, educational attainment, family and socio-economic circumstances, occupational history, parenting and social attitudes (Elliott and Shepherd 2006; Wadsworth et al. 2003). Follow-up sweeps were undertaken at 5, 10 and 16, and at 26 (a postal survey), 30 and 34. We look at whether the characteristics of the cohort members' neighbourhood at 16 is related to their likelihood of becoming a parent before the age of 20, after controlling for other factors. We are interested in age 16 data as this is a key transition point into adulthood, being the age at which young people can legally leave schooling.

In line with most studies on this topic, we confine our enquiry to teenage mothers, rather than fathers. We use retrospective fertility data collected at age 30 and 34 to construct 6,065 fertility histories for female cohort members (some of the approximately 8,500 original female members did not continue with the study). In total at age 16 years, 88% of records had a valid postcode[1], enabling us to match to neighbourhood data, and more had a local authority recorded, although not a postcode. Because of the additional computational difficulty of mapping data for Scotland and Wales, we limit the sample to those living in England. Sample size is also reduced when we include only those cohort members for whom all relevant background information was collected at relevant ages. While one of the great advantages of using this data source is the wide range of data available to control for individual and family background characteristics, some of the data are missing in some years. To deal with this, we adopted multiple imputation techniques (see Goldstein 2009 for an example using

[1]For this chapter we also returned to the original survey paper copies to re-transcribe mistranscribed postcodes in order to improve the quality of the data.

Table 6.2 Sample size reductions

Total in BCS70 at birth	17,000 (approx.)
Total females	8,500 (approx.)
Total females for whom fertility information is known	6,065
Total females for whom fertility information is known and local authority recorded, and for whom relevant information is known or can be imputed	4,865
Total females for whom fertility information is known and postcode recorded, and for whom relevant information is known or can be imputed	3,631

the cohort studies; Royston 2004; Royston 2005), facilitated by using the same measurements from earlier time points, as well as other predictors, to estimate values for missing data across 21 replicate sets. After multiple imputation, we are able to include 4,865 female cohort members for analyses at large scale geographies (local authority level and above) and 3,631 for analysis using smaller geographies. Eight percent of the English cohort members gave birth before their 20th birthday. We show these sample size reductions in Table 6.2 to illustrate the sample size problem with enquiries like this. By their nature, disadvantageous outcomes are experienced by a minority of the population, which means that very large sample sizes are needed to model predictors with any degree of methodological sophistication.

To analyse place effects, postcodes were matched[2] to Census boundaries using GIS software, enabling Census data to be matched to the individual records. Note that the age 16 sweep of the survey does not coincide with a Census. The closest Censuses occurred in 1981 and 1991. The inability to match data in corresponding years is a common problem working with any UK data prior to 2000 when much better inter-censal data became available. In this case, we decided to use the 1981 Census data to represent 1986 neighbourhood characteristics. Rather than using general measures (such as overall poverty or deprivation rates) we selected Census variables which reflected as closely as possible the theoretical propositions identified in the literature. Unusually, we were also able to draw on attitudinal data from the survey itself from parents and children about educational expectations and the desire to have children. Table 6.3 shows the measures we use, and also where there were no measures available. It demonstrates that this data source enables testing of opportunity cost and value mechanisms much better than it does social networks or other characteristics of neighbourhood.

We were also able to control for a wide range of other characteristics (other than neighbourhood ones) that are known to effect teenage parenthood, using data from the survey itself.[3] Based on previous literature including systematic reviews by Harden et al. (2006) and Imamura et al. (2007), and other studies including those

[2]Since postcoded data cannot be released for confidentiality reasons, the matching process was done by staff at the Centre for Longitudinal Studies, to whom we are immensely grateful.

[3]In theory it should also be possible to minimise 'selection effects' – the notion that people with certain characteristics select into certain neighbourhoods - by constructing residential histories linked to other events such as unemployment, loss of income or change in health status, although we do not do so in this chapter.

Table 6.3 Place measures used

Mechanism	Relevant measures of place	Actual measures used
Opportunity cost	Census measures of labour market opportunities	(i) Industrial structure (% males in each industrial sector) (ii) % economically active young people 25–34 not in employment (see note a) (iii) % economically active adults seeking work
	School quality indicators	No measures available for school quality
Social class, values and social exclusion	Social class composition of neighbourhood	% households headed by each social class group (Census)
	Attitudes to early parenthood and abortion	No measures available
	Measures of labour market expectation and participation	(i) Parental and child expectations (BCS 70 – see note b) (ii) % of married women with children 0–4 who are employed (Census)
	Levels of educational participation in local areas	(i) % of women aged 16–24 who are students (Census) (ii) % of children disliking school or playing truant (BCS70)
Characteristics of social networks	Whether other young people are married and have children	% of women aged 16–19 and 20–24 who are married.
	Attitudinal variables reflecting neighbourliness and social participation and control	No measures available
	Measures of intergenerational network	No measures available
Other neighbourhood factors	Including Census data on housing, migration data, intercensal change (e.g. in employment), local data on service provision	% households in social housing Other census measures e.g. migration could be explored in the future

Note: (a) "not in employment" in the Census includes students, so data for the 16–24 age group is likely to reflect educational participation rather than non-employment. For this reason we use older age groups. (b) Attitudinal data available in BCS70 includes parental and children's expectations of further education and parental attitudes to maternal employment, "high value of children", "a better life for women", and "child independence". We derive these data from the average values in sub-region. Sub-regions are a bespoke geography used in this chapter to represent contiguous old counties; we describe their derivation later in this chapter

that have used data from the cohort studies (Hobcraft 2008; Hobcraft and Kiernan 2001; Kiernan 1997; Sigle-Rushton 2004), we developed a set of individual predictors including:

- *Educational Expectation Measures*: expectations reflective of leaving school at the minimum age, leaving school at 18, entering higher education and being uncertain about the future educational trajectories.

- *Socioeconomic Factors*: measuring having a father in a manual social class (Ermisch and Pevalin 2003), living in social housing (Hawkes et al. 2004), and living on state supported benefits (Harden et al. 2006).
- *Educational Measures*: Early maths and reading ability, dislike of school and truancy/school attendance.
- *Behavioural and Philoprogenitive Measures*: Behavioural tendencies at age 16, and philoprogenitive tendencies (as measured by the importance to a cohort member of having their own family in the future, collected at age 16).
- *Home Learning Environment and Demographic Measures*: Family structure at age 16 and age of the cohort member's mother at birth, as well as parental interest in children's education.

Of course, it might well be argued that some of these measures are influenced by people's home neighbourhoods. If this is the case, the effect is likely to underestimate neighbourhood influences, although selection effects, as discussed later, have the opposite effect.

We model the probability of becoming a teenage mother versus not becoming a teenage mother using binary logistic models. In these models we include all the family level and individual level predictors listed above, as well as the specific characteristics of place shown in Table 6.3. We do this for several different geographies, an approach we explain as the chapter unfolds. For the higher geographies that we test (labour market and sub-region) we also adopt a multi-level modelling approach (Rabe-Hesketh and Skrondal 2008; Rasbash et al. 2004; Snijders and Bosker 1999) to establish how much of the variation in the probability of becoming a teenage mother is accounted for by place (simply the difference between living in one place rather than another, rather than the specific characteristics of the places). At our smallest geography, the electoral ward, such an approach is inappropriate, since fewer than 7% of the sample lived in a ward where another four or more cohort members were present, and less than half of the sample lived in a ward with even one other cohort member present. For wards, therefore, we do not attempt to establish place effects *per se*, but concentrate on the effect of specific ward characteristics, as suggested in Table 6.3.

Initial Results and Further Geographic Investigations

To begin with, we adopt what might be regarded as a traditional approach to neighbourhood geography. We matched our survey data to the existing Census geography most closely approximating to 'neighbourhood': the electoral ward. Wards in England are typically of around 5,000 people in size. Those which contained cohort members in our sample had an average population of 8,140, although some wards contained as many as 38,000 in the population and some as few as 390.

This reveals some apparent neighbourhood effects (Table 6.4). Young women in wards with higher proportions of manual social classes were more likely to become teenage mothers, as were those in wards with high marriage rates among young

Table 6.4 Odds of becoming a teenage mother for BCS70 women for selected ward level characteristics controlling for individual level factors (see Table 6.5 for full controls)

Neighbourhood characteristic	Standardised proportion of females aged 16–24 in further or higher education	Standardised proportion of males in social class IV & V	Standardised proportion of women aged 16–19 married
Odds ratio (t-statistic)	0.768** (−3.04)	1.180* (1.97)	1.210* (2.31)

$*p<0.05$; $**p<0.01$

women, and those in wards with a low proportion of young women progressing to further education. These results suggest the influence of values around fertility. The latter two results were robust to the social class structure of the ward, suggesting that there can be localised patterns of education and marriage among working class communities: local formations of class. An increase of one standard deviation above the mean in the proportion of female students was associated with an odds of 0.77 (23% reduction) of becoming a teenage mother. The higher the proportion of married women in a ward, the higher the odds of becoming a teenage parent – an increase of one standard deviation above the mean of the proportion of married women across all wards resulted in a 21% increased odds of teenage motherhood.

If we were to take these results at face value without any further investigation, we might draw the conclusion that that at the neighbourhood level, localised values influence fertility patterns. Some policy makers would interpret this as an imperative to encourage young women to pursue education and perhaps to defer marriage. Others (in the spirit of the "Moving to Opportunity" demonstration programme see Goering et al. 2003),would argue that individual interests would be best served by facilitating moves away from wards with low female education and high marriage rates. However, what if wards do not represent a meaningful geography but simply a convenient one? Variations in ward characteristics may reflect differences between bigger geographical units, such as cities, labour markets or sub-regions, at which differences in the propensity to become a parent, marry early or continue in education are structured by historic patterns of industry and employment, religion or culture. This would not mean that ward differences are unimportant, merely that we would not look to this geographical unit to provide explanation or action.

It might also mean that we could miss significant place effects. If for example, very weak labour demand within a labour market area deterred young women from entering the labour market, this effect would not necessarily be influenced by the behaviour of others in the immediate neighbourhood: people might be equally deterred whether they lived in a high marriage or low marriage rate neighbourhood. The absence of ward differences in this case might lead us to conclude that there was no labour market effect, but ward is the wrong geography to capture labour market effects.

To address some of these difficulties, we therefore develop our analysis by testing for place effects at higher geographies. To test for wider labour market effects we first use another existing geography – the Travel to Work Area (TTWA). These are areas in which the majority of the working age population actually worked in the area. The minimum size is 3,500 and the largest accounts for several million workers. In 1981, in UK wide terms 334 travel to work areas existed, although the number has since reduced as commuting distance increased. In our data, 228 English TTWAs are represented. TTWAs are commonly used as a standard labour market geography and reflect the geography of work opportunities better than, for example, administrative geographies such as local authorities. However, they are by no means perfect in reflecting the work horizons of many youth. Firstly, for large cities with well developed public transport systems, TTWAs become very large, governed by commuting into the centre. London, for example is one TTWA, although it would be very unusual for someone in the outer South West of the city to travel across it to the far North East. More localised commuting flows become subsumed in the definition process by in-out flows. Second, we know that many young people with low skills or from manual social class backgrounds will have very limited 'travel-to-work' areas, partly because of the cost of transport relative to wages and partly because of traditional expectations of local employment (Green and White 2007). Third, it might be argued that some TTWAs are too small for our purposes. Attitudes to work and fertility might be influenced, for example, not by the fact of being in a particular labour market around a large town, but by wider cultural and industrial heritages – being in the peripheral and rural agricultural East of the country, for example, rather than being in a particular town and its hinterland. In our sample the average number of cohort members per TTWA was 17 with the smallest containing just one and the largest 245.

To tackle these issues, we experimented with two approaches. One was to develop bespoke geographies, in other words to divide the country into bounded smaller units, not necessarily of the same size, that represent clusters of areas which were similar in terms of the variables we were interested in. This approach seeks to create real named places, ideally with enough cohort members in them to enable testing of any one against any other. Initially we constructed a sub-regional geography by combining contiguous local authorities in order to build 29 areas which had sufficient cohort members within them to provide robust sample sizes for statistical testing, while reflecting as closely as possible real geographical and topographical divisions (according to our existing knowledge). These areas averaged 211 cohort members each in our sample, with up to 378 in the largest (Outer London). We then attempted a number of more complex approaches to the 'bespoke geography' question, building up from wards by joining contiguous wards with similar characteristics to create clusters – for example clusters of high manufacturing wards, which we also named (for example 'Pennine Textile Belt'). These attempts did not produce a convincingly better geography than the existing ones. A principal difficulty is that a contiguous boundary approach can lead to long thin areas covering many miles. Although the areas at each end have more in common with each other on single variables than they do with the areas around, it becomes hard to argue that they represent in any sense a geography of shared identity. We therefore did not persist with the clusters.

Our second approach was more intuitive: to maintain the existing geographies (i.e. wards), but to take into account their surroundings by weighting the ward values on any variable according to the values of the surrounding wards. For the purposes of this chapter we adopted a simple approach to this, following work by Rae (2009). Using GIS software we identified each ward's first order neighbours (those with which it shares a boundary) and calculated a mean neighbour score for the variables we were interested in. We then inserted these scores into our regression model, thus addressing the question "does it matter what kind of neighbourhoods surround a person's own neighbourhood?".

Results

To summarise, these geographical investigations left us with two geographies to test in addition to ward (Sub-region and TTWA) and a new set of variables (nearest neighbour scores) at the ward level. We present the results as odds ratios in Table 6.5. These show the odds of experiencing teenage motherhood (versus not experiencing it) for one characteristic relative to another, or for continuous measurements for each additional unit increase. A value over one indicates a higher likelihood of teenage motherhood, and a value under one that the likelihood of teenage motherhood is reduced. Beneath the values in Table 6.5, the t-values are displayed in brackets. We show the full output in Table 6.5 to allow the reader to view the effect of all covariates on the probability of teenage motherhood. Model 2 in Table 6.5 shows that low parental education expectations, residence in social housing, residence in the North of England, having parents with little interest in education, family receipt of unemployment or sickness benefits, being aggressive and having a younger mother are all predictive of teenage motherhood; other factors are also included in the model. For example, living in social housing is associated with a 70% increased probability of teenage motherhood (OR: 1.702) compared to living in owner occupied housing. For simplicity, we do not show the full models again after Table 6.5 for reasons of parsimony, only the effect of the neighbourhood factors[4].

Sub-Regional Results

At the sub-regional level, an empty model (Model 1) with no individual predictors suggests that only a moderate amount of the variance (3%) in teenage parenthood can be attributed to the sub-regional level. The introduction of explanatory variables

[4]While there are no large changes in the patterns of the individual and family level predictors with the addition of other neighbourhood factors, some variation does occur. Full output is available on request from d.kneale@ioe.ac.uk/dylankneale@ilc.org.uk

Table 6.5 Models showing odds of becoming a teenage mother with significant sub-regional level predictors of teenage motherhood

	Model 1			Model 2			Model 3			Model 4			Model 5		
	Coeff.	Std err.	Sig	Coeff.	Std err.	Sig	Coeff.	Std err.	Sig	Coeff.	Std err.	Sig	Coeff.	Std err.	Sig
Parental education expectations age 16 (base: leave at the minimum)															
Leave at 18				0.749	−1.850	BS	0.752	−1.820	BS	0.749	−1.850	BS	0.750	−1.840	BS
Higher education				0.815	−0.950		0.817	−0.940		0.811	−0.970		0.813	−0.960	
Can't say				0.854	−0.460		0.856	−0.460		0.857	−0.440		0.859	−0.440	
Cohort member education expectations age 16 (base: leave at the minimum)															
Leave at 18				0.911	−0.400		0.914	−0.390		0.908	−0.420		0.915	−0.380	
Higher education				0.599	−1.410		0.600	−1.420		0.596	−1.420		0.600	−1.420	
Can't say				0.850	−0.950		0.848	−0.960		0.849	−0.950		0.851	−0.940	
Standardised reading score age 10				0.828	−1.630		0.829	−1.620		0.823	−1.700		0.828	−1.640	
Standardised maths score age 10				1.014	−0.150		1.013	−0.130		1.020	−0.200		1.015	−0.150	
Dislike of school age 16 (base: don't like school)															
Like school				0.846	−0.960		0.845	−0.970		0.841	−1.000		0.849	−0.940	
Somewhat like school				0.748	−1.630		0.746	−1.640		0.745	−1.650		0.746	−1.630	
Tenure age 16 (base: owner occupation)															
Social housing				1.706	−4.000	**	1.702	−3.980	**	1.711	−4.040	**	1.701	−3.960	**
Private rented and other				1.632	−1.850		1.652	−1.890		1.660	−1.910		1.674	−1.930	
Behavioural score: aggression				1.229	−3.570	**	1.226	−3.520	**	1.224	−3.520	**	1.226	−3.550	**

(continued)

Table 6.5 (continued)

	Model 1			Model 2			Model 3			Model 4			Model 5		
	Coeff.	Std err.	Sig	Coeff.	Std err.	Sig	Coeff.	Std err.	Sig	Coeff.	Std err.	Sig	Coeff.	Std err.	Sig
Philoprogenitive tendencies (base: children matter very much to adult life)															
Children matter somewhat				0.929	−0.480		0.925	−0.500		0.931	−0.460		0.926	−0.500	
Children don't matter				0.807	−1.210		0.806	−1.220		0.809	−1.200		0.808	−1.190	
Mum's age first birth				0.898	−6.000	***	0.898	−5.990	***	0.899	−6.000	***	0.899	−5.950	***
Family receipt of unemployment benefits				1.412	−2.090	*	1.412	−2.100	*	1.422	−2.120	*	1.399	−2.040	
Mother's interest in child's education age 16 (baseline: healthy interest)															
Some interest				1.239	−1.480		1.235	−1.460		1.239	−1.480		1.234	−1.450	
Unhealthy interest				1.624	−2.500	*	1.639	−2.540	*	1.624	−2.500	*	1.655	−2.600	*
Can't say				1.220	−1.140		1.217	−1.130		1.215	−1.120		1.212	−1.110	
Father's social class (base: professional and managerial)															
Lower than professional and managerial				1.207	−0.840		1.204	−0.830		1.209	−0.860		1.204	−0.840	
Not working or no father				1.276	−0.790		1.275	−0.790		1.284	−0.820		1.261	−0.770	
Parental structure age 16 (baseline: two natural parents)															
Reconstituted family				1.101	−0.590		1.110	−0.640		1.101	−0.590		1.114	−0.670	
Lone parent family				1.178	−0.590		1.181	−0.600		1.181	−0.600		1.194	−0.640	
School days missed age 10				1.004	−0.430		1.003	−0.410		1.003	−0.340		1.003	−0.410	

Standardised vocabulary score age 5		0.931	−1.620		0.932	−1.590		0.931	−1.630		0.931	−1.610	
Region in the north of England		1.640	−3.910	**	1.390	−2.120	*	1.365	−2.210	*	1.354	−2.160	*
Standardised proportion of males employed in manufacturing in sub-region					1.142	−1.730	BS						
Standardised proportion of married women working in sub-region								1.200	−2.610	*			
Standardised proportion of households in privately rented housing in sub-region											0.780	−2.530	*
ρ (Intraclass correlation)	0.032	0.006			0.004			0.003			0.002		
N in model	4,865	4,865			4,865			4,865			4,865		

BS borderline significant ($p<0.08$)
$*p<0.05$; $**p<0.01$; $***p<0.001$

on the individual and family level reduced this effect even further to less than 2% (not shown) and less than 1% with the addition of region (Model 2). When we look at the specific characteristics of sub-regions (Models 2–5) we find that some of these are significant, namely:

- Proportions of men employed in manufacturing in the sub-region (more manufacturing increased the probability of teen motherhood – borderline significant). The sub-regional employment rate (not shown) was, however, not significant, suggesting that industrial structure rather than current strength of the labour market is important.
- The proportion of married women in full-time employment[5] (more married women working predicting lower odds of teenage motherhood).

Both of these findings point to the influence of opportunity costs on fertility decisions. Interestingly attitudinal variables (philoprogenitive tendencies and educational expectations) were not significant, even in simple bivariate models. Greater proportions of privately rented housing were also generally associated with a greater likelihood of occurrence. The latter could be taken as an indication of transience and possibly lower levels of community ties, although it is hard to isolate the mechanisms at work and for this reason we decided not to pursue this avenue of enquiry further. We looked for non-linear patterns in the data through initially dividing the variables into quartiles, and also through spline analyses, although we found no evidence for non-linear effects at this level.

In terms of targeting sub-regions, these results suggest that those sub-regions with an industrial structure based around manufacturing, with high levels of transient housing tenure and low levels of women who combine family life and employment, are likely to be those areas with elevated levels of teenage parenthood. In this case, these characteristics do amount to 'area effects' to a certain extent, representing relationships between individual behaviour and area characteristics that are not explained by individual attributes. However, they remain relatively weak predictors at this level next to individual characteristics. Again, this may be a problem of scale. We originally constructed sub-regional areas for the purposes of creating areas large enough to examine attitudinal variables (derived from the cohort data itself). However, they may be too small to capture large scale cultural influences. For example, Joshi and Hawkes (2006) found that motherhood in Wales occurred earlier than in the remainder of the UK, and found that this correlated with different intergenerational and childcare structures. On the other hand, the sub-regions may be too large to capture labour market effects with any great precision. For example, our sub-regional classification groups parts of deprived and industrial Lancashire with parts of rural and wealthy Cheshire. Therefore, we do not explore area effects at the sub-regional level any

[5] This variable actually reflects the proportion of married women with children who are Economically Active. However, we often refer to this as married women working.

further, but take forward the notion that manufacturing and married women in employment may be important features of an area and move to explore these characteristics at smaller geographies.

Travel to Work Area and Ward Nearest Neighbour Analyses

Examination of lower level geographies reveals some rather different results. Surprisingly, the TTWA geography yielded few significant effects. There were significant differences between TTWAs. However the amount of variance accounted for by changes in TTWA were small in magnitude and the intraclass correlation coefficient suggested that only 5% of the shared variance in teenage motherhood was due to cohort members being resident in the same travel to work areas. This halved with the addition of family and individual covariates into the model. These coefficients were of similar order of magnitude as for the sub-regional level. However none of the variables reflecting area characteristics that we found to be significant at the sub-regional level or at the ward level were significant at the TTWA level. The proportion of workers in different industries did not significantly predict teenage parenthood (not shown), and neither did the proportion of married women with children who were economically active, or the patterns of post-compulsory educational uptake among young women (although the latter did achieve borderline statistical significance ($p<0.08$)). The absence of significant labour market effects was surprising given that the TTWA level was designed to reflect the local labour market.

Using the ward nearest neighbour scores, the proportion of men working in manufacturing was also not a significant predictor of teenage parenthood. However, this may not be of great surprise when we consider that the ward level is unlikely to capture labour market characteristics. We repeated the analyses by splitting the distributions of the proportion in manufacturing by ward and contiguous ward into quartiles to examine non-linearities but were unsuccessful in finding any statistically significant results. Finally, we divided both ward and ward nearest neighbour measures into quartiles and grouped both middle quartiles into a 'middle' category with a remaining 'high' category and 'low' category, and created groups based on the differences; again there was little evidence that this was significant. The evidence indicates that manufacturing seems to be an important predictor at higher level geographies, although is not a significant explanatory component at a lower level. As was the case for the models with manufacturing, the proportion of married women with children under 15 who were economically active (as a proportion of married women with children under 15) was a weak predictor using the ward nearest neighbour score. We went through the same process as was the case for manufacturing to test for non-linear effects, and found little evidence to support the hypothesis that the proportion of married women working is important at the small area level (not shown).

We also tested for the effect of the proportion of young women in full-time education using the ward nearest neighbour scores. The results are shown in Fig. 6.1 which shows

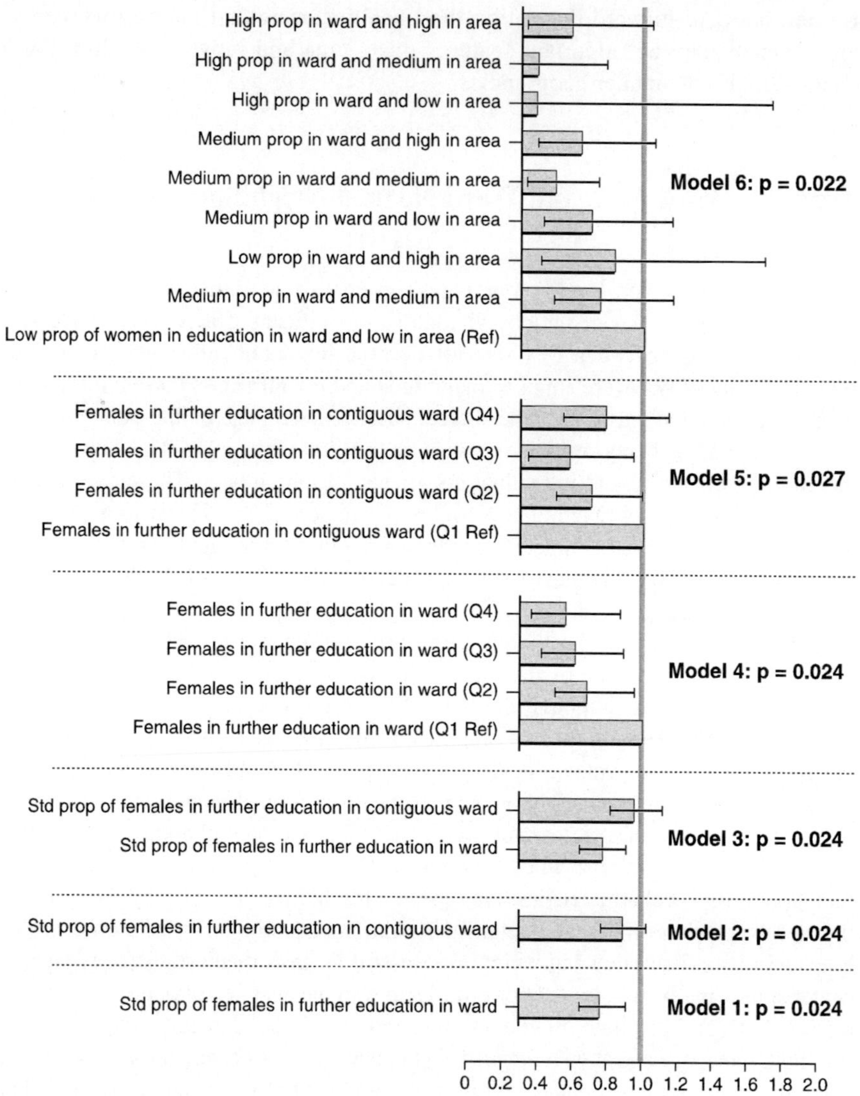

Fig. 6.1 Models showing odds of becoming a teenage mother with area level proportion of young women in further education and travel to work area neighbourhood effect (see notes). *Notes*: Chart shows odds ratios for the proportion of females 16–24 in further education based on ward and contiguous ward values from models with full controls. Odds ratios in Models 1–3 represent the change in the probability of teenage motherhood for a one standard deviation increase in the proportion of females 16–24 in further education. Models 4–6 represent the change in the odds of teenage motherhood relative to the baseline category (Ref). *Error bars* represent the confidence interval. Rho values (ρ) represent the intraclass correlation coefficient for Travel to Work Area. N for all models: 3,631

the odds ratios for area characteristics from models with full individual, family and regional controls as. The vertical line indicates an odds ratio of 1 which indicates that the characteristic has no effect in either raising or lowering the probability of becoming a teenage mother – for variables that are treated as categorical (including those that represent the quartile of a characteristic), the baseline category is set to one and all other effects are interpreted relative to this baseline category. Although at the ward level the proportion of young women in full-time education was a significant predictor of becoming a teenage mother, this was not the case for the ward nearest neighbour score, the odds of the latter being closer to one (Model 2, Fig. 6.1). However, when we divide both the individual ward score and the variables into quartiles reflecting their respective distributions, we find that a possible reason for the non-significant result in the case of the contiguous ward value may lie in the non-linear effect it appears to exert (Model 5, Fig. 6.1). Relative to living in an area that had the lowest quartile of the proportion of women in further education, those in either of the middle quartiles thereafter were significantly less likely to become teenage mothers, although there was no significant advantage in terms of a reduction of risk through living in an area with a proportion in the highest quartile.

When we explore whether the ward effect differs by the value of contiguous wards through creating a variable depicting the relationship between the ward and contiguous value, we find only limited evidence that the pattern in contiguous area matters beyond the pattern in the cohort member's own ward (Model 6, Fig. 6.1; we also looked at interaction terms and found similar results). For those living in wards with low levels of post-compulsory participation in education, having contiguous areas with high levels was most protective against teenage parenthood; for those in wards which themselves had high levels, the protective effect attenuated and even reversed slightly. However, when we explicitly test the effect of contiguous areas stratified by our different ward categories, the differences were no longer significant with the addition of individual level covariates.

A final step was to combine the evidence from sub-regional, ward and ward nearest neighbour models in the same model. When we combine the results from the sub-regional models and the ward and contiguous ward models we find that only the variable reflecting the proportion of young women in further education in the ward and nearest neighbourhoods remains significant (Fig. 6.2). Holding all other factors constant, we see that relative to women who were living in wards with a low proportion of women in further education in the ward and in contiguous wards, those living in wards with a 'medium' level (quartile 2 and 3) in their own ward and in contiguous wards were around half as likely to become teenage mothers. Those in wards with a high level of women in full-time education but where the contiguous wards had a medium level had an even lower odds of becoming a teenage mother compared to the baseline (OR: 0.4), where the odds ratio bar being much lower than one and the error bars not crossing the red line (which represents no effect). The results for the effect of males employed in manufacturing and married women in employment in the sub-region were no longer significant in the combined model; this is shown by the odds ratio bars being closer to one, with the error bars crossing the vertical line.

In summary, we find that some variables seem to be important at the most local (ward) level – in particular the social class composition of the ward, the marriage

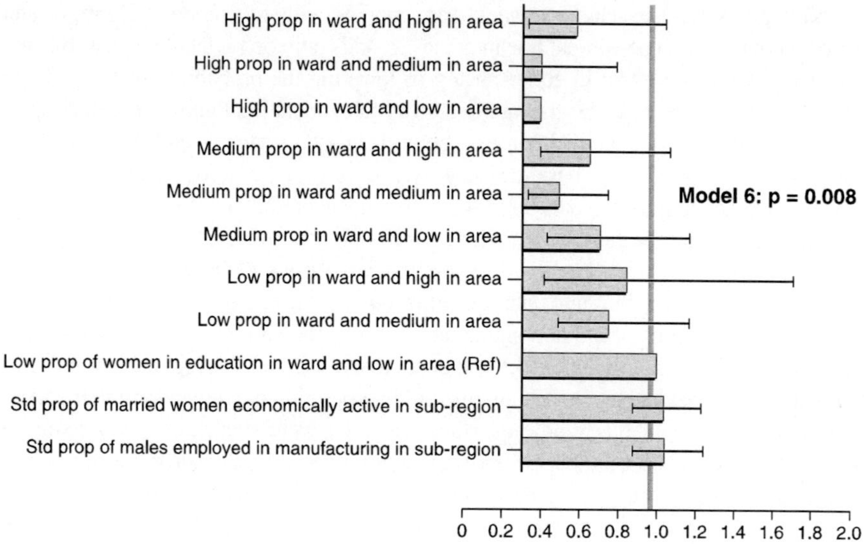

Fig. 6.2 Annotated output from model showing odds of becoming a teenage mother with the sub-regional effect, selected sub-regional characteristics and the proportion of women in further education (see text). *Notes*: Chart shows odds ratios for the neighbourhood characteristics based on ward, contiguous ward and sub-regional values from models with full controls. Odds ratios represent the change in the probability of teenage motherhood for a one standard deviation increase in the proportion of sub regional characteristic or the change in the odds of teenage motherhood relative to the baseline category (Ref). *Error bars* represent the confidence interval. Rho values (ρ) represent the intraclass correlation coefficient for Sub-region Area. N for the model: 3,631

rate and the proportion of young women progressing to further education. These suggest differential values around gender, education and fertility. These results develop the existing literature for the UK. Sloggett and Joshi (1998), using a deprivation score at the ward level rather than specific variables, found no neighbourhood effect. Growing up in areas which have a high proportion of women in further education may mean a higher exposure to positive role models and opportunities. Our study appears to suggest that only specific components of deprivation matter, and in terms of our final models, the results suggest that it is a lack of positive role models and a deprivation of opportunities that form the crucial elements of neighbourhood deprivation for teenage parenthood.

At the sub-regional level, industrial structure was seen to be important, with higher proportions in manufacturing being important as well as the proportion of married women in full time employment. This suggests that opportunity costs are also influential, although looking for these at the most local level would mean that they are missed. Interestingly, both our attempts to refine geographies so that they better reflect labour markets yield very few results. There is some evidence of non-linear effects when nearest neighbours of individual wards are considered, but these are not conclusive. Here our results contradict earlier UK work (Ermisch and

Pevalin 2003) which did find labour market effects at the travel to work area level, although this study looked at the strength of the contemporaneous labour market rather than its historical structure. Looking at the strength of the labour market was less suitable for our study, where a 5 year lag appears between labour market measurement and age 16, than looking at industrial structure, which can be argued to be less fluid. However, Ermisch and Pevalin (2003) also used a more limited set of controls, opening up the suggestion that our stronger set of controls for individual factors has eliminated what might have appeared to be neighbourhood effects.

In common with other studies, all the neighbourhood effects we find are small in magnitude. While some of the limitations of our approach are discussed earlier and are made transparent throughout, and others are discussed in the next section, it is important to emphasise that all of our results are 'gross' of selection effects. In other words, we have taken no measures to control for the fact that some effects that governed the selection of the parents of cohort members into neighbourhoods may also govern the cohort member's propensity to experience a teenage birth, so that when modelling the influence of neighbourhood type, the estimate may be inflated (see Galster et al. 2007; Harding 2003 for further description). However, examining this is beyond the scope of the current chapter, where we maintain focus on issues of mechanisms and scale.

Discussion and Conclusion

We approached this enquiry from two critiques of much existing neighbourhood effects research. One was that some studies do not originate in the existing literature about the topic in hand and thus have no theoretical basis. The other was that geographical units are sometimes used without any particular logic or justification. Just because a certain level of geography exists and can be matched to the data source does not mean that it has any theoretical relevance or that conclusions about the scale of action can be drawn from it.

Using teenage parenthood in the UK as an example and drawing on a rich longitudinal data source, we have attempted to design a study that explicitly tackles and attempts to overcome some of these difficulties, in order to make the process of enquiry into neighbourhood effects rather more transparent than is often the case in academic journal articles.

Drawing on the existing teen pregnancy literature enabled us to identify three hypotheses about what influences early fertility (in short, opportunity costs, values and networks) and how place might have an effect in each of these. This enabled us to identify measures and scales that could test these. One important conclusion from this process was that the social processes leading to early parenthood cannot entirely be separated from one another. For example, class identity or parental trajectories or social networks may influence calculations of opportunity costs. Once place is introduced, the interactions become even more complex, since the same characteristics of place that influence one mechanism may also influence another. This rather suggests that although in principle a theory-driven approach that identifies and tests specific mechanisms is the right one, in practice it may be impossible ever to do this adequately with quantitative methods and data. Much greater integration of qualitative and quantitative approaches is needed to get to the answer: a solution that is not facilitated by the strong disciplinary boundaries that still exist in most UK universities. That said, our own attempt to test specific place mechanisms at the scales at which they might be expected to operate does shed more light than previous studies that have used more generic measures of place. We discover some evidence of values-related place effects at the neighbourhood level (the social class composition of the ward, the marriage rate and the proportion of young women progressing to further education) and labour market structural affects at the sub-regional level. Attitudes to education and childbearing, which we could only test at the sub-regional level because of sample size (these were variables derived from the survey) were not significant at this level, which may suggest either that they are not influential, or that they operate at smaller geographies.

In addition to testing at different spatial levels for mechanisms that are hypothesised to work at these levels, we have also attempted to go beyond the use of existing geographies and develop bespoke geographical units that would better reflect the spaces over which the social processes in which we are interested might operate. We suggested that there are essentially two approaches that might be adopted. One is to draw new boundaries, creating new units, for example around clusters of contiguous similar areas. In this case, this approach proved problematic both for conceptual and technical reasons, although there may be other social processes where bespoke geographies are obviously appropriate and easier to construct (for example measuring education outcomes by the characteristics of school catchment areas), and indeed there are examples of this kind of work (for example Gibbons 2002). A second approach is to work with existing geographies but consider the characteristics of neighbouring units, either by simply creating a new variable to capture this or by weighting the scores of existing units according to what is around them. In our view there is scope for the neighbourhood effects field to develop a lot further in the ways it theorises and measures geography. In scratching the surface in this chapter we have merely aimed to illustrate some of the issues not to solve them. Interestingly, our own ward nearest neighbour measurements yielded few significant results, which tends to suggest that place effects on values around fertility do indeed operate at a relatively small spatial scale.

Perhaps most importantly, we have attempted to show that robust investigation of neighbourhood effects demands deliberate testing of specific mechanisms at specific scales. This is particularly important given that policy interventions may be made at different scales. Results may indicate that a mechanism is or is not important, or they may indicate that it does not operate at that particular scale. While there remain considerable conceptual and technical hurdles (not least data availability), it seems important to us that the neighbourhood effects field, including both quantitative and qualitative researchers, moves towards more explicit and transparent considerations of geography in order to make a stronger contribution to knowledge of place effects.

Acknowledgements We would like to acknowledge the assistance of staff from the Centre for Longitudinal Studies (Institute of Education) in obtaining the area data used: Jon Johnson, Peter Shepherd, Rachel Rosenberg, Robert Browne and Denise Brown. We would also like to acknowledge the assistance of Richard Wiggins from the Department of Quantitative Social Science (Institute of Education) in helping to make this project a reality. Finally, we would like to acknowledge the helpful feedback received from attendees to the Neighbourhood Effects seminar in St Andrews where we first presented this work, many of whom have also contributed to this volume.

References

Afable-Munsuz, A., Speizer, I., Magnus, J. H., & Kendall, C. (2005). A positive orientation toward early motherhood is associated with unintended pregnancy among New Orleans youth. *Maternal and Child Health Journal, 10*(3), 265–276.

Becker, G. S. (1991). *A Treatise on the family*. Cambridge: Harvard University Press.

Bell, J., Clisby, S., Craig, G., Stanley, N., Measor, L., & Petrie, S. (2004). *Living on the edge: Sexual behaviour and young parenthood in seaside and rural areas*. London: Department of Health.

Browning, C. R., Leventhal, T., & Brooks-Gunn, J. (2005). Sexual initiation in early adolescence: The nexus of parental and community control. *American Sociological Review, 70*(5), 758–778.

Burchardt, T., Le Grand, J., & Piachaud, D. (2002). Introduction. In J. Hills, J. Le Grand, & D. Piachaud (Eds.), *Understanding social exclusion*. Oxford: Oxford University Press.

Casterline, J. B. (2001). Diffusion processes and fertility transition. In J. B. Casterline (Ed.), *Diffusion processes and fertility transition*. Washington, DC: National Academy Press.

Cleveland, H. H., & Gilson, M. (2004). The effects of neighborhood proportion of single-parent families and mother–adolescent relationships on adolescents' number of sexual partners. *Journal of Youth and Adolescence, 33*(4), 319–329.

Crane, J. (1991). The epidemic theory of ghettos and neighborhood effects on dropping out and teenage childbearing. *American Journal of Sociology, 96*(5), 1226–1259.

Del Bono, E. (2004). *Pre-marital fertility and labour market opportunities: Evidence from the 1970 British Cohort Study*. Oxford: Department of Economics, University of Oxford.

Driscoll, A. K., Sugland, B. W., Manlove, J., & Papillo, A. R. (2005). Community opportunity, perceptions of opportunity, and the odds of an adolescent birth. *Youth & Society, 37*(1), 33–61.

Elliott, J., & Shepherd, P. (2006). Cohort profile: 1970 British Birth Cohort (BCS70). *International Journal of Epidemiology, 35*(4), 836–843.

Ermisch, J., & Pevalin, D. (2003). *Who has a Child as a Teenager?* (No. 30). Colchester: Institute for Social and Economic Research: University of Essex.

Galster, G. (2011). The mechanism(s) of neighbourhood effects: Theory, evidence, and policy implications. In M. van Ham, D. Manley, N. Bailey, L. Simpson, & D. Maclennan (Eds.), *Neighbourhood effects research: New perspectives*. Dordrecht: Springer.

Galster, G., Marcotte, D. E., Mandell, M., Wolman, H., & Augustine, N. (2007). The influence of neighborhood poverty during childhood on fertility, education and earnings outcomes. *Housing Studies, 22*(5), 723–751.

Gibbons, S. (2002). *Geography, resources and primary school performance*. London: Centre for the Economics of Education, London School of Economics and Political Science.

Goering, J., Fein, J. D., & Richardson, T. M. (2003). What have we learned about housing mobility and poverty deconcentration? In J. Goering & J. D. Fein (Eds.), *Choosing a better life?* Washington, DC: The Urban Institute Press.

Goldstein, H. (2009). Handling attrition and non-response in longitudinal data. *Longitudinal & Life Course Studies, 1*(1), 63–72.

Green, A. E., & White, R. F. (2007). *Attachment to place, social networks, mobility and prospects of young people*. York: Joseph Rowntree Foundation.

Harden, A., Brunton, G., Fletcher, A., Oakley, A., Burchett, H., & Backhans, M. (2006). *Young people, pregnancy and social exclusion: A systematic synthesis of research evidence to identify effective, appropriate and promising approaches for prevention and support*. London: The Evidence for Policy and Practice Information and Co-ordinating Centre (EPPI-Centre), Social Science Research Unit, Institute of Education, University of London.

Harding, D. J. (2003). Counterfactual models of neighbourhood effects: The effect of neighbourhood poverty on dropping out and teenage pregnancy. *American Journal of Sociology, 109*(3), 676–719.

Haveman, R., Wolfe, B., Peterson, E., & Wilson, K. (1997). *Do teens make rational childbearing choices?: Family, neighborhood, and net benefit determinants of teen nonmarital childbearing*. Madison: Institute for Research on Poverty, University of Wisconsin-Madison.

Hawkes, D., Joshi, H., & Ward, K. (2004). *Working Paper 6 – Unequal entry into motherhood and unequal starts in life: Evidence from the first survey of the UK Millennium Cohort* (Working Paper). London: Centre for Longitudinal Studies, Institute of Education, University of London.

Hobcraft, J. (2008). The timing and partnership context of becoming a parent: Cohort and gender commonalities and differences in childhood antecedents. *Demographic Research, 19*(34), 1281–1322.

Hobcraft, J., & Kiernan, K. (2001). Childhood poverty, early motherhood and adult exclusion. *British Journal of Sociology, 52*(3), 495–517.

Holmlund, H. (2005). Estimating long-term consequences of teenage childbearing – An examination of the siblings approach. *Journal of Human Resources, 40*(3), 716–743.

Imamura, M., Tucker, J., Hannaford, P., Oliveira da Silva, M., Astin, M., Wyness, L., Bloemenkamp, K. W. M., Jahn, A., Karro, H., Olsen, J., & Temmermen, M. (2007). Factors associated with teenage pregnancy in the European Union countries: A systematic review. *European Journal of Public Health, 17*(6), 630–636.

Joshi, H. (1998). The opportunity costs of childbearing: More than mothers' business. *Journal of Population Economics, 11*(1), 161–183.

Joshi, H. (2002). Production, reproduction, and education: Women, children and work in a British perspective. *Population and Development Review, 28*(3), 445–474.

Joshi, H., & Hawkes, D. (2006). *Early and late entry to motherhood in wales: Evidence of socio-economic inequalities in the first survey of the UK Millennium Cohort*. London: Centre for Longitudinal Studies, Institute of Education, University of London.

Kiernan, K. E. (1997). Becoming a young parent: A longitudinal study of associated factors. *The British Journal of Sociology, 48*(3), 406–428.

Kiernan, K., & Hobcraft, J. (1997). Parental divorce during childhood: Age at first intercourse, partnership and parenthood. *Population Studies, 51*(1), 41–55.

Kneale, D. (2010). Pushy parents make for later grandparents: Parents' educational expectations and their children's fertility among two British cohorts. *Longitudinal & Life Course Studies, 1*(2), 137–154.

Lappegard, T., & Ronsen, M. (2005). The multifaceted impact of education on entry into motherhood. *European Journal of Population, 21*(1), 31–49.

Lesthaeghe, R., & Neels, K. (2002). From the first to the second demographic transition: An interpretation of the spatial continuity of demographic innovation in France, Belgium and Switzerland. *European Journal of Population, 18*(4), 325–360.

Lupton, R. (2003). *'Neighbourhood Effects': Can we measures them and does it matter?* London: Centre for Analysis of Social Exclusion, London School of Economics.

Lupton, R., & Kneale, D. (2010). *Are there neighbourhood effects on teenage parenthood in the UK, and does it matter for policy?: A review of theory and evidence*. London: Centre for the Analysis of Social Exclusion, London School of Economics.

McCulloch, A. (2001). Teenage childbearing in Great Britain and the spatial concentration of poverty households. *Journal of Epidemiology and Community Health, 55*(1), 16–23.

Rabe-Hesketh, S., & Skrondal, A. (2008). *Multilevel and longitudinal modeling using stata*. College Station: Stata Press.

Rae, A. (2009). Isolated entities or integrated neighbourhoods? An alternative approach to the measurement of deprivation. *Urban Studies, 46*(9), 1859–1878.

Rasbash, J., Steele, F., Browne, W., & Prosser, B. (2004). *A user's guide to MLwiN version 2.1*. London: Centre for Multilevel Modelling, Institute of Education.

Rendall, M. S., & Smallwood, S. (2003). Higher qualifications, first birth timing and further childbearing in England and Wales. *Population Trends, 111*, 18–27.

Rendall, M. S., Couet, C., Lappegard, T., Robert-Bobee, I., Ronsen, M., & Smallwood, S. (2005). First birth by age and education in Britain, France and Norway. *Population Trends, 121*, 27–34.

Royston, P. (2004). Multiple imputation of missing values. *The STATA Journal, 4*(3), 227–241.

Royston, P. (2005). Multiple imputation of missing values: Update of ice. *The STATA Journal, 5*(4), 527–536.

Seltzer, J. A., Bachrach, C. A., Bianchi, S. M., Bledsoe, C. H., Casper, L. M., Chase-Lansdale, P. L., DiPrete, T. A., Hotz, J. A., Morgan, S. P., Sanders, S. G., & Thomas, D. (2005). Explaining family change and variation: Challenges for family demographers. *Journal of Marriage and Family, 67*(4), 908–925.

Sigle-Rushton, W. (2004). *Intergenerational and life-course transmission of social exclusion in the 1970 British Cohort Study*. London: London School of Economics.

Sloggett, A., & Joshi, H. (1998). Deprivation indicators as predictors of life events 1981–1992 based on the UK ONS longitudinal study. *Journal of Epidemiology and Community Health, 52*(4), 228–233.

Smith, S., & Ratcliffe, A. (2009). Women's education and childbearing: A growing divide. In J. Stillwell, E. Coast, & D. Kneale (Eds.), *Fertility, living arrangements, migration and care*. Dordrecht: Springer.

Snijders, T. A. B., & Bosker, R. J. (1999). *Multilevel analysis*. London: Sage Publications.

South, S. J., & Crowder, K. D. (1999). Neighborhood effects on family formation: Concentrated poverty and beyond. *American Sociological Review, 64*(1), 113–132.

Wadsworth, M., Ferri, E., & Bynner, J. (2003). Changing Britain. In E. Ferri, J. Bynner, & M. Wadsworth (Eds.), *Changing Britain, changing lives; Three generations at the turn of the century*. London: Bedford Way Papers: Institute of Education, University of London.

Chapter 7
Neighbourhood Effects, Housing Tenure and Individual Employment Outcomes

David Manley and Maarten van Ham

Introduction

Many academics and policy makers believe in neighbourhood effects: the idea that living in a poor neighbourhood can severely reduce an individual's life chances with respect to their health outcomes, educational achievement and labour market performance (for an overview of the literature see Ellen and Turner 1997; Dietz 2002; Galster 2011). The literature suggests a wide range of theoretical pathways by which the neighbourhood context can influence individual outcomes (see Durlauf 2004). These pathways include a lack of positive role models, negative socialisation, a physical disconnection from job-finding networks, a culture of dysfunctional values and disconnection from mainstream society, discrimination by institutions and employers from outside the neighbourhood, access to low quality public services, and an exposure to high levels of criminal behaviour.

There is a large body of qualitative and quantitative research showing evidence of negative neighbourhood effects of living in deprived neighbourhoods. Qualitative research has contributed greatly to the development of theory and hypotheses and has mostly found evidence in favour of the neighbourhood effects hypothesis. But by its very nature, qualitative research does not produce generalisable outcomes as very specific cases are studied. The body of quantitative studies is inconclusive with regard to the question whether neighbourhood effects exist, and if they do, which of the causal pathways are the most important. Much of the quantitative work has been criticised for failing to address a series of econometric problems – most importantly

D. Manley (✉)
Centre for Housing Research, School of Geography and Geosciences,
University of St Andrews, Fife, Scotland, UK
e-mail: d.manley@st-andrews.ac.uk

M. van Ham
OTB Research Institute for the Built Environment, Delft University of Technology,
Delft, The Netherlands
e-mail: m.vanham@tudelft.nl

M. van Ham et al. (eds.), *Neighbourhood Effects Research: New Perspectives*,
DOI 10.1007/978-94-007-2309-2_7, © Springer Science+Business Media B.V. 2012

selection bias – related to the identification of causal relationships (Durlauf 2004; Cheshire 2007; van Ham and Manley 2010).

There is no doubt that neighbourhood poverty and individual disadvantage are strongly correlated, but it is much less certain that there is a *causal* relationship between the two. The main challenge in the study of neighbourhood effects is identifying causal pathways, directly relating neighbourhood context to individual outcomes (Cheshire 2011). It has been argued that the apparent relationship between the neighbourhood context and individual outcomes may not be causal, and might merely be a reflection of the ability of different groups in society to select a good neighbourhood in which to live (van Ham and Manley 2010; and see the chapter by Hedman and van Ham 2011 in this volume). It is highly problematic to identify causal neighbourhood effects while using observational data (Durlauf 2004) as in such data households are normally not allocated to neighbourhoods randomly. Even in (quasi) experimental data, for instance from the Gautreaux Project in Chicago or the wider Moving To Opportunity Programs (Rosebaum 1995; Katz et al. 2001; Ludwig et al. 2001; Goering et al. 2002) allocation to neighbourhoods is not completely random. Studies using observational and experimental data reach conflicting conclusions, with some studies positively identifying neighbourhood effects, while other report no (causal) effects. The question whether the effects found are selection effects or causal effects is highly relevant in the context of formulating policy responses to concentrations of poverty and associated problems. This chapter aims to give more insight into the potential effects of selection bias on estimates of neighbourhood effects. The chapter builds on studies by Oreopoulos (2003) and van Ham and Manley (2010) which analysed neighbourhood effects separately for social renters and homeowners. The use of separate models is motivated by the fact that the entry route for social renters into a dwelling and therefore into a neighbourhood is very different to the entry route of owner occupiers. Homeowners choose their dwelling and neighbourhood based on preferences, resources (income and ability to get a mortgage) and what is available in the market. The more resources a household can use for housing, the larger the choice set. As a result, the choice of neighbourhood is strongly related to the labour market status and potential of households. This is not the case for households in social renting where the choice set is limited to those properties that the household qualifies for within the administrative structure. In the social sector, households are allocated a dwelling by a housing officer, and although this process is not completely random, it can be argued that the allocation mechanism is largely independent from the labour market situation and potential of the household and approaches a quasi-experimental setting. This was especially true for 1991, the first data point in our analysis, when choice-based letting was not yet widely used in Scotland. Given the different selection mechanisms of owner-occupiers and social renters, it can be argued that associations between neighbourhood characteristics and individual outcomes for social renters are more likely to be free of selection bias – and therefore more likely to be real causal neighbourhood effects – than any such associations for owner occupiers.

In this chapter we explore the tenure split approach by testing the hypothesis that living in a neighbourhood with high levels of unemployment has a negative effect on individual labour market outcomes. Based on the theoretical literature it would be expected that those living in neighbourhoods with high levels of unemployment are less likely to be in contact with positive role models, are more likely to have unemployed friends and neighbours which might cause negative socialisation, are less likely to know people who can help finding a job, and are more likely to be stigmatised by outsiders. This study will investigate the following two questions: To what extent does living in a neighbourhood with a high level of unemployment reduce an individual's chances of having work in 2001 if they were unemployed in 1991?; To what extent does living in a neighbourhood with a high level of unemployment reduce an individual's chances of having work in 2001 if they were employed in 1991? We used unique data from the Scottish Longitudinal Study (SLS) which is a 5.3% anonymised sample of the Scottish population linked through time by matching census records from 1991 and 2001. Using this data, we are able to link 1991 neighbourhood characteristics to 2001 labour market outcomes. The data is not without limitations, but it is one of the best longitudinal datasets available to study neighbourhood effects.

Background

Theoretical Considerations

There is a large and growing literature investigating how the neighbourhood context can influence individual life chances (see for literature reviews Ellen and Turner 1997; Friedrichs 1998; Leventhal and Brooks-Gunn 2000; Dietz 2002; Sampson et al. 2002; Durlauf 2004; Bolster et al. 2007). Galster (2011, in this volume) posited 15 different mechanisms through which the neighbourhood context can influence individual level outcomes. We summarise these mechanisms into two categories: internal neighbourhood mechanisms and external neighbourhood mechanisms.

The first group of mechanisms are internal to the neighbourhood and the best known example of such a mechanism is derived from the work of Wilson (1987) who documented the outcomes of individuals living in high poverty neighbourhoods in Chicago during the 1980s. He concluded that increasing concentrations of poverty in large public housing projects were creating a negative environment for residents, which was directly leading to further disadvantage and increased the propensity of unemployment. This is a so-called social-interactive mechanism (see Galster 2011 this volume): if a neighbourhood environment lacks individuals with higher levels of education or employment other residents may lower their expectations about what they could achieve, or accept unemployment as a norm (see Manski 2000; Blume and Durlauf 2001; Brock and Durlauf 2001; Bolster et al. 2007). Wilson's (1987) study is viewed by many as the starting point for much of the current neighbourhood effects research and the conclusions of his ethnographic research in Chicago are widely cited in the neighbourhood effects literature. However, Small and Feldman (2011, in this volume) questioned whether the theoretical pathways which have been developed in Chicago can be translated to other cities or national contexts as Chicago is an atypical rust-belt city in the USA.

The second group of mechanisms are external to the neighbourhood. Examples of these include the spatial mismatch hypothesis and stigmatisation based on neighbourhood reputations. The spatial mismatch hypothesis (Kain 1968) suggests that individuals living in inner city neighbourhoods are unable to find employment because employment opportunities are inaccessible from the locations in which they live. Gobillon and colleagues (2005) identified seven mechanisms related to the spatial mismatch, four of which relate to the accessibility of employment for workers and include commuting costs, information access, incentive to search for work over large distances and high costs of searching beyond the immediate neighbourhood. As with the work of Wilson (1987, 1991) the spatial mismatch hypothesis was developed in the context of Afro-American workers in the USA. However, research by van Ham (2002) and Houston (2001, 2005) has shown that the hypothesis can also be applied to other national contexts such as the Netherlands and the United Kingdom. The second external mechanism relates to neighbourhood reputations. Employers, banks, and other external agents tend to form opinions on the abilities and suitability of individuals based on the reputation of the neighbourhoods in which they live. Research has shown that individuals living in certain neighbourhoods are excluded from finance (see Aalbers 2009) reducing their ability to obtain loans or mortgages for purchasing a car or a house. In terms of labour market outcomes, research has shown that neighbourhood reputations can detrimentally affect an individual's chances of getting a job (see Dean and Hastings 2000), because employers refuse to hire residents from certain neighbourhoods (see Wilson 1991; Wacquant 1993; Permentier et al. 2007). Stigma can become a structural barrier to gaining or keeping employment when it is institutionalised.

Methodological Considerations

Studies consistently find that people living in deprived neighbourhoods are less likely than people in affluent neighbourhoods to do well in life. However, this does not necessarily mean that living in deprived neighbourhoods causes people to do less well. A major challenge in the empirical investigation of neighbourhood effects is the identification of causal relationships. Many studies which claim to have found causal neighbourhood effects are likely to have only found correlations between neighbourhood characteristics and individual outcomes, without clear evidence of the direction of causation. The literature distinguishes several econometric problems in the investigation of neighbourhood effects (see Moffitt 2001) which are summarized in Table 7.1.

The simultaneity problem, also referred to as Manski's reflection problem (Manski 1993), arises when a researcher tries to infer whether the average behaviour in some group influences the behaviour of the individuals that make up the group.

Table 7.1 Econometric problems in neighbourhood effects research

Problem	Example
Simultaneity problem	In neighbourhoods with a high overall level of unemployment individuals are de facto more likely to be unemployed
Omitted-context-variables	Neighbourhood context information that is omitted: for instance neighbourhood ethnicity and deprivation are frequently correlated. Omission of deprivation could result in ethnicity being falsely identified as the driver of individual outcomes
Endogenous membership	Characteristics of the individual that are either not collected or not included in analysis but are related to the outcome of interest

Problems with simultaneous causation may arise because the contextual conditions themselves may be the result of respondents' behaviour (endogenous effect). For example, if we are interested in whether a high level of unemployment in a neighbourhood causes residents to be more likely to be unemployed, an econometric problem arises because unemployed individuals in the neighbourhood contribute to the overall neighbourhood unemployment level. One solution is to relate past neighbourhood context (unemployment levels) to current (unemployment) outcomes. A second issue is the omitted-context-variables problem, also called the correlated unobservables problem. This problem refers to the omission of important context characteristics from a regression model which are correlated with included variables (at the neighbourhood level). If important variables are omitted, researchers might draw the wrong conclusions from the estimated effects of context variables which are included. A clear example of this is apparent in the racial proxy hypothesis (Harris 1999) whereby it appears that the proportion of ethnic minority groups in a neighbourhood is the cause of out migration by the native population, where-as in reality the out migration is caused by correlated neighbourhood deprivation. A third problem is the endogenous membership problem. This problem also involves omitted variables, but this time relating to the individual. The core of this problem is self-selection into and out of neighbourhoods. Sorting into neighbourhoods is not based on a random process and if unobserved individual characteristics are correlated with both the location decision and the dependent variable, endogeneity occurs. In most studies it is likely that selective mobility into neighbourhoods leads to biased estimates of neighbourhood effects.

Hedman and van Ham (2011, this volume) suggest that neighbourhood mobility is highly structured and neighbourhood selection is strongly related to individual characteristics: individuals tend to move into neighbourhoods with populations which are similar to themselves. Individuals with greater financial resources will, all other things being equal, enter a neighbourhood in which the income of other residents is also relatively high. Those with a high income avoid neighbourhoods with a low average income or high levels of crime, anti-social behaviour or poor access to services. Hedman and colleagues (2010) used Swedish data on moves over a 10 year period to show that neighbourhood self-selection is a key determinant of neighbourhood composition. As a result, it is likely that much previous evidence of neighbourhood effects is at least in part, attributable to selective mobility into and out of neighbourhoods. Cheshire (2007, p. 2) succinctly summed up the problem by asking the question: "do poor people live in poor neighbourhoods because living in affluent ones costs too much? Or does living in a poor neighbourhood make poor people significantly poorer?"

The gold standard in avoiding selection bias is the use of (quasi) experimental data in which households are randomly assigned to neighbourhoods. There have been several poverty deconcentration programs in the US since the 1970s which operated a (quasi) experimental design. The best known are the Chicago Housing Association Gautreaux, the Moving to Opportunity (MTO) and Hope VI programs. Whilst the operational details of the programs differ, the overall idea behind the programs was similar: households living in concentrations of poverty in large scale housing projects were offered a number of different 'treatments' which included relocation from their current poor neighbourhood to a more affluent one using housing vouchers to access the private rental market, counselling for moving from welfare to work, and relocation in a regenerated neighbourhood. The idea behind the programs was that households who received 'treatment' would do better as a result of their move than they would have done had they remained living in their original neighbourhoods. Theoretically, participation in the relocation schemes was random with open selection criteria for households wishing to participate. As such, the relocation programs should provide an ideal test of whether or not neighbourhood characteristics affect the outcomes of individual life courses.

Some have criticised the findings of the large randomised trials discussed above, and have urged caution regarding the interpretation of research findings of these trials (see Moffitt 2001; Clark 2008). In all programs only a small proportion of the households living in concentrations of poverty were given the means to move to more affluent suburban locations. In conjunction with the moves, households were offered counselling and support to assist with the move and finding employment and other opportunities in their new neighbourhoods. However, the main criticism relates to the selection into the programs and the support received. In all programs there were a number of criteria that residents had to fulfil in order to qualify for participation. For instance, in the Gautreaux program residents had to nominate themselves, and were not accepted if they had "more than four children, large debts or unacceptable housekeeping" (Rosenbaum 1994, p. 4). In addition the managers and counsellors of the program identified the families they felt were more likely to succeed and placed them in the better neighbourhoods, leaving the less suitable locations for less deserving families (Goering et al. 2002). This process placed self and institutional selection criteria on program participation neither of which are independent of the outcomes that the program sought to improve. Of the three projects, the MTO comes closest to an experimental design, with the fewest constraints on recruitment, although the selection was still far from random. It should also be noted that there is evidence that, although households changed their residential location, many tended to maintain contacts and use their old networks for education and employment opportunities and even return to their original neighbourhoods rather than integrate into their new neighbourhood locations and networks (Boyd et al. 2006). So although theoretically experimental designs are ideal for the study of neighbourhood effects, in practice they are very expensive and difficult to execute.

An additional challenge in neighbourhood effects research is the identification of the most appropriate spatial scale at which to measure neighbourhood characteristics. The meaning of neighbourhoods is highly contested (see Galster 2001; Flowerdew et al. 2008), although the issue of scale is frequently omitted from discussions in the empirical literature. Theoretical contributions highlight that the scale at which the neighbourhood is conceptualised is an important component of the neighbourhood effects thesis. Important questions relating to scale and neighbourhood boundaries are often not asked because administrative units are used as proxies for neighbourhoods driven by the availability of data. This is a problem because spatial scale should be driven by the mechanism and hypothesis under investigation. For example, testing hypotheses on the effect of neighbourhood reputation or neighbourhood stigma might require larger neighbourhood units than studies testing hypotheses on peer group effects. If models searching for neighbourhood effects incorrectly specify the spatial scale of neighbourhoods then it is possible that the modelled outcomes are unable to identify any effects simply because they do not operate at the scale chosen for the analysis (Manley et al. 2006).

Neighbourhood Effects and Labour Market Outcomes

Musterd and Andersson (2005) used data from the Netherlands to investigate the impact of neighbourhood context on unemployment. They found that employed individuals living in neighbourhoods with a high proportion of people in receipt of welfare benefits were more likely to experience unemployment than individuals in neighbourhoods with a lower proportion of people in receipt of benefits. Their study gives some support to the hypothesis of negative socialisation as a source of neighbourhood effects. Repeating the analysis with Swedish data, Musterd and Andersson (2006) found similar results. After controlling for a range of individual characteristics they found that the probability of an individual remaining unemployed increased as the proportion of neighbourhood unemployment increased beyond a neighbourhood unemployment threshold of 16%, the probability of remaining out of work did not increase further. Using data from the 1991 Population Census of Great Britain, including information the individual level Sample of Anonymised Records, Clark and Drinkwater (2002) studied neighbourhood effects on employment outcomes for ethnic minorities in England and Wales. They reported that employment outcomes for ethnic minorities are related to the ethnic composition of the neighbourhood in which they live. Ethnic minority individuals living in ethnic enclaves are at a greater risk of experiencing unemployment compared to ethnic individuals in less ethnically concentrated areas. However, although Clark and Drinkwater were able to use individual data in the analysis, they note that they were unable to control for self-selection into areas with differing levels of ethnic concentration and that this may be behind some of the correlations found.

The quasi-experimental studies discussed above have produced a wealth of neighbourhood effects research on a wide range of outcomes (see for instance: Venkatesh, et al. 2004 for Gautreaux; Elhassan et al. 1999 for MTO; Ciseros and Engdahl 2009 for HOPE VI). Popkin and colleagues (1993) investigated outcomes for the Gautreaux programs and assessed how well black women from inner city housing projects performed in the labour market after moving to suburban neighbourhoods in Chicago. They found that residents who moved to the suburbs had an increased probability of finding employment, even when that individual had experienced long term unemployment in the past. However, although the probability of employment was higher, no differences were found in the wages of working women in the inner city and the suburb as the types of employment were similar in both locations. In a follow-up paper Mendenhall and colleagues (2006) concluded that the outcomes observed by Popkin and colleagues (1993) were maintained in the longer term: women who had relocated to suburban locations as a result of the Gautreaux program spent less time out of work, and were less likely to claim welfare assistance.

Assessing outcomes for the more extensive HOPE VI project, Popkin and Cunningham (2009) reported a mixed picture. Using data from the HOPE VI panel study they showed that the program resulted in dramatic improvements in the levels of well-being, including reduction in fear of crime and violence, for those residents who had moved to different neighbourhoods using housing vouchers to rent in the private market. The well-being outcomes contrast with the employment outcomes, where "there were no changes in employment or self-sufficiency for private market renters, the few HOPE VI movers, or those who remained in traditional public housing" (Popkin and Cunningham 2009, p. 197), with unemployment remaining at 48% throughout the panel period.

The third of the major deconcentration programs, Moving to Opportunity (MTO), was initially assessed at 4 and 7 years after the randomised neighbourhood reassignment of participants. The interim study found that although there were improvements in satisfaction, perceptions of neighbourhood safety, and participant's mental health, there were no positive benefits for participation in the labour market. Moreover, for the male youth cohort there were reports of negative effects on behavioural outcomes despite the moves to neighbourhoods with lower concentrations of poverty (Orr et al. 2003). In a separate study, King, Liebman and Katz (2007) found similar labour market results to those of Orr and colleagues (2003) for individuals who had moved through the MTO program. The fact that the employment status of many relocated individuals did not improve over time, even when relocated to significantly better neighbourhoods in terms of violence, crime, unemployment and housing quality, raises doubt over the neighbourhood effects hypothesis in relation to employment outcomes.

Unfortunately, experimental and quasi-experimental data is rarely available, so a number of authors have attempted to use alternative methods to address the problem of selection bias while using observational data (see Oreopoulos 2003; Bolster et al.

2007; van Ham and Manley 2010). Oreopoulos (2003) used administrative data from Toronto to investigate labour market outcomes of adults who were assigned into various social housing projects during their childhood. The households the children lived in were assigned to dwellings in neighbourhoods which varied in levels of crime, unemployment and poverty. It can be argued that the selection mechanism was largely independent from the characteristics of the child and that the data is therefore quasi-experimental. Oreopoulos did not find any evidence of neighbourhood effects for adults who had grown up in social housing. However, for the control group, consisting of individuals in private housing in the same neighbourhoods, significant 'neighbourhood effects' were found. Oreopoulos concluded that those in private housing self-selected into neighbourhoods and that the correlations found were most likely selection effects and not causal neighbourhood effects. Similar results were found by Bolster and colleagues (2007), using data from the British Household Panel Survey (BHPS). They found small neighbourhood effects for those living in the private housing but no effects for those in social housing, although they do not explicitly discuss this outcome in their paper. Van Ham and Manley (2010) investigated the effect of living in deprived neighbourhoods and mixed tenure neighbourhoods on labour market outcomes using data from the Scottish Longitudinal Study. They found that living in a deprived neighbourhood is correlated with employment outcomes for those living in private housing, but not for those in social housing. They also concluded that self-selection into deprived neighbourhoods by homeowners with poor labour market prospects most likely caused the correlations found.

This chapter builds on the work of van Ham and Manley (2010). Instead of using a composite measure of neighbourhood deprivation we test a more specific hypothesis based on the effects of living in neighbourhoods with high levels of unemployment (see Wilson 1987; Musterd and Andersson 2005). As discussed in the introduction, according to the neighbourhood effect hypothesis it can be expected that those living in neighbourhoods with high levels of unemployment are less likely to be in contact with positive role models, are more likely to have unemployed friends and neighbours (which can lead to negative socialisation), are less likely to know people who can help finding a job, and are more likely to be stigmatised by outsiders. Thus, living in a neighbourhood with a high level of unemployment might make it harder for individuals out of work to get a job, and for those in employment to keep their job.

Data and Methods

We used individual level longitudinal data from the Scottish Longitudinal Study (SLS). The SLS is an anonymised 5.3% sample of the Scottish population with matched census records from 1991 to 2001 (Boyle et al. 2008). The sample gives approximately 274,000 SLS members available for analysis. The SLS members are geocoded which allows researchers to link individual records to neighbourhood characteristics at a low geographic scale. The longitudinal structure of the data is highly appropriate for neighbourhood effects research as it enables researchers to follow individuals and investigate the effects of 1991 characteristics on their 2001 outcomes.

For the analysis presented in this chapter, the SLS population is restricted to include only those individuals aged between 15 and 50 years old in 1991, and only those individuals who were available for paid employment in both 1991 and 2001. This means that those who were students, retired or permanently ill in 1991 were excluded from the analysis. For employed individuals, part time and full time work are coded as employed with no distinction made between the two categories. We have also included individuals who had secured a job but not yet started it as employed. A restriction of the data is that we have no information on what an individual's employment situation was between the data points in 1991 and 2001. So, for example, if an individual was unemployed for the 1991 and 2001 Census days but had employment for the whole of the period in-between they would appear identical in our employment variable as an individual who had been unemployed through the whole time. Although this is a serious limitation of the data, we feel that size of the data set, the low level geocoding, and the longitudinal nature of the data outweigh the problems posed by the lack of information between the time points.

The outcome variables used in this study measure employment status in 2001. In the first set of models we investigated the probability of having a job in 2001 for those individuals who were unemployed in 1991. Those individuals who remained unemployed in 2001 were coded 0 and those who had a job in 2001 were coded 1. In the second set of models we investigated the probability that those individuals who were employed in 1991 are unemployed in 2001. Thus, from those who were employed in 1991 the outcome of still being employed in 2001 is coded 0, while those individuals who were unemployed in 2001 are coded 1. Since both dependent variables are binary we have used logistic regression models, with a correction for the clustering of individuals in neighbourhoods.

Quantitative studies using secondary data rely on administrative spatial units when making neighbourhoods operational (see Galster 2001; Manley et al. 2006). In many neighbourhood effects studies there is a mismatch between the spatial level at which the theoretical causal mechanism is thought to operate and the spatial level at which neighbourhoods have been made operational. Many studies use relatively large neighbourhoods because lower level geocoding was not available in the data used. In this study we investigate the effect of neighbourhood unemployment levels on individual employment outcomes. The literature identifies several causal mechanisms through which the neighbourhood context can have an effect on employment outcomes and these mechanisms can operate on different spatial scales. For instance, (negative) role model effects can be expected to operate on a relatively low spatial scale. Direct neighbours are probably more important than those living a few blocks away. Local networks through which people might find employment can also be expected to operate on a relatively low spatial level as these networks often need face-to-face interaction. Stigmatisation of neighbourhood residents by outsiders based on the reputation of the neighbourhood is likely to operate at the level of larger neighbourhood units. Because of the variety of spatial scales at which causal mechanisms might be at play we use two definitions of neighbourhoods (see also Graham et al. 2009; van Ham and Manley 2010). The first neighbourhood scale uses Output Areas (OAs), which contain around 119 people on average. The second neighbourhood scale uses Continuous Areas Through Time (CATTs) which are much larger and contain around 503 people on average.

Two neighbourhood characteristics are included in the models, both of which are measured in 1991 at the OA and CATT level. We measured neighbourhood characteristics in 1991 to minimise problems with reversed causality: in our design, individuals lived in their neighbourhoods prior to any change in their labour market status. Although we cannot be certain of identifying causal pathways this way, we can be relatively confident that any neighbourhood effects we observe are more than merely correlations. The main neighbourhood level characteristic in the models is the percentage of unemployed individuals in the neighbourhood in 1991. This is calculated by dividing the number of 16–64 year old people who are looking for employment by the total number of people available for work in that age group. The neighbourhood effects literature suggests that, when unemployment levels reach a certain critical level (threshold), it is more likely that individuals will be affected by negative socialization and negative role models. It has also been suggested that neighbourhoods with high levels of unemployment are more likely to suffer from negative external reputations and that individuals searching for work whilst living in such neighbourhoods are less likely to find work because employers are reluctant to employ them. Neighbourhood unemployment is categorized into 5 groups: 0–1.9%; 1.9–3.6%; 3.64–6.0%; 6.0–10.1%, and; 10.1–54.9% (for reference the national average level of unemployment recorded in the 1991 Census for Scotland was 6.2%). We have used categories for the variable to account for the possibility that any relationship is non-linear.

The second neighbourhood characteristic included in the models is an urban-rural classification which serves as a proxy for access to job opportunities (see also van Ham 2002). The urban–rural classification is based on population size and access to concentrations of population (Scottish Executive, 2004) and measured in six categories: (i) cities (over 125,000 people); (ii) urban areas (10,000–125,000 people); (iii) small towns (3,000–10,000 people or within 30 min from towns with 10,000 people or more); (iv) remote towns (3,000–10,000 people over 30 min from settlement of over 10,000); (v) accessible rural (less than 3,000 people and within 30 min from places with over 10,000 people); and (vi) remote (settlements with under 3,000 people, over 30 min from places with over 10,000 people).

We also included a range of individual level control variables in our models. These include dichotomized variables for gender, ethnicity, limiting long term illness, household status, housing tenures and educational outcomes. An individual's age is included as a continuous variable. To minimise causality issues, all control variables are measured for 1991. Summary statistics of all variables can be found in Table 7.2.

Table 7.2 Variable summary statistics

	Unemployed in 1991		Employed in 1991	
	$N = 3{,}639$		$N = 60{,}048$	
	OA	CATT	OA	CATT
Percentage employed in 2001	71.7%		97.5%	
Neighbourhood level variables				
Neighbourhood unemployment 1991 (ref = 0–1.9%)	8.4	6.8	23.5	22.2
1.9–3.5%	13.5	12.3	25.5	21.6
3.6–5.9%	19.5	18.1	21.2	21.1
5.9–10.0%	28.8	24.2	17.9	19.5
10.1–54.9%	29.8	38.7	12.0	15.7
Urban-rural classification 1991 (reference = remote)[a]	4.4	4.4	5.2	5.2
Accessible rural areas	10.2	10.2	13.5	13.5
Remote towns	1.9	1.9	2.7	2.7
Small towns	10.0	10.0	10.7	10.7
Urban areas	31.2	31.2	31.2	31.2
Cities	42.3	42.3	36.7	36.7
Individual and household level variables				
Qualifications 1991 (reference = none)		85.3		76.2
Less than degree		2.6		10.6
Degree or better		2.6		8.4
Not stated		5.2		2.2
Tenure 1991 (reference = owners)		34.7		69.5
Social renter		58.1		24.3
Private renter		7.2		6.2
Age (average age in 1991)		28.9 years		32.9 years
Female (reference = male)		33.2		44.3
Ethnic (reference = non ethnic)		0.9		0.7
Partner works 1991 (reference = not work)		55.9		77.1
Change in health (reference = no LLTI)		88.5		92.6
LLTI 91 & 01		1.1		0.8
LLTI 91		2.1		0.9
LLTI 01		8.3		5.7
Change in household type (reference = couple)		55.4		73.9
91 & 01 single		9.4		4.8
91 single/01 couple		4.4		3.7
91 couple/01 single		30.8		17.6
Change in presence of children (reference = children)		24.4		25.6
91/01 no children		41.1		33.1
91 no child/01 child		14.9		16.7
91 child/01 no child		19.6		24.6
Moved (reference = not moved)		69.5		62.1

[a]The Urban-rural classification is the same for CATTs
Source: Calculations done by the authors using data from the SLS

Modelling Results

Table 7.3 reports models of the probability that those who were unemployed in 1991 also had a job in 2001. Models 1 and 2 include neighbourhood characteristics on the CATT level and Models 3 and 4 include neighbourhood characteristics measured at the Output Area level. Model 1 shows that the probability of having a job in 2001 decreases as the level of unemployment in the 1991 neighbourhood of residence increases. Those living in neighbourhoods with more than 10% unemployment are the least likely to have a job in 2001. Although we use longitudinal data, we cannot be certain that the effects found are causal effects. A major cause of potential bias is the self-selection of residents into neighbourhoods prior to 1991.

Model 2 also controls for a large range of individual and household characteristics and includes a job access proxy in the form of an urban to rural classification scheme. Including all these control variables in the model reduces the size of the neighbourhood unemployment parameters, but the general pattern stays the same. The model results show that unemployed women in 1991 are much more likely than unemployed men to have a job in 2001. We did not find an effect for ethnicity. One possible explanation is that the number of individuals in the data who belong to ethnic minority groups is very low and that we combined all individuals who were indentified in the Census as non-white in one category. Differences in labour market performance between ethnic minorities might therefore average out. With increasing age, unemployed people are less likely to have a job 10 years later.

Individuals who were unemployed in 1991 and who are single in both 1991 and 2001 are the least likely to get a job in 2001. This is probably an example of reversed causation where unemployed people are also less likely than employed people to find a partner. Also those who split up from their partner between 1991 and 2001 have a reduced probability of being employed in 2001. People without children in both years are less likely than others to have a job in 2001. Again, this is probably a case of reversed causality. As expected, level of education is an important predictor of 2001 employment. Having at least basic school level qualifications or better, compared with individuals without qualifications, significantly improves an individual's chances of having a job in 2001. The effects of individual level education are relatively large compared with the effects of the neighbourhood level characteristics. Finally, social renters and private renters are less likely to have a job in 2001 compared with homeowners. Model 3 and 4 in Table 7.3 include neighbourhood level variables at the Output Area level. As explained before, Output Areas are much smaller spatial units than CATTs. Interestingly, the overall results for the Output Area level analyses are very similar to the CATT area analyses. Most other research using more than one spatial level in their modelling reported stronger effects at the lower geographies.

Table 7.4 reports models of the probability that those who were employed in 1991 are unemployed in 2001. Again, Models 5 and 6 include neighbourhood characteristics for the CATT neighbourhoods and Models 7 and 8 include characteristics for neighbourhoods represented by Output Areas. Model 5 shows a strong

Table 7.3 Logistic regression of the probability of employment in 2001 for those unemployed in 1991

	CATT level						OA level					
	Model 1			Model 2			Model 3			Model 4		
	Coef.	S.e.[b]	Sig	Coef.	S.e.[b]	Sig	Coef.	S.e.[b]	Sig	Coef.	S.e.[b]	Sig
Neighbourhood level variables												
Neighbourhood unemployment 1991 (reference = 0–1.9%)												
1.9–3.5%	–0.353	0.249		–0.237	0.278		–0.534	0.294	*	–0.381	0.320	
3.6–5.9%	–0.741	0.230	***	–0.486	0.264	*	–0.550	0.278	**	–0.348	0.302	
6.0–10.0%	–1.094	0.223	***	–0.762	0.257	***	–1.083	0.265	***	–0.717	0.292	***
10.1–54.9%	–1.278	0.218	***	–0.882	0.253	***	–1.347	0.260	***	–0.865	0.291	***
Urban-rural classification 1991 (reference = cities)												
Urban areas				–0.008	0.107					–0.009	0.107	
Small towns				–0.003	0.159					0.056	0.154	
Remote towns				0.275	0.318					0.318	0.321	
Rural areas				0.076	0.171					0.122	0.169	
Remote areas				–0.209	0.220					–0.175	0.222	
Individual and household level variables												
Female (reference = male)				0.848	0.112	***				0.842	0.110	***
Ethnic (reference = non ethnic)				–0.170	0.475					–0.195	0.477	
Age in years (1991)				–0.015	0.005	***				–0.015	0.005	***
Change in household type (reference = couple)												
91 & 01 single				–1.105	0.145	***				–1.092	0.147	***
91 single/01 couple				0.327	0.257					0.348	0.258	
91 couple/01 single				–0.744	0.105	***				–0.737	0.103	***

Change in presence of children (reference = children)												
91/01 no children				–0.346	0.124	***				–0.348	0.125	***
91 no child/01 child				0.184	0.166					0.188	0.168	
91 child/01 no child				0.010	0.147					0.004	0.145	
Qualifications 1991 (reference = none)												
Less than degree				0.633	0.315	**				0.638	0.318	**
More than degree				0.979	0.316	***				0.962	0.318	***
None stated				–0.317	0.203					–0.323	0.204	
Tenure 1991 (reference = owners)												
Social renter				–0.633	0.113	***				–0.640	0.114	***
Private renter				–0.669	0.176	***				–0.656	0.179	***
Change in health (reference = no LLTI)												
LLTI 91 & 01				–0.530	0.397					–0.538	0.397	
LLTI 91				–0.010	0.289					0.002	0.285	
LLTI 01				–0.825	0.142	***				–0.830	0.141	***
Constant	2.255	0.209		3.212	0.314	***	2.326	0.254	***	3.197	0.347	***
Initial log pseudo LL		–2045.941										
Log pseudo likelihood		–2001.816			–1629.250			–2002.017			–1614.122	
Wald		76.000			359.240			77.840			370.530	
Number of observations		3,639			3,639			3,639			3,639	

*$p < 0.10$; **$p < 0.05$; ***$p < 0.01$

[a]Robust standard errors adjusted for clustering at the CATT/OA level

Source: Calculations done by the authors using data from the SLS

Table 7.4 Logistic regression of the probability of unemployment in 2001 for those employed in 1991

	CATT level						OA level					
	Model 5			Model 6			Model 7			Model 8		
	Coef.	S.e.[a]	Sig	Coef.	S.e.[a]	Sig	Coef.	S.e.[a]	Sig	Coef.	S.e.[a]	Sig
Neighbourhood level variables												
Neighbourhood unemployment 1991 (reference = 0–1.9%)												
1.9–3.5%	0.157	0.083	*	0.105	0.088		–0.002	0.081		–0.044	0.086	
3.6–5.9%	0.296	0.079	***	0.135	0.084		0.303	0.077	***	0.166	0.082	**
6.0–10.0%	0.561	0.077	***	0.320	0.084	***	0.485	0.076	***	0.249	0.082	***
10.1–54.9%	0.915	0.076	***	0.487	0.088	***	0.889	0.074	***	0.431	0.086	***
Urban–rural classification 1991 (reference = cities)												
Urban areas				0.066	0.061					0.057	0.060	
Small towns				0.054	0.092					0.026	0.089	
Remote towns				0.205	0.139					0.177	0.139	
Rural areas				0.103	0.084					0.072	0.083	
Remote areas				0.117	0.118					0.087	0.120	
Individual and household level variables												
Female (reference = male)				–0.618	0.054	***				–0.617	0.054	***
Ethnic (reference = non ethnic)				0.683	0.241	***				0.689	0.244	***
Age in years (1991)				0.009	0.003	***				0.009	0.003	***
Change in household type (reference = couple)												
91 & 01 single				0.852	0.092	***				0.854	0.092	***
91 single/01 couple				–0.076	0.158					–0.079	0.159	
91 couple/01 single				0.848	0.059	***				0.849	0.059	***

Change in presence of children (reference = children)												
91/01 No children				0.193	0.068	***				0.190	0.070	***
91 No child/01 child				−0.078	0.093					−0.075	0.093	
91 Child/01 no child				0.143	0.073	*				0.141	0.074	*
Qualifications 1991 (reference = none)												
Less than degree				−0.403	0.099	***				−0.403	0.100	***
More than degree				−0.697	0.120	***				−0.702	0.119	***
None stated				0.157	0.152					0.157	0.151	
Tenure 1991 (reference = owners)												
Social Renter				0.524	0.058	***				0.513	0.058	***
Private Renter				0.355	0.101	***				0.343	0.102	***
Change in health (reference = no LLTI)												
LLTI 91 & 01				1.047	0.175	***				1.051	0.174	***
LLTI 91				−0.321	0.289					−0.320	0.288	
LLTI 01				0.951	0.075	***				0.946	0.073	***
Constant	−3.826	0.059	***	−4.381	0.140	***	−3.771	0.057	***	−4.316	0.138	***
Initial log pseudo LL		−8678.380										
Log pseudo likelihood		−8580.202			−7358.909			−8573.306			−7358.936	
Wald		190.850			988.240			213.170			1013.390	
Number of observations		60,048			60,048			60,048			60,048	

*$p < 0.10$; **$p < 0.05$; ***$p < 0.01$

[a]Robust standard errors adjusted for clustering at the CATT level

Source: Calculations done by the authors using data from the SLS correlation between neighbourhood unemployment levels and the probability of being unemployed in 2001. The higher the neighbourhood unemployment levels, the less likely employed people are to still have a job in 2001. Adding control variables in Model 6 does not alter the overall pattern of the main variable of interest, but the parameters are much smaller and not all significant.

Model 6 also includes a range of control variables. Females are less likely to be out of work than males. Those belonging to ethnic minority groups are more likely than non-ethnic minority individuals to be out of employment in 2001. With increasing age the probability of being out of employment in 2001 increases. Those who were single in both 1991 and 2001 and those who became single between the two Census years are the most likely to be unemployment in 2001. Those without children in both Census years are more likely than those with kids to be out of employment in 2001. Having qualifications greatly reduces the probability of becoming unemployed. Those with degrees are the least likely to be out of employment. Private renters and especially social renters are much more likely than home owners to be out of employment in 2001. A possible explanation is selection into tenures where those with better employment prospects are the most likely to be homeowners in the first place. As expected, individuals with poor health in both Census years, along with those whose health deteriorates between 1991 and 2001, are more likely to be out of work than those with good health. Again we repeat the models including Output Area level variables. Model 7 shows that the results of the models including Output Area level neighbourhood characteristics are very similar to the results of the models including CATT level neighbourhood characteristics.

Stopping the analysis here could lead to the confirmation of the neighbourhood effects hypothesis as we have found significant 'effects' of neighbourhood unemployment levels on individual employment outcomes. However, earlier in this chapter we have argued that modelling employment outcomes separately for social renters and homeowners can provide more insight in whether the correlations found are indeed causal effects (see Oreopoulos 2003; van Ham and Manley 2010). Tables 7.5 and 7.6 presents tenure split models including neighbourhood characteristics at the CATT level (the results at the OA level were very similar). Models 9–12 (Table 7.5) estimate the probability of having a job in 2001 for those who were out of work in 1991. Models 9 and 10 only include social renters and models 11 and 12 only include owner occupiers. The main difference between the models for social renters and the models for owner occupiers is that we found hardly any significant effects of neighbourhood unemployment levels for social renters, while for owner occupiers the significant effects seen previously in Table 7.3 are still present. For social renters we only found a significant effect on employment outcomes of living in neighbourhoods with the highest levels of unemployment. Models 13–16 (Table 7.6) estimate the probability that those with a job in 1991 are out of employment in 2001. Models 13 and 14 only include social renters and models 15 and 16 only include owner occupiers. Again we find that there are more significant effects of living in a high unemployment neighbourhood on labour market outcomes for owner occupiers than for social renters. The overall pattern in Tables 7.4 and 7.5 is that we find relatively strong correlations between neighbourhood unemployment

Table 7.5 CATT logistic regression for entering employment (Models 9–12) split by tenure

	Entering employment											
	Social renters						Owner occupiers					
	Model 9			Model 10			Model 11			Model 12		
	Coef.	S.e.[a]	Sig	Coef.	S.e.[a]	Sig	Coef.	S.e.[a]	Sig	Coef.	S.e.[a]	Sig
Neighbourhood level variables												
Neighbourhood unemployment 1991 (reference = 0–1.88)												
1.9–3.5%	−0.351	0.452		−0.517	0.438		0.049	0.347		0.050	0.382	
3.6–5.9%	−0.198	0.434		−0.269	0.417		−0.656	0.303	**	−0.620	0.348	*
6.0–10.0%	−0.493	0.421		−0.597	0.398		−0.895	0.299	***	−0.891	0.338	***
10.1–54.9%	−0.623	0.415		−0.763	0.392	*	−0.901	0.309	***	−0.749	0.347	**
Urban-rural classification 1991 (reference = cities)												
Small towns				−0.082	0.190					0.210	0.348	
Remote towns				0.415	0.368					0.771	1.092	
Rural areas				0.140	0.238					−0.036	0.305	
Remote areas				−0.474	0.339					−0.182	0.347	
Individual and household level variables												
Female (reference = male)				0.676	0.241	***				1.232	0.403	***
Ethnic (reference = non ethnic)												
				1.588	0.852	*				−0.871	0.611	
Age in years (1991)				−0.014	0.007	*				−0.002	0.011	
Change in household type (reference = couple)												
91 & 01 Single				−1.235	0.175	***				−1.052	0.382	***
91 Single/01 couple				0.131	0.286					0.744	1.116	
91 Couple/01 single				−0.735	0.129	***				−0.810	0.207	***

(continued)

Table 7.5 (continued)

	Entering employment											
	Social renters						Owner occupiers					
	Model 9			Model 10			Model 11			Model 12		
	Coef.	S.e.[a]	Sig	Coef.	S.e.[a]	Sig	Coef.	S.e.[a]	Sig	Coef.	S.e.[a]	Sig
Change in presence of children (reference = children)												
91/01 No children				−0.314	0.153	**				−1.120	0.353	***
91 No child/01 child				0.179	0.195					−0.638	0.447	
91 Child/01 no child				0.036	0.173					−0.599	0.379	
Qualifications 1991 (reference = none)												
Less than degree				0.959	0.583					0.571	0.506	
More than degree				0.849	0.606					0.832	0.426	*
None stated				−0.295	0.247					−0.223	0.379	
Change in health (reference = no LLTI)												
LLTI 91 & 01				−0.452	0.446					−1.058	0.863	
LLTI 91				−0.332	0.306					–	–	
LLTI 01				−0.811	0.176	***				−1.062	0.264	***
Constant	1.488	0.410	***	2.236	0.489	***	2.516	0.257	***	3.165	0.574	***
Log pseudo LL		−1302.673			−1052.958			−511.288			−413.752	
Wald		10.120			198.770			20.020			106.170	
Number of obs.		1,992						1,270				

*$p < 0.10$; **$p < 0.05$; ***$p < 0.01$

[a]Robust standard errors adjusted for clustering at the CATT level

Source: Calculations done by the authors using data from the SLS

Table 7.6 CATT logistic regression for loosing employment (Models 13–16)

	Losing employment											
	Social renters						Owner occupiers					
	Model 13			Model 14			Model 15			Model 16		
	Coef.	S.e.[a]	Sig	Coef.	S.e.[a]	Sig	Coef.	S.e.[a]	Sig	Coef.	S.e.[a]	Sig
Neighbourhood level variables												
Neighbourhood unemployment 1991 (reference = 0–1.88)												
1.9–3.5%	–0.168	0.201		–0.203	0.210		0.188	0.098	*	0.179	0.103	*
3.6–5.9%	0.061	0.181		–0.017	0.197		0.229	0.097	***	0.167	0.102	
6.0–10.0%	0.319	0.167	*	0.274	0.182		0.335	0.102	***	0.283	0.107	***
10.1–54.9%	0.476	0.162	***	0.359	0.180	**	0.566	0.113	***	0.491	0.122	***
Urban-rural classification 1991 (reference = cities)												
Urban areas				0.073	0.097					0.060	0.083	
Small towns				–0.029	0.154					0.110	0.117	
Remote towns				0.184	0.196					0.237	0.215	
Rural areas				0.158	0.156					0.065	0.111	
Remote areas				0.178	0.218					0.104	0.166	
Individual and household level variables												
Female (reference = male)				–1.202	0.147	***				–0.671	0.114	***
Ethnic (reference = non ethnic)		0.098		0.749						0.882	0.262	***
Age in years (1991)				0.009	0.005	*				0.010	0.005	**
Change in household type (reference = couple)												
91 & 01 single				0.978	0.045	***				0.498	0.150	***
91 Single/01 couple				0.084	–0.161					–0.155	0.230	

(continued)

Table 7.6 (continued)

	Losing employment											
	Social renters						Owner occupiers					
	Model 13			Model 14			Model 15			Model 16		
	Coef.	S.e.[a]	Sig	Coef.	S.e.[a]	Sig	Coef.	S.e.[a]	Sig	Coef.	S.e.[a]	Sig
91 Couple/01 single				0.894	−0.003	***				0.775	0.083	***
Change in presence of children (reference = children)												
91/01 No children				0.978	0.141	***				0.446	0.097	***
91 No child/01 child				0.084	0.253					0.135	0.129	
91 Child/01 No child				0.894	0.094	***				0.378	0.102	***
Qualifications 1991 (reference = none)												
Less than degree				−0.511	0.265	*				−0.397	0.116	***
More than degree				−0.245	0.367					−0.758	0.138	***
None stated				0.366	0.230					−0.187	0.244	
Change in health (reference = no LLTI)												
LLTI 91 & 01				1.029	0.260	***				1.122	0.244	***
LLTI 91				−1.596	0.717	**				0.428	0.313	
LLTI 01				0.937	0.118	***				0.985	0.100	***
Constant	−3.129	0.151	***	−3.140	0.287	***	−3.957	0.070	***	−4.401	0.223	***
Log pseudo LL		−3110.544			−2542.160			−4817.586			−4261.940	
Wald		31.750			399.000			27.580			457.460	
Number of obs.		14,833						41,825				

*$p < 0.10$; **$p < 0.05$; ***$p < 0.01$

[a]Robust standard errors adjusted for clustering at the CATT level

Source: Calculations done by the authors using data from the SLS

levels and employment outcomes for owner occupiers, but not for social renters. It is unlikely that neighbourhood effects only influence owner occupiers and that social renters in the same neighbourhoods are immune to neighbourhood effects. As a result, alternative explanations must be explored, including the possibility that the effects of selection bias are stronger for owner occupiers than for social renters.

Discussion

In this chapter we investigated the hypothesis that living in neighbourhoods with a high level of unemployment can negatively affect the labour market prospects of neighbourhood residents. Several theoretical causal pathways were suggested through which the neighbourhood context could influence individual outcomes: those living in neighbourhoods with high levels of unemployment are less likely to be in contact with positive role models, are more likely to have unemployed friends and neighbours which might cause negative socialisation, are less likely to know people who can help finding a job, and are more likely to be stigmatised by outsiders. We used longitudinal data to study the effect of 1991 neighbourhood characteristics on 2001 labour market outcomes. As expected, we found a strong correlation between neighbourhood unemployment levels and individual labour market outcomes 10 years later, even after controlling for a range of individual and household characteristics. These results could lead to the conclusion that there are strong causal neighbourhood effects at play.

However, the data used from the Scottish Longitudinal Study did not allow us to control our results for self-selection of individuals into neighbourhoods prior to 1991, a process that is likely to be correlated with individual level labour market outcomes. To overcome this restriction and to gain greater insight into potential selection mechanisms we estimated tenure split models which showed significant effects of neighbourhood unemployment levels on labour market outcomes for owner occupiers, but not for social renters. Previously, similar effects were found by Oreopoulos (2003) and van Ham and Manley (2010). Intuitively, one would expect negative neighbourhood effects for social renters and not for owner occupiers. It was suggested that the effects found are related to the differences in mechanisms through which social renters and owner occupiers 'select' their neighbourhoods. In Scotland, social renters in the early 1990s and before were largely randomly allocated a dwelling by a housing officer, without the option to express any choice with regard to dwelling or neighbourhood. We do acknowledge however that allocation was not completely random as ethnicity, household size, and household structure all played a role in the allocation process (Duke 1970; Simpson 1981; Henderson and Karn 1984; Clapham and Kintrea 1984; Malpass and Murie 1994; Peach 1996; Somerville 2001; Sarre et al. 1989). As a result of the allocation process in social housing, it is reasonable to argue that selection bias is less likely to influence model outcomes for social renters than for owner occupiers. Owner occupiers had greater freedom in choosing where they wanted to live within the restrictions of their own budget and the constraints of the local housing market. Mortgage providers play an important indirect role in neighbourhood selection by homeowners as they determine how much an applicant can borrow and therefore which houses in which neighbourhoods are affordable (see Aalbers 2009). Mortgage providers look at individual and household income, but also at job security and potential career development. Those with the least secure jobs get the lowest mortgages and therefore select themselves into the cheapest neighbourhoods, often neighbourhoods with high unemployment levels. Thus, labour market outcomes are also driving the selection of neighbourhoods by households in the owner-occupied sector.

The tenure split models suggest that the correlations found between neighbourhood unemployment levels and individual unemployment are, at least in part, the result of selection bias. To enable neighbourhood effects research to move forward, it is necessary to incorporate the neighbourhood selection process into models of neighbourhood effects. Such an approach requires richer longitudinal data than currently available in most datasets. Where it is not possible to model neighbourhood selection before modelling the impact of neighbourhood characteristics on individual outcomes, an approach such as the one explored in this chapter is recommended.

Moving beyond the question of whether or not correlations between neighbourhood characteristics and individual outcomes are the result of causal pathways or selection effects, it is important to remember that concentrations of unemployment and poverty in neighbourhoods are real. The question of whether and how to tackle these concentrations of poverty is more than an academic and methodological question. It is also a question of social and spatial justice (Smith 1994; Soja 2010). The lack of a causal pathway between neighbourhood context and individual outcomes does little to solve the problems of uneven neighbourhood resources. Nevertheless, the realisation that individual outcomes, in the case of this chapter individual unemployment, are not exacerbated by living in concentrations of unemployment is important. Policy prescriptions, such as reducing the concentration of social housing to deconcentrate poverty, and by default unemployment, will most likely not by itself lead to significant improvements in individual outcomes. This argument does not mean that neighbourhoods with high concentrations of poverty or other perceived social problems should not be invested in or offered regeneration. Rather, it is necessary to recognise the limitations of such policies with respect to the impact they will have on individuals and the limited potential they have to improve individual outcomes. The empirical results shown in this chapter highlight the importance of, amongst other things, an individual's educational achievement as a means through which their chances of employment increase significantly. As a result, policies that specifically target individuals are more likely to offer real outcomes and tangible changes in individual life courses.

Acknowledgments The help provided by staff of the Longitudinal Studies Centre–Scotland (LSCS) is acknowledged. The LSCS is supported by the ESRC/JISC, the Scottish Funding Council, the Chief Scientist's Office and the Scottish Executive. The authors alone are responsible for the interpretation of the data. Census output is Crown copyright and is reproduced with the permission of the Controller of HMSO and the Queen's Printer for Scotland.

References

Aalbers, M. B. (2009). Redlining. In N. Thrift & R. Kitchin (Eds.), *International encyclopaedia of human geography* (Vol. X, pp. 117–124). Oxford: Elsevier.

Blume, L., & Durlauf, S. (2001). The interactions-based approach to socioeconomic behaviour. In S. Durlauf & H. P. Young (Eds.), *Social dynamics* (pp. 15–44). Cambridge: MIT Press.

Bolster, A., Burgess, S., Johnston, R., Jones, K., Propper, C., & Sarker, R. (2007). Neighbourhoods, households and income dynamics: A semi-parametric investigation of neighbourhood effects. *Journal of Economic Geography, 7*(1), 1–38.

Boyd, M., Edin, K., Duncan, G., & Clampet-Lundquist, S. (2006). The durability of the Gautreaux Two Residential Mobility Program: A qualitative analysis of who stays and who moves from low-poverty neighbourhoods. Population Association of American, Annual Meeting, Los Angeles, 30th March, 2006. http://paa2006.princeton.edu/download.aspx?submissionId=61724. Accessed 3 June 2010.

Boyle, P. J., Feijten, P., Feng, Z., Hattersley, L., Huang, Z., Nolan, J., & Raab, G. (2008). Cohort profile: The Scottish Longitudinal Study (SLS). *International Journal of Epidemiology, 38*(2), 385–392.

Brock, W., & Durlauf, S. (2001). Interactions-based models. In J. Heckman & E. Leamer (Eds.), *Handbook of econometrics* (Vol. 5, pp. 3297–3380). Amsterdam: Elsevier Science B.V.

Cheshire, P. (2007). *Are mixed communities the answer to segregation and poverty?* York: Joseph Rowntree Foundation.

Cheshire, P. (2011). Policies for mixed communities: Still looking for evidence? In M. van Ham, D. Manley, N. Bailey, L. Simpson, & D. Maclennan (Eds.), *Neighbourhood effects research: New perspectives*. Dordrecht: Springer.

Ciseros, H. G., & Engdahl, L. (Eds.). (2009). *From despair to hope: HOPE VI and the new promise of public housing in America's cities*. Washington DC: The Brookings Institution.

Clapham, D., & Kintrea, K. (1984). Allocation systems and housing choice. *Urban Studies, 21*(3), 261–269.

Clark, W. A. V. (2008). Re-examining the moving to opportunity study and its contribution to changing the distribution of poverty and ethnic concentration. *Demography, 45*(3), 515–535.

Clark, K., & Drinkwater, S. (2002). Enclaves, neighbourhood effects and employment outcomes: Ethnic minorities in England and Wales. *Journal of Population Economics, 15*(1), 5–29.

Dean, J., & Hasting, A. (2000). *Challenging images: Housing estates, stigma and regeneration*. Bristol: Policy Press.

Dietz, R. D. (2002). The estimation of neighborhood effects in the social sciences: An interdisciplinary approach. *Social Science Research, 31*(4), 539–575.

Duke, C. (1970). *Colour and rehousing: A study of redevelopment in Leeds*. London: Institute of Race Relations.

Durlauf, S. (2004). Neighborhood effects. In J. V. Henderson & J. F. Thisse (Eds.), *Handbook of regional and urban economics* (Cities and Geography, Vol. 4, pp. 2173–2242). Amsterdam: Elsevier Science B.V.

Elhassan, H., Feins, J., Goering, J., Holin, M. J., Kraft, J., & McInnis, D. (1999). *Moving to opportunity for fair housing demonstration program: Current status and initial findings*. Washington DC: U.S. Department of Housing and Urban Development.

Ellen, I. G., & Turner, M. A. (1997). Does neighbourhood matter? Assessing recent evidence. *Housing Policy Debate, 8*(4), 833–866.

Flowerdew, R., Manley, D., & Sabel, C. (2008). Neighbourhood effects on health: Does it matter where you draw the boundaries? *Social Science and Medicine, 66*(6), 1241–1255.

Friedrichs, J. (1998). Do poor neighborhoods make their residents poorer? Context effects of poverty neighborhoods on their residents. In H. Andress (Ed.), *Empirical poverty research in a comparative perspective* (pp. 77–99). Aldershot: Ashgate.

Galster, G. (2001). On the nature of neighbourhood. *Urban Studies, 38*(12), 2111–2124.

Galster, G. (2011). The mechanism(s) of neighbourhood effects: Theory, evidence, and policy implications. In M. van Ham, D. Manley, N. Bailey, L. Simpson, & D. Maclennan (Eds.), *Neighbourhood effects research: New perspectives*. Dordrecht: Springer.

Gobillon, L., Selod, H., & Zenou, Y. (2005). *The mechanisms of spatial mismatch*. Research Unit Working Papers 0510, Laboratoire d'Economie Appliquee, INRA.

Goering, J., Feins, J. D., & Richardson, T. M. (2002). A cross-site analysis of initial moving to opportunity demonstration results. *Journal of Housing Research, 13*(1), 1–30.

Graham, E., Manley, D., Hiscock, R., Boyle, P., & Doherty, J. (2009). Mixing housing tenures: Is it good for social well-being? *Urban Studies, 46*(1), 139–165.

Harris, D. R. (1999). "Property values drop when blacks move in, because...": Racial and socio-economic determinants of neighborhood desirability. *American Sociological Review, 64*(3), 461–479.

Hedman, L., & van Ham, M. (2011). The impacts of residential mobility on measurements of neighbourhood effects. In M. van Ham, D. Manley, N. Bailey, L. Simpson, & D. Maclennan (Eds.), *Neighbourhood effects research: New perspectives*. Dordrecht: Springer.

Hedman, L., van Ham, M., & Manley, D. (2010). *Neighbourhood reproduction through neighbourhood choice*. Paper presented at meeting of the European Network for Housing Research, Istanbul, Turkey, July 2010.

Henderson, J., & Karn, V. (1984). Race, class and the allocation of public housing in Britain. *Urban Studies, 21*(1), 115–128.

Houston, D. S. (2001). Testing the spatial mismatch hypothesis in the United Kingdom using evidence from firm relocations. *European Research in Regional Science, 11*(1), 134–151.

Houston, D. S. (2005). Employability, skills mismatch and spatial mismatch in metropolitan labour markets. *Urban Studies, 42*(2), 221–243.

Kain, J. F. (1968). Housing segregation, negro employment, and metropolitan decentralization. *Quarterly Journal of Economics, 82*(1), 175–197.

Katz, L., Kling, J., & Liebman, J. (2001). Moving to opportunity in Boston: Early results of a randomized mobility experiment. *Quarterly Journal of Economics, 116*(2), 607–654.

King, J. R., Liebman, J. B., & Katz, L. F. (2007). Experimental analysis of neighborhood effects. *Econometrica, 75*(1), 83–119.

Leventhal, T., & Brooks-Gunn, J. (2000). A randomized study of neighborhood effects on low-income children's educational outcomes. *Developmental Psychology, 40*(4), 488–507.

Ludwig, J., Duncan, G. J., & Hirschfield, P. (2001). Urban poverty and juvenile crime: Evidence from a randomized housing-mobility experiment. *Quarterly Journal of Economics, 116*(2), 655–680.

Malpass, P., & Murie, A. (1994). *Housing policy and practice* (4th ed.). London: MacMillan Press.

Manley, D., Flowerdew, R., & Steel, D. (2006). Scales, levels and processes: Studying spatial patterns of British census variables. *Computers, Environment and Urban Systems, 30*(2), 143–160.

Manski, C. (1993). Identification of endogenous social effects: The reflection problem. *Review of Economic Studies, 60*(3), 531–542.

Manski, C. (2000). Economics analysis of social interactions. *Journal of Economic Perspectives, 14*(1), 115–136.

Mendenhall, R., Duncan, G. J., & DeLuca, S. A. (2006). Neighborhood resources, racial segregation, and economic mobility: Results from the Gautreaux program. *Social Science Research, 35*(6), 892–923.

Moffitt, R. (2001). Policy interventions, low-level equilibria, and social interactions. In S. Durlauf & H. Young (Eds.), *Social dynamics* (pp. 45–82). London: MIT Press.

Musterd, S., & Andersson, R. (2005). Housing mix, social mix and social opportunities. *Urban Affairs Review, 40*(6), 761–790.

Musterd, S., & Andersson, R. (2006). Employment, social mobility and neighbourhood effects: The case of Sweden. *International Journal of Urban and Regional Research, 30*(1), 120–140.

Oreopoulos, P. (2003). The long-run consequences of living in a poor neighborhood. *Quarterly Journal of Economics, 118*(4), 1533–1575.

Orr, L., Feins, J. D., Jacob, R., Beecroft, E., Sanbonmatsu, L., Katz, L. F., Liebman, J. B., & Kling, J. R. (2003). *Moving to opportunity interim impacts evaluation*. Washington, DC: US Department of Housing and Urban Development, Office of Policy Development and Research.
Peach, C. (1996). Does Britain have ghettos? *Transactions of the Institute of British Geographers, 21*(1), 216–235.
Permentier, M., van Ham, M., & Bolt, G. (2007). Behavioural responses to neighbourhood reputations. *Journal of Housing and the Built Environment, 22*(2), 199–213.
Popkin, S. J., & Cunningham, M. K. (2009). Has HOPE VI transformed residents' lives. In H. L. Cisneros & L. Engdahl (Eds.), *From despair to hope: HOPE VI and the new promise of public housing in America's cities* (pp. 191–205). Washington, DC: Brookings Institute Press.
Popkin, S. J., Rosenbaum, J. E., & Meaden, P. M. (1993). Labor market experiences of low-income black women in middle-class suburbs: Evidence from a survey of Gautreaux program participants. *Journal of Policy Analysis and Management, 12*(3), 556–573.
Rosebaum, J. (1995). Changing the geography of opportunity by expanding residential choice: Lessons from the gatreaux program. *Housing Policy Debate, 6*(1), 231–269.
Rosenbaum, J. (1994). *Housing mobility strategies for changing the geography of opportunity*. Evanston: Institute for Policy Research, Northwestern University.
Sampson, R., Morenoff, J., & Gannon-Rowley, T. (2002). Assessing 'neighbourhood effects': Social processes and new directions in research. *Annual Review of Sociology, 28*, 443–478.
Sarre, P., Phillips, D., & Skellington, R. (1989). *Ethnic minority housing: Explanations and policies*. Aldershot: Avebury.
Simpson, A. (1981). *Stacking the decks*. Nottingham: Nottingham Community Relations Council.
Small, M., & Feldman, J. (2011). Ethnographic evidence and neighbourhood effects: Strong and weak approaches to testing propositions from the field. In M. van Ham, D. Manley, N. Bailey, L. Simpson, & D. Maclennan (Eds.), *Neighbourhood effects research: New perspectives*. Dordrecht: Springer.
Smith, D. M. (1994). *Geography and social justice*. Oxford: Blackwell.
Soja, E. W. (2010). *Seeking spatial justice*. Minneapolis: University of Minnesota Press.
Somerville, P. (2001). Allocating housing or letting people choose? In D. Cowan & A. Marsh (Eds.), *Two steps forward: Housing policy into the new millennium* (pp. 113–132). Bristol: Policy Press.
van Ham, M. (2002). *Job access, workplace mobility, and occupational achievement*. Delft: Eburon.
van Ham, M., & Manley, D. (2010). The effect of neighbourhood housing tenure mix on labour market outcomes: A longitudinal investigation of neighbourhood effects. *Journal of Economic Geography, 10*(2), 257–282.
Venkatesh, S. A., Celimli, I., Miller, D., Murphy, A., & Turner, B. (2004). *Chicago public housing transformation: A research report*. New York: Center for Urban Research, Columbia University.
Wacquant, L. J. D. (1993). Urban outcasts: Stigma and division in the black American ghetto and the French periphery. *International Journal of Urban and Regional Research, 17*(3), 366–383.
Wilson, W. J. (1987). *The truly disadvantaged*. Chicago: University of Chicago Press.
Wilson, W. J. (1991). Another look at the truly disadvantaged. *Political Science Quarterly, 106*(4), 639–656.

Chapter 8
Neighbourhood Social Capital and Individual Mental Health

Gindo Tampubolon

Introduction

It has been claimed that neighbourhood effects have been found to be important for a wide range of outcomes, including schooling, housing and health (Durlauf 2004). Many of the empirical works reviewed by Durlauf claim to present evidence of the importance of neighbourhood effects and identified some of the underlying causal mechanisms such as peer group effects and (lack of) information effects. Durlauf also noted that a significant part of the existing body of evidence does not deal with identification problems or selection problems, thus potentially weakening claims about causal neighbourhood effects. (Econometric) identification of causal mechanisms is the main challenge in neighbourhood effects research and in recent years major advances have been made in this field. Durlauf (2004) also noted that a tighter link between empirical work and substantive theory (human capital theory or housing demand theory or health capital theory) is needed in order to transform the promise of neighbourhood effects research into real advances. Such advances will spur more fruitful theoretical works and more relevant policy input. For policy responses to assumed neighbourhood effects it is crucial to identify causal pathways which link neighbourhood characteristics with individual level outcomes.

Relatively recent but intense efforts focusing on neighbourhood effects originate from within the body of literature on public health and social epidemiology. There is a fast growing body of literature which focuses on neighbourhood effects on individual health outcomes such as obesity, mental health, physical health and health-related quality of life. This literature has identified two neighbourhood attributes potentially relevant for individual health: physical/environmental deprivation of the neighbourhood and neighbourhood social capital.

G. Tampubolon (✉)
Institute for Social Change, The University of Manchester, Manchester, UK
e-mail: tampubolon@manchester.ac.uk

M. van Ham et al. (eds.), *Neighbourhood Effects Research: New Perspectives*,
DOI 10.1007/978-94-007-2309-2_8, © Springer Science+Business Media B.V. 2012

The claim that neighbourhood social capital matters seems intuitive; yet supporting evidence remains elusive. Studies in the US show that neighbourhood social capital correlates with individual health (Kawachi et al. 1997, 1999; Subramanian et al. 2005; Viswanath et al. 2006; Farquhar et al. 2005; Perry et al. 2008). In the UK however comparable evidence is difficult to find. The few existing studies of social capital and health in the UK failed to find a general association between social capital and health outcomes (Duncan et al. 1993; Sloggett and Joshi 1998; Mohan et al. 2005; Propper et al. 2005; Stafford et al. 2008). The nearest to find a negative effect of neighbourhood social capital on individual mental health is a study by Stafford et al. (2008: 304). They report a negative association "between social capital and common mental disorders which was limited to economically 'stressed' residents and neighbourhoods." Studies from other countries such as New Zealand and Sweden have failed to settle the issue (Blakely et al. 2006; Islam et al. 2006). Kawachi and Berkman (2003) have identified some of the mechanisms which are thought to link neighbourhood social capital and individual health (see the next section for an overview).

Despite the identification of potential mechanisms, studies on social capital and health fail to connect with a theoretical model of health production, particularly the Grossman health model (Grossman 1972a, b), thereby depriving them of formal grounding. In this influential model of health demand, Grossman posits that health, like human capital, is produced using various market and non-market inputs. Initially endowed with health stock at birth, individuals maintain or produced the level of health desired by consuming various inputs including time, medical care, housing, exercise, education and other goods.

Conversely, studies in health economics which follow Grossman's model largely ignore the potential of neighbourhood social capital in influencing individual health decisions. How neighbourhood social capital produces a better quality of life through health benefits for neighbourhood residents is left unspecified. Another potential shortcoming of studies on neighbourhood social capital and individual health outcomes is that they often rely on respondents' reports of their neighbourhood social environment. The assessment of social capital was obtained from the same respondents whose health outcomes were measured and this raises a potential reflection problem (Manski 1993), which might prevent the identification of causal effects. Also, the level of spatial aggregation to define 'neighbourhoods' has varied across previous studies. For example, many studies in the UK, admittedly by necessity rather than by design, use the administrative units of wards as a proxy of 'neighbourhoods' – which many consider to be too large and heterogeneous for studying the impact of neighbourhood social environments on health outcomes (e.g. Mohan et al. 2005). In 2009 the population of wards ranged from 90 in Walbrook to 32,373 in Sparkbrook with a mean of 5,945 (Office of National Statistics www.statistics.gov.uk/statbase/ Accessed 29 October 2010). Finally, rarely does a study on the effects of neighbourhood social capital on health outcomes use a widely validated health instrument.

This chapter contributes to the literature on neighbourhood effects and health outcomes by proposing an extension of the influential Grossman model of health

(Grossman 1972a, b), by explicitly including interactions with the neighbourhood context. The extended model elaborates on social interactions and their effects on individual decisions, particularly health maintenance and health risk decisions. I shall draw upon the Blume-Brock-Durlauf social interaction model which will be discussed below (Blume 1993; Brock 1993; Durlauf 1997; Brock and Durlauf 2001a, b; Durlauf 2002; Blume and Durlauf 2005) to study the effect of social capital on mental health, using data from the Welsh Health Survey 2007 (WHS) and the Living in Wales 2007 (LiW) survey.

Instruments or exclusion restrictions that are theoretically motivated within the extended Grossman model are readily obtained from other studies in public health, epidemiology, and economics using the Grossman model within a neighbourhood context. For this study neighbourhoods are defined as lower super output areas (LSOA), a geography purposefully designed for social research, with a mean population of about 1,500. For comparison, wards have a mean population of about 2,500 people in Wales and a general practice or 'primary care doctor practice' has a catchment area with a mean population of 5,600 (Department of Health 2006). LSOAs are thus a finer scale for delineating neighbourhood for the purpose of health research. Moreover, this standardised geography enables independent measures of neighbourhood social capital and neighbourhood deprivation, obtained from administrative sources, to be linked to the data.

This study uses a widely validated instrument of health related quality of life, Short Form-36 (SF36), to measure mental health (Ware 2004; Wilkin et al. 1992). SF36 is the most frequently used measure of generic health status across the world (Bowling 2005: 63). It consists of 36 item health status questions and has been widely psychometrically validated. The items measure eight health dimensions including physical functioning, social functioning, role limitations due to physical problems, role limitations due to emotional problems, mental health, energy/vitality, pain and general health perception. Two summary scores are derived from these eight dimensions: the physical component summary and the mental component summary. It is the latter summary that is used here.

Neighbourhood Social Capital and Health

The concept 'social capital' is the result of a crystallisation of ideas that have been around since researchers began to examine systematically the relationships between society, especially neighbourhoods, and individual outcomes. A definition that will suffice for our purpose comes from Putnam (1993: 167): "social networks and norms and trust" residing in a neighbourhood. It is obvious that social networks, norms and trust grow out of and circulate in social interactions; see also the discussion by Woolcock (1998). The literature on models of social interactions will be one of the main sources of econometric modelling ideas drawn upon in this study.

Recent works in social epidemiology have attempted to be more specific about how social capital in the neighbourhood can influence health and well-being (Berkman and

Kawachi 2000; Kawachi and Berkman 2003). Kawachi and Berkman (2003) write about three mechanisms linking neighbourhood social capital and individual health. First, more cohesive neighbourhoods are better equipped to disseminate information and mobilize collective action, for example, to prevent fast food outlets to open in a neighbourhood. Second, more cohesive neighbourhoods are better equipped to enforce and maintain social norms, and hence to maintain residents' sense of health. However, it is now also recognised that social norms can influence health in negative ways, as shown in the case of obesity (Christakis and Fowler 2007). The third mechanism is indirect; collective efficacy and informal control in preventing crime and violence reduce environmental stresses suffered by residents in their day to day activities and increases the take up of health maintenance behaviour such as physical exercise. Finally, Marmot et al. (2010: 136) note that high levels of neighbourhood social capital also enable communities to be more responsive to national and local organisations that seek involvement and engagement at the local level. The above overview of mechanisms reminds us that social processes remain to an important extent rooted in places.

The Grossman Model of Health and Its Extensions to Neighbourhood Effects

The formal model of neighbourhood social capital and individual health outcomes developed in this chapter draws on the Grossman model (Grossman 1972b; see also Grossman 1972a). In the Grossman model, health is produced using various market and non-market inputs. Initially endowed with health stock at birth, individuals produced the level of health desired by consuming various inputs including medical care, housing, exercise, education and other goods. Following the notation of Case and Deaton (2005), assume there is an instantaneous felicity function to represent the utility of consumption for an individual where t is age, δ_t is consumption, and H_t is the stock of health. Health is produced according to:

$$H_{t+1} = \theta m_t + (1 - \delta_t) H_t \tag{8.1}$$

where m_t is the decisions and behaviours for maintenance of health (including medical care bought and health behaviours like regular physical exercise (m_t^+), and smoking (m_t^-), positive and negative behaviours respectively), θ is the efficiency or conversion factor which is affected by education (and other socioeconomic status indicators) and δ is the rate of health deterioration at t. People maximise a life cycle welfare function:

$$U = \sum_0^T (1+\rho)^t v(c_t, H_t) \tag{8.2}$$

where ρ expresses time preference, and T is the length of life. The welfare function is optimized subject to full wealth constraints incorporating both wealth and time limits:

$$\sum_0^T \frac{c_t}{(1+r)^t} + \sum_0^T \frac{p_m m_t}{(1+r)^t} = W_0 + \sum_0^T \frac{y_t(H_t)}{(1+r)^t} \tag{8.3}$$

where r is the market rate of interest, p_m is the price of medical care and other health behaviours, W_0 is initial assets, and $y_t(H_t)$ is earnings, itself a function of health.

Optimising the welfare function subject to the constraints (3) and the changes in health stock (1) gives insights into the role of education and inequalities in health. These have been widely tested empirically by assuming functional forms of the elements of the theory (often of Cobb-Douglas form). Wagstaff (1986) provides some example assumptions which enable empirical estimation. On estimation, Van Doorslaer (1987) recommends a focus on the health production function to avoid problems when estimating the health demand function. Equations for health production function and for health maintenance suitable for estimation are:

$$H = H(M, W, X, \mu_h) \tag{8.4}$$

and

$$M = M(W, Y, \mu_m) \tag{8.5}$$

where W is wealth, X and Y include age, education and other exogeneous variables; and the μ's are residuals.

This is emphatically a recursive or triangular system as M, in turn, enters the health production function. Maintaining or neglecting health is affected by various determinants including access to wealth and individual resources; in turn, health maintenance ultimately affect individual health stock or health status. This system is also known as multiprocess system. Recently, for example, Balia and Jones (2008) [1] estimated a similar recursive system of health maintenance behaviour, health outcomes and mortality. Their recursive structure is intuitively and formally in this order: health maintenance, health outcome, mortality.

I propose an extension broadening the formal model to include neighbourhood effects. This extension acts as a bridge between the economics of health and epidemiology and public health. In the Grossman model, demand for the maintenance of health, M, is narrowly defined for each individual. However, if we construe maintenance to include the general maintenance of health and the avoidance of health risks then we are in a position to include neighbourhood effects. As explained earlier, the neighbourhood context can be expected to influence individual health outcomes. The inclusion of neighbourhood effects have the potential to better explain health outcomes, and offer scope for policy intervention.

[1] The published version dropped citation to Grossman and introduced a typographic error compared to the working paper version.

Statistical Mechanics of Social Interactions, Social Capital and Health

The theoretical justification for including broader actions, specifically neighbours' actions, on a resident's individual health is grounded in works on social interaction and its identification (Blume 1993; Brock 1993; Manski 1993; Durlauf 1997; Young 1998; Becker and Murphy 2000; Manski 2000; Brock and Durlauf 2001a, b; Glaeser and Scheinkman 2001; Durlauf 2002; Glaeser et al. 2002; Glaeser and Scheinkman 2003; Cutler and Glaeser 2005; Durlauf and Fafchamps 2005; Blume and Durlauf 2005).

Blume, Brock and Durlauf in a series of papers cited earlier draw upon statistical mechanics to understand the process of social interactions and how individual choices within them give rise to interesting aggregate behaviours.[2] In our context, social interactions facilitate the various forms of social capital which give rise to aggregate or widespread health behaviours such as jogging in the neighbourhood or smoking.

I follow closely Durlauf (1997) and Brock and Durlauf (2001b) which consider a binary choice setting.[3] This setting allows all parameters to be given their structural interpretation and facilitates econometric identification. Other works (Brock and Durlauf 2001a; Durlauf 2002) discuss identification in a linear-in-means setting as discussed below. Each individual is set in a population N where social interactions are present. Each individual resident chooses a binary action m_i with support $\{-1,1\}$. This support, instead of the usual $\{0,1\}$, is common in a social interactions model and shows its provenance in statistical mechanics. There the support is typically 'spin up' and 'spin down' and the aggregate behaviour of the 'population' of interest is typically macroscopic magnetization. Intuitively, these spin directions map onto jogging or not while macroscopic magnetization maps onto neighbourhood health activity.

Individual utility $V(m_i)$ is assumed to consist of three terms: private utility associated with a choice, $u(m_i)$; social utility associated with the choice, $S(.,.)$; and a random utility term which is independently and identically distributed, $\varepsilon(m_i)$; in the following equation:

$$V(m_i) = u(m_i) + S(m_i, \mu_i^e(m_{\setminus i})) + \varepsilon(m_i). \tag{8.6}$$

The term $\mu_i^e(m_{\setminus i})$ denotes the conditional probability that resident i puts on the choice of others at the time of making his or her own decision. In case of indiscriminate or total strategic complementarity, this social utility depends solely on $\overline{w}_i^e = (N-1)^{-1}\sum_{i \neq j} w_{i,j}^e$, where $w_{i,j}^e$ denotes the subjective expected value from the perspective of resident i of resident j choice.

---

[2] The closely related field of spatial statistics which is interested in *spatial* interactions also draws upon the same statistical mechanics literature, see Ripley (1990).

[3] Their model parallels the probability structure of the so-called Curie-Weiss model in statistical mechanics (Brock and Durlauf 2001b: 240). They refer to Ellis (1985, chap. 4) though Parisi (1988: 24) and Baxter (1982: 39) give more accessible accounts of Ising model with mean field which result in similar aggregate behaviour of magnetization m^*.

Brock and Durlauf assume parametric forms for the social utility term and the probability density of the random utility term.[4] They consider forms of social utility which exhibit indiscriminate strategic complementarity, as above, and are constant. The social utility then obeys $\frac{\partial S(m_i, \overline{w}_i^e)}{\partial m_i \partial \overline{w}_i^e} = J > 0$. These forms allow capture of the degree of dependence across residents' choices in a single parameter. With the constant degree of dependence, two forms of social utility suggest themselves. First, $S(m_i, \overline{w}_i^e) = Jm_i\overline{w}_i^e$ which exhibits proportional spillovers (strength of dependence). Second, $S(m_i, \overline{w}_i^e) = -\frac{J}{2}(m_i - \overline{w}_i^e)^2$ which exhibits conforming or restraining norms. The latter penalises deviations from the mean more strongly than the former. Additionally, the two forms differ in levels.

With ε assumed to be independent and extreme-value (Gumbel) distributed, the differences in the errors become logistically distributed. This widely used assumption in discrete choice literature, see e.g. Maddala (1983), allows a direct link between the theoretical model and its econometric estimation. To derive an equilibrium condition, we assume that decisions are made in noncooperative fashion, that is, each resident makes a choice without strategic communication or coordination. It follows from the extreme-value distribution assumption that:

$$\text{Prob}(m_i) = \frac{exp(\beta(u(m_i) + Jm_i\overline{w}_i^e))}{\sum_{n_i \in \{-1,1\}} exp(\beta(u(n_i) + Jn_i\overline{w}_i^e))}. \tag{8.7}$$

The parameter gives the extent to which the deterministic components of utility determine actual choice. Because of independence, the joint probability over all choices is:

$$\text{Prob(m)} = \frac{exp(\beta(\sum_1^N (u(m_i) + Jm_i\overline{w}_i^e)))}{\sum_{n_1 \in \{-1,1\}} \cdots \sum_{n_N \in \{-1,1\}} exp(\beta(\sum_1^N (u(n_i) + Jn_i\overline{w}_i^e)))}. \tag{8.8}$$

In the absence of a social interaction effect, $J = 0$, the probability above is proportional to logistic density; in its presence, $J \neq 0$, it captures interaction influence on behaviours in the neighbourhood. They then linearise the private utility $u(m_i) = hm_i + k$ with a further inspiration from statistical mechanics.[5] With this linearization, and using the definition of hyperbolic functions, the expectation becomes:

$$\text{E}(m_i) = tanh(\beta(h + J(N-1)^{-1}\sum_{i \neq j} m_{i,j}^e)). \tag{8.9}$$

---

[4] Physicists, instead, start with the working assumption that the coordinates and momenta in the equation of motion, at equlibria, follow the canonical distribution given by the so-called Boltzmann formula. See Parisi (1988: 2, Eq. 1.5) or Baxter (1982: 8, Eq. 1.4.1).

[5] Again see Parisi (1988: 2) on h the magnetic field and k the Boltzmann coefficient.

Furthermore, self-consistent and symmetric beliefs of residents (no residents are privileged compared to any other resident) give $\mathrm{E}(m_i) = \mathrm{E}(m_j) \forall i, j$. Together with the last equation, these guarantee there exists at least one expected choice level m^* [1, Proposition 1]:

$$m^* = tanh(\beta(h + Jm^*)) \tag{8.10}$$

Demonstrating the existence of equilibrium is one thing; achieving identification is another. Identification has always been a fraught issue in social interaction models. Manski (1995) and Durlauf (2002) have done a lot of work on deriving the conditions necessary for identification in linear and non-linear models of social interaction. Manski (2000: 129) lists a number of possibilities for identification including time lags and spatial lags of individual behaviours, non-linear models such as Brock and Durlauf's above, or other non-linearities (such as median neighbourhood behaviour), and the use of instrumental variables which affect the outcomes of a subset of the neighbours. The last possibility is the most relevant here. Durlauf (2002: 468, proposition 3) demonstrates that two or more instruments are needed to estimate the effect of neighbourhood social capital on an individual outcome; see also Brock and Durlauf (2001a) on linear-in-means model identification.

In sum, social interaction models lay the foundation for understanding the effects of social interaction in neighbourhoods on individual behaviour. With suitable instruments the effect of social capital, facilitated by social interaction on individual health, can be estimated. In fact, the formal model shows that ignoring social interaction may lead to an under-specified model, as leaving out social interaction effectively assumes such interactions to be negligible, $J = 0$, and omits any possibility of it being beneficial or harmful, $J \neq 0$.

Somewhat more prosaically, the effects of social interaction on health can be illustrated with an example on obesity. Food portions in America have increased over the last three decades (Nielsen and Popkin 2003). Finishing your meal and leaving an empty plate while dining out with friends, can be seen as an effect of social interaction influencing health behaviour in a negative way, m_t^-. What one orders to begin with ("Just a salad for me." Or "The full monty, please") and what one finishes are not unrelated to what everyone else around the table orders or how much they eat. This mechanism can be extended to the neighbourhood social context over time. For instance, Christakis and Fowler (2007) suggest that in Framingham, part of the greater Boston area, a network of friends acted as the conduit of an acceptable norm of body weight. Operating over 30 years, interactions in this network of friends led to an increase in obesity through their social interactions. The authors were careful to account for individual socio-demographic factors and other place-based factors. Across the Atlantic, Tampubolon et al. (2009) found, using data from Wales, that friendly neighbours and neighbourhoods can also lead to an increase in obesity. They also separated out the effects of individual sociodemographic and geographic factors in a multilevel multiprocess model which simultaneously explained consumption, physical exercise and obesity.

Both these empirical studies go some way towards revising the notion that social capital is always or primarily associated with positive benefits as read by Durlauf and Fafchamps (2005).[6]

Glaeser and Scheinkman (2003: 352) show that the so-called moderate social influence condition holds. It means the effect of one's action on one-self must be greater than the induced effect through social interaction on one's neighbours. Again, using obesity as an illustration: jogging, a health maintenance behaviour, by an individual should improve the individual's body mass composition. Ceteris paribus, this improvement should be greater than induced improvement in the body mass composition of the neighbours. Some neighbours were inspired to take up jogging while others were not. Alternatively, consider smoking, a well-known health risk. Smoking by an individual harms the individual's health. This deleterious effect should be more severe for that individual than the harm induced in the health of the neighbours through either passive smoking or through social interaction or social norm effects. The cases of excessive drinking and social drinking work similarly. In these cases, the moderate social influence condition is satisfied. Because social interaction can produce discrete multiple equilibria in health behaviours, it is not surprising to observe that different neighbourhoods in greater Boston (for instance, Framingham versus Backbay) possess different obesity rates. The discreteness, and hence the possibility of estimating them, is guaranteed by the moderate social influence condition.

Notably, this moderate social influence condition is consistent with the basic tenet of epidemiology or public health research in the form of 'population strategy'. In the words of Rose (1992: 135) "A 10% lowering of the population's levels of blood cholesterol can be expected to reduce coronary heart disease by 20–30%, and such a reduction of a condition that now kills one-quarter of the population would be a benefit indeed. A reduction of one-third in the nation's salt intake, ... might also reduce by up to one-half the number of people requiring drug treatment for hypertension." It is well known that neighbourhood effects on health behaviour are usually much smaller, often an order of magnitude smaller, than the effects of individual characteristics (in individual level regression or multilevel regression models). The threshold for effect magnitude in a public health setting can be lower than that in a clinical setting. An intervention bringing a 2% decrease in the average population body mass index is already considered important though a larger effect by an order of magnitude is perhaps needed for a clinically obese individual.

[6] In this connection, none other than Brock and Durlauf (2001a, p. 166) would welcome such empirical studies. "... this hardly means that these literatures [under-theorised empirical studies in the sense below] are incapable of providing useful insights. In this respect, we find arguments to the effect that because an empirical relationship has been established without justification for auxiliary assumptions such as linearity, exogeneity of certain variables, etc., one can ignore it, to be far overstated. In our view, empirical work establishes greater or lesser degrees of plausibility for different claims about the world and therefore the value of any study should not be reduced to a dichotomy between full acceptance or total rejection of its conclusions. Hence the determination of the plausibility of any exclusion restriction is a matter of degree and dependent on its specific context."

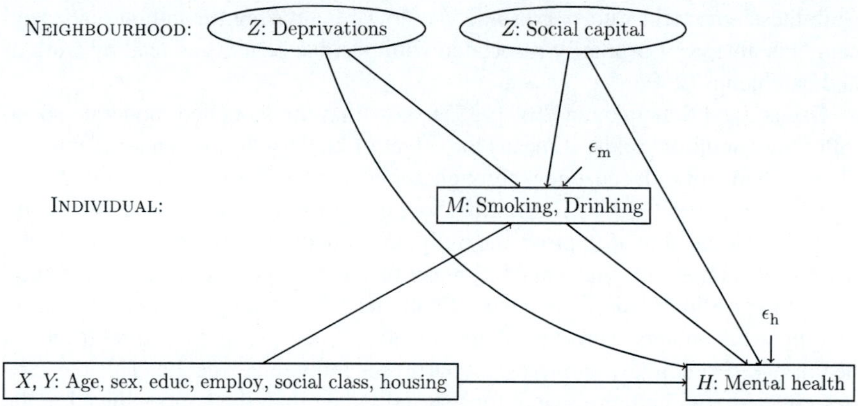

Fig. 8.1 Health maintenance (*M*) and production (*H*) in their individual and neighbourhood contexts

This lower threshold for population or higher sensitivity is accepted because one bears in mind that the ultimate effect is for the whole population and not confined to a single individual.

In parallel to theoretically recognising the importance of social interaction, it is practically acknowledged that built (physical) and social features of neighbourhood can induce benefits as well as pose risks to health (Srinivasan et al. 2003). In sum, the recursive system (Eqs. 8.4 and 8.5) incorporating insights from social interaction (Eq. 8.10) is modified by including neighbourhood effects. These include effects such as neighbourhood social capital and neighbourhood deprivation (to capture lack of leisure space for social interactions), Z , in the health production function. This is estimated as a reduced form using instrumental variable estimation.

The extended model can also be presented as in Fig. 8.1 where it is depicted that processes determining health are not circumscribed entirely within the individual but are also affected by neighbourhood social capital and deprivation. By implication, although this extended model is conceived to explain mental health, its application is broad and encompasses other health outcomes such as obesity. The demonstration below shows promising ways of examining how individual and neighbourhood factors bring about healthy outcomes.

Instruments for Estimating Neighbourhood Effects

The moderate social influence condition is not a constructive condition; it does not show how to estimate the effect of individual and neighbourhood factors. In the absence of a randomised experiment moving residents from one neighbourhood to another, instrumental variable estimation is deemed the most appropriate technique to avoid biased estimates. Instruments, v, must satisfy both exclusion restriction $E(v, \varepsilon) = 0$, and relevance condition $E(v, Z) >> 0$. It is well known that the exclusion restriction is essentially untestable due to unobserved ε hence a strong theory like the extended Grossman model is needed; whereas the strength of the correlation is routinely judged using a rule of thumb of F statistics greater than ten (Angrist and Pischke 2009; Cameron and Trivedi 2005).

Two instruments are proposed: ethnic diversity and length of stay in the neighbourhood. Neither the original Grossman model nor the proposed extension has any role for neighbourhood ethnic diversity, hence $E(diversity, \varepsilon) = 0$. Ethnic diversity as an instrument thus satisfies the exclusion restriction. Furthermore, Putnam (2007) demonstrates that ethnic diversity can erode social capital. This motivates the instrument's relevance, a test of which is provided below. The second instrument, length of stay in the neighbourhood acts as a proxy for individual attachment to the neighbourhood. Thus, the longer an individual stays in the neighbourhood the greater the intensity of any effect. Transient residents may not be affected one way or another by changes in neighbourhood ethnic diversity or social capital; long-time residents are. In summary, neighbourhood ethnic diversity and average length of stay in the neighbourhood are the instruments.

Data

The Welsh Assembly Government provided two independent surveys: Welsh Health Survey 2007 (WHS) and the Living in Wales 2007 (LiW) survey. The WHS selected a random sample of postcode sectors from the Post Office Postcode Address File. The sample was stratified by the 22 unitary authorities within Wales and 30 addresses were selected in each of them. Health measurements were requested by health care professionals for adults and all children aged between 2 and 15 years old living at the selected addresses. Written consent, in English and Welsh, for these measurements was obtained in advance. Interviewers, who speak English or Welsh, carried out the interviews and measurements according to a standardised written protocol (Fuller and Heeks 2008). More than four in five (82.1%) of adults selected responded to the survey. Further details are available in the WHS technical report (Fuller and Heeks 2008).

The neighbourhood here is defined as the lower super output area (LSOA), a geography purposefully designed for social research (The Cabinet Office 2009; The Office for National Statistics 2004; the Office of the Deputy Prime Minister 2005). Such a definition of a neighbourhood compares favourably with other studies using larger or more heterogeneous areas as proxies for neighbourhoods. Neighbourhood and individual variables were selected to conform to the extended Grossman model. The 2005 Index of Multiple Deprivation for Wales (WIMD) provided a measure of neighbourhood deprivation and was also used as a proxy measure for (the lack of) access to various facilities in neighbourhoods. Neighbourhood social capital measures capture the 'trust' and 'network' social capital available in neighbourhoods. The LiW survey collected information on trust, sense of community and friendliness of neighbours with the following questions:

- Would you say that you trust 'most of the people in the neighbourhood', 'many', 'a few', or 'do not trust people in the neighbourhood'?
- What do you like most about living in this neighbourhood? What else? Options include: 'I feel like I belong to this neighbourhood', 'The friendships and associations I have with other people in my neighbourhood mean a lot to me'.

The information for these questions was averaged for each neighbourhood to construct neighbourhood social capital measures. The instrument of ethnic diversity is constructed using the Herfindahl index scaled to range between 0 and 1 as is common in the literature on ethnic diversity and social capital (Putnam 2007; Letki 2008). The average length of residency is constructed from the LiW survey accordingly since respondent was asked how long someone has been resident in the neighbourhood.

Linking the Welsh Health Survey and Living in Wales Survey

Neighbourhood social capital information from the LiW is linked to the WHS using the LSOA code assigned to each respondent. Data for a total of 1,152 neighbourhoods was matched to 13,917 respondents. In our data there was an average of approximately 19 residents per neighbourhood, with a minimum of 1 and a maximum of 56. Some respondents did not provide sociodemographic information required by the extended model, hence they were excluded from the analysis. The final dataset included 13,557 respondents with information on health, sex, social class, education, and tenure, plus neighbourhood information such as social capital and deprivation.

Results

Table 8.1 gives some summary statistics for the sample used. The Table shows that the data are gender balanced, but that the older age groups are overrepresented. Trust is quite abundant since residents tend to trust many people around them. Residents tend to agree with the statement that local friendships mean a lot to them and with the statement that they belong to the neighbourhood (from the potential categories of completely agree, agree, indifferent, and completely disagree).

The results of the instrumental variable estimation are given in Table 8.2. I elaborate on the neighbourhood deprivation and social capital effects first. Over and above individual determinants and behaviours, neighbourhood effects matter sizably and are significant at 10%. Neighbourhood deprivation reduces mental health quality. However, two forms of neighbourhood social capital more than compensate

Table 8.1 Summary statistics for the sample

Variable	Mean/mode*
SF36 physical summary	48.0
Women	54%
Age (5 year group)*	55–59,75+
Employed	47%
Unemployed	1.4%
Professional	35%
Intermediate	19%
Degree educated	15%
Tenure own	78%
Tenure private	7.4%
Neighbourhood deprivation: WIMD 2005	20.88
Trust people in the neighbourhood	2.2 (Many)
Local friendships mean a lot	1.0 (Agree)
I feel like I belong to this neighbourhood	1.1 (Agree)

Table 8.2 Neighbourhood social capital and individual mental health (SF36)

	β	*p*	β	*p*	β	*p*
Individual						
Female	−1.959	0.000	−1.977	0.000	−1.957	0.000
Age	−0.578	0.000	−0.535	0.000	−0.574	0.000
Age^2	0.054	0.000	0.051	0.000	0.054	0.000
Class: professional	1.002	0.000	1.171	0.000	1.046	0.000
Class: intermediate	1.004	0.001	1.170	0.000	1.083	0.000
Degree educated	0.091	0.715	0.062	0.814	0.086	0.729
Tenure: owner	3.098	0.000	3.174	0.000	3.035	0.000
Tenure: private tenant	1.120	0.056	1.416	0.024	1.103	0.057
Last year subj. health	−3.574	0.000	−3.552	0.000	−3.556	0.000
Alcohol consumption	0.530	0.000	0.529	0.000	0.537	0.000
Smoking	1.008	0.000	0.990	0.000	1.024	0.000
Neighbourhood						
Deprivation	−0.021	0.277	−0.040	0.001	−0.040	0.000
Trust	1.415	0.098				
Friendly place			6.660	0.105		
Belong to neighbourhd					1.118	0.065
Constant	53.807	0.000	54.066	0.000	55.840	0.000
$J_{statistics}$	1.002	0.317	0.001	0.979	0.573	0.449
$F_{statistics}$	12.491		3.217		31.636	

for this deleterious effect. Living in a trusting neighbourhood (compared to living in a less trusting neighbourhood) independent of whether the resident is trusting of other people, increases the resident's mental health by 1.4 points. To gain a sense of magnitude, SF36 (the Short Form Heath Survey frequently used to assess medical outcomes) is constructed to have a mean of 50 and a standard deviation of ten (Bowling 2005: 65). The second largest benefit for mental health outcomes is related to a sense of belonging where it improves mental health by 1.1 point. The generous level of significance is perhaps excused by the overall significance of two forms of social capital as well as the inefficiency of the estimator. Furthermore, given the predominantly null findings in the literature (Duncan et al. 1993; Sloggett and Joshi 1998; Mohan et al. 2005; Propper et al. 2005; Stafford et al. 2008), the overall pattern of significant effects of different forms of social capital is encouraging.

Tests of strength and relevance for the instruments (F , Hansen J and its p value) confirm the usefulness of the instruments in identifying the effects of social capital. In this context, one should not read too much into the substance of the relationship between the instruments and social capital (as captured in the implicit 'first stage' regression). There is nothing inevitable nor immutable about the relationship between ethnic diversity and residence length on the one hand and social capital on the other. For contrasting views about this, see Putnam (2007) and Letki (2008).

Individual Effects

The model in Table 8.2 also includes a range of control variables. The results show that overall men are more likely to report that they are healthier in comparison with women. As age increases people report that they are less healthy. The results show clear health inequality between occupational classes. Manual workers (compared to the intermediate and professional workers) tend to be less healthy. Another measure of socioeconomic status, education, appears not to stratify mental health in the population. Homeowners and private renters report better mental health than those living in social housing such that homeowners' health is a full three points better that that of social renters. Housing tenure is the second best predict or of mental health, an unsurprising result. Wealth, represented through housing is well known to improve health since it enables access to healthier foods and more active leisure activities along with other advantages. However, reverse causality is also likely to play a role as healthy people are more likely to be home owners.

Last years subjective health condition (i.e. the individual's assessment of their general health in the previous year) is the strongest predictor of mental health. If mental health was poor, then the current state of health is also likely to be poor. Respondents who smoke and drink alcohol report better mental health than non-smokers and non-drinkers (Lasser et al. 2000; Hughes et al. 1986). There is a sizeable literature on these behaviours which discusses these behaviours as somehow mentally 'comforting'. For instance, Lasser et al. (2000) elaborates on the relationship between smoking and mental health. Notably, the sizes of the effects are comparable to those of social capital. In other words, a similar improvement in mental health can be gained by smoking/drinking (generally accepted as a health risk) as is obtained by living in a trusting neighbourhood.

Discussion and Conclusion

This chapter contributes to the literature on neighbourhood social capital and health by extending the Grossman health model by explicitly including interactions between individuals within a neighbourhood context. The extended model draws upon the Blume-Brock-Durlauf social interaction model and includes social interactions and their effects on individual mental health. Compared to recent studies on neighbourhood social capital and health in developed countries such as Sweden, New Zealand and England (Blakely et al. 2006; Islam et al. 2006; Duncan et al. 1993; Sloggett and Joshi 1998; Mohan et al. 2005; Propper et al. 2005; Stafford et al. 2008) the study reported in this chapter finds contradicting evidence with neighbourhood social capital generally being beneficial to individual mental health.

The evidence presented here is obtained using a combination of an extended theoretical model and an instrumental variable (IV) method for causal estimation. The extended theoretical model allows causal effects of neighbourhood social capital on health to be estimated. It achieves this by motivating the strong instruments of ethnic diversity and length of residence in the neighbourhood which help to recover the effect of neighbourhood social capital on individual health related quality of life. Various aspects of neighbourhood social capital, such as social cohesion aspects (trust, a sense of belonging) are effective in improving individual health. Each of these aspects of social capital is shown to more than compensate for the deleterious effect of overall neighbourhood deprivation. These causal effects help to identify entry routes for public health interventions involving the neighbourhood as well as the individual and could include, for instance, interventions to make neighbourhood spaces friendlier for interaction.

Given that the effect of neighbourhood social capital on individual health has so far proved elusive in other industrial countries, why is Wales different? It might be tempting to explain this result in the commonly accepted argument of egalitarian society (Islam et al. 2006). In highly unequal societies, neighbourhood social capital tends to be effective to fill in the vacuum of needed health services that are not provided by the state or other organisations. Yet this is not the case with Wales since the UK National Health Service provides such services.

The extended Grossman health production function combined with independent neighbourhood social capital measures may have uncovered the elusive effect of neighbourhood social capital. Previous studies have not benefited from recent methodological development nor have the fortune of access to independent data (e.g. neighbourhood data are often derived from the same individual sample). Mohan et al. (2005) for instance desired for the latter to address their null finding on the effect of social capital.

This study indicates that the extended Grossman model is applicable in settings other than health quality of life such as obesity (Tampubolon et al. 2009) and it facilitates the tracing of the mechanisms by which neighbourhood effects improve individual health. The last words should probably go to Geoffrey Rose, the eminent public health educator. Despite the difficulties, anticipated by prominent economists (Arrow 2000; Dasgupta 2000; Solow 2000), facing researchers setting out to examine the effects of social interactions in the neighbourhood on individual health, one should not be disheartened. Ultimately, as Rose (1992: 161) insisted, "The primary determinants of disease are mainly economic and social, and therefore its remedies must also be economic and social. Medicine and politics cannot and should not be kept apart."

Acknowledgments I thank the Welsh Assembly Government – Llywodraeth Cynulliad Cymru – for funding this work. Data used in this analysis are provided by the Government under contract designed to protect the anonymity of respondents and neighbourhoods. Ichiro Kawachi and Subu Subramanian have been especially generous with their ideas for which I am in debt. I am also grateful to the participants of the seminar series "Neighbourhood Effects" and in particular to Maarten van Ham and David Manley who organised the series.

References

Angrist, J. D., & Pischke, J. S. (2009). *Mostly harmless econometrics: An Empiricist's companion*. Princeton: Princeton University Press.

Arrow, K. J. (2000). Observations on social capital. In I. Serageldin & P. Dasgupta (Eds.), *Social capital: A multifaceted perspective* (pp. 3–5). Washington, DC: The World Bank.

Balia, S., & Jones, A. M. (2008). Mortality, lifestyle and socio-economic status. *Journal of Health Economics, 27*(1), 1–26.

Baxter, R. J. (1982). *Exactly solved models in statistical mechanics*. London: Academic Press.

Becker, G. S., & Murphy, K. M. (2000). *Social economics: Market behavior in a social environment*. Cambridge: Harvard University Press.

Berkman, L. F., & Kawachi, I. (2000). *Social epidemiology*. Oxford: Oxford University Press.

Blakely, T., Atkinson, J., Ivory, V., Collings, S., Wilton, J., & Howden-Chapman, P. (2006). No association of neighbourhood volunteerism with mortality in New Zealand: A national multilevel cohort study. *International Journal of Epidemiology, 35*(4), 981–989.

Blume, L. E. (1993). The statistical mechanics of strategic interaction. *Games and Economic Behavior, 5*(3), 387–424.

Blume, L. E., & Durlauf, S. N. (2005). *Identifying social interactions: A review*. Mimeo. Cornell University.

Bowling, A. (2005). *Measuring health: A review of quality life measurement scales*. Maidenhead: Open University Press.

Brock, W. A. (1993). Pathways to randomness in the economy: Emergent nonlinearity and chaos in economics and finance. *Estudios Economicos, 8*(1), 3–55.

Brock, W. A., & Durlauf, S. N. (2001a). Interactions-based models. In J. J. Heckman & E. Leamer (Eds.), *Handbook of econometrics, Vol 5* (pp. 3297–3800). Amsterdam: North Holland.

Brock, W. A., & Durlauf, S. N. (2001b). Discrete choice with social interactions. *The Review of Economic Studies, 68*(2), 235–260.

Cameron, A. C., & Trivedi, P. K. (2005). *Microeconometrics: Methods and applications*. Cambridge: Cambridge University Press.

Case, A., & Deaton, A. (2005). Broken down by work and sex: How our health declines. In D. A. Wise (Ed.), *Analyses in the economics of aging* (pp. 185–212). Chicago: University of Chicago Press.

Christakis, N. A., & Fowler, J. H. (2007). The spread of obesity in a large social network over 32 years. *The New England Journal of Medicine, 357*(4), 370–379.

Cutler, D., & Glaeser, E. L. (2005). *What explains differences in smoking, drinking and other health-related behaviors?* Discussion Paper 2060. Harvard: Harvard Institute of Economic Research.

Dasgupta, P. (2000). Economic progress and the idea of social capital. In I. Serageldin & P. Dasgupta (Eds.), *Social capital: A multifaceted perspective* (pp. 325–424). Washington, DC: The World Bank.

Department of Health. (2006). *National quality & outcomes framework statistics for England 2005/2006. Technical report*. London: National Health Service Information Centre. http://www.ic.nhs.uk. Accessed 19 May 2010.

Duncan, C., Jones, K., & Moon, G. (1993). Do places matter? A multi-level analysis of regional variations in health-related behaviour in Britain. *Social Science & Medicine, 37*(6), 725–733.

Durlauf, S. N. (1997). Statistical mechanics approaches to socioeconomic behavior. In W. B. Arthur, S. N. Durlauf, & D. A. Lane (Eds.), *The economy as an evolving complex system II* (A proceedings volume in the Santa Fe studies in the sciences of complexity, pp. 81–104). Reading: Addison Wesley.

Durlauf, S. N. (2002). On the empirics of social capital. *Economic Journal, 112*(483), F459–F479.

Durlauf, S. N. (2004). Neighborhood effects. In J. V. Henderson & J. F. Thisse (Eds.), *Handbook of regional and urban economics* (Vol. 4, pp. 2173–2242). Amsterdam: Elsevier.

Durlauf, S. N., & Fafchamps, M. (2005). Social capital. In P. Aghion & S. N. Durlauf (Eds.), *Handbook of economic growth* (Vol. 1, pp. 1639–1699). Amsterdam: North Holland.

Ellis, R. S. (1985). *Entropy, large deviations, and statistical mechanics*. Berlin: Springer.

Farquhar, S. A., Michael, Y. L., & Wiggins, N. (2005). Building on leadership and social capital to create change in 2 urban communities. *American Journal of Public Health, 95*(4), 596–601.

Fuller, E., & Heeks, F. (2008). *Welsh health survey 2007 Technical Report: Prepared for the Welsh Assembly Government*. Technical report. National Centre for Social Research.

Glaeser, E. L., & Scheinkman, J. A. (2001). Measuring social interactions. In S. N. Durlauf & H. P. Young (Eds.), *Social dynamics* (pp. 83–131). Cambridge: MIT Press.

Glaeser, E. L., & Scheinkman, J. A. (2003). Nonmarket interactions. In M. Dewatripont, L. P. Hansen, & S. J. Turnovsky (Eds.), *Advances in economics and econometrics: Theory and applications, eighth world congress volume I* (pp. 339–369). Cambridge: Cambridge University Press.

Glaeser, E. L., Laibson, D., & Sacerdote, B. (2002). An economic approach to social capital. *Economic Journal, 112*, F437–F458.

Grossman, M. (1972a). On the concept of health capital and the demand for health. *The Journal of Political Economy, 80*(2), 223–255.

Grossman, M. (1972b). The Demand for health: A Theoretical and empirical investigation. *National Bureau of Economic Research*. Occasional Paper 119.

Hughes, J. R., Hatsukami, D. K., Mitchell, J. E., & Dahlgren, L. A. (1986). Prevalence of smoking among psychiatric outpatients. *American Journal of Psychiatry, 143*(8), 993–997.

Islam, M. K., Merlo, J., Kawachi, I., Lindström, M., Burström, K., & Gerdtham, U. G. (2006). Does it really matter where you live? A panel data multilevel analysis of Swedish municipality-level social capital on individual health related quality of life. *Health Economics, Policy and Law, 1*(3), 209–235.

Kawachi, I., & Berkman, L. F. (Eds.). (2003). *Neighbourhoods and health*. Oxford: Oxford University Press.

Kawachi, I., Lochner, K., Kennedy, B. P., & Prothrow-Stith, D. (1997). Social capital, income inequality, and mortality. *American Journal of Public Health, 87*(9), 1491–1498.

Kawachi, I., Kennedy, B. P., & Glass, R. (1999). Social capital and self-rated health: A contextual analysis. *American Journal of Public Health, 89*(8), 1187–1193.

Lasser, K., Boyd, J. W., Woolhandler, S., Himmelstein, D. U., McCormick, D., & Bor, D. H. (2000). Smoking and mental illness: A population-based prevalence study. *Journal of the American Medical Association, 284*(20), 2606–2610.

Letki, N. (2008). Does diversity erode social cohesion? Social capital and race in British neigbourhoods. *Political Studies, 56*(1), 99–126.

Maddala, G. S. (1983). *Limited-dependent and qualitative variables in econometrics* (Econometric society monographs). Cambridge: Cambridge University Press.

Manski, C. F. (1993). Identication of endogenous social effects: The reflection problem. *Review of Economic Studies, 60*(3), 531–542.

Manski, C. F. (1995). *Identication problems in the social sciences*. Cambridge: Harvard University Press.

Manski, C. F. (2000). Economic analysis of social interactions. *Journal of Economic Perspectives, 14*(3), 115–136.

Marmot, M., Atkinson, T., Bell, J., Black, C., Broadfoot, P., Cumberlege, J. I., Diamond, I., Ham, C., Meacher, M., & Mulgan, G. (2010). *Fair society, healthy lives: The Marmot review, strategic review of health inequalities in England post-2010*. Technical report, London: The Marmot Review.

Mohan, J., Twigg, L., Barnard, S., & Jones, K. (2005). Social capital, geography and health: A small-area analysis for England. *Social Science & Medicine, 60*(6), 1267–1283.

Nielsen, S. J., & Popkin, B. M. (2003). Patterns and trends in food portion sizes, 1977–1998. *Journal of the American Medical Association, 289*(4), 450–453.

Office of the Deputy Prime Minister. (2005). *Neigbourhood statistics: Geography policy*. http://www.neighbourhood.statistics.gov.uk/HTMLDocs/images/GeographyPolicy_tcm97-51009.pdf. Accessed 19 May 2009.

Parisi, G. (1988). *Statistical field theory*. Redwood City: Addison Wesley.

Perry, M., Williams, R. L., Wallerstein, N., & Waitzkin, H. (2008). Social capital and health care experiences among low-income individuals. *American Journal of Public Health, 98*(2), 330–336.

Propper, C., Jones, K., Bolster, A., Burgess, S., Johnston, R., & Sarker, R. (2005). Local neighbourhood and mental health: Evidence from the UK. *Social Science & Medicine, 61*(10), 2065–2083.

Putnam, R. D. (1993). *Making democracy work: Civic traditions in modern Italy*. Princeton: Princeton University Press.

Putnam, R. D. (2007). E Pluribus Unum: Diversity and community in the twenty-frst century. The 2006 Johan skytte prize lecture. *Scandinavian Political Studies, 30*(2), 137–174.

Ripley, B. D. (1990). Gibbsian interaction models. In D. A. Grith (Ed.), *Spatial statistics: Past, present, and future* (pp. 3–53). Ann Arbor: Institute of Mathematical Geography.

Rose, G. (1992). *Rose's Strategy of preventive medicine*. Oxford: Oxford University Press.

Sloggett, A., & Joshi, H. (1998). Deprivation indicators as predictors of life events 1981–1992 based on the UK ONS longitudinal study. *Journal of Epidemiology and Community Health, 52*(4), 228–233.

Solow, R. M. (2000). Notes on social capital and economic performance. In I. Serageldin & P. Dasgupta (Eds.), *Social capital: A multifaceted perspective* (pp. 6–9). Washington, DC: The World Bank.

Srinivasan, S., O'Fallon, L. R., & Drearry, A. (2003). Creating healthy communities, healthy homes, healthy people: Initiating a research agenda on the built environment and public health. *American Journal of Public Health, 93*(9), 1446–1450.

Stafford, M., De Silva, M. J., Stansfeld, S., & Marmot, M. (2008). Neighbourhood social capital and common mental disorder: Testing the link in a general population sample. *Health & Place, 14*(3), 394–405.

Subramanian, S. V., Kim, D., & Kawachi, I. (2005). Covariation in the socioeconomic determinants of self rated health and happiness: A multivariate multilevel analysis of individuals and communities in the USA. *Journal of Epidemiology and Community Health, 59*(8), 664–669.

Tampubolon, G., Kawachi, I., & Subramanian, S. (2009). *Neighbourhood social capital and individual obesity: Evidence from a national sample*. Available from author.

The Cabinet Office. (2009). *Report of Policy Action Team 18: Better information*, April 2000. London: The Stationery Office.

Van Doorslaer, E. K. (1987). *Health, knowledge and the demand for medical care: An econometric analysis*. PhD thesis, Rijkuniversiteit Limburg te Maastricht, Maastricht.

Viswanath, K., Steele, W. R., & Finnegan, J. R., Jr. (2006). Social capital and health: Civic engagement, community size, and recall of health messages. *American Journal of Public Health, 96*(8), 1456–1461.

Wagstaff, A. (1986). The demand for health: Some new empirical evidence. *Journal of Health Economics, 5*(3), 195–233.

Ware, J. E. (2004). SF-36 health survey update. In M. E. Maruish (Ed.), *The use of psychological testing for treatment planning and outcomes assessment* (pp. 693–718). Mahwah: Lawrence Erlbaum Associates.

Wilkin, D., Hallam, L., & Doggett, M. A. (1992). *Measures of need and outcome for primary health care*. Oxford: Oxford University Press.

Woolcock, M. (1998). Social capital and economic development: Toward a theoretical synthesis and policy framework. *Theory and Society, 27*(2), 151–208.

Young, H. P. (1998). *Individual strategy and social structure: An evolutionary theory of institutions*. Princeton: Princeton University Press.

Chapter 9
The Notable and the Null: Using Mixed Methods to Understand the Diverse Impacts of Residential Mobility Programs*

Stefanie DeLuca, Greg J. Duncan, Micere Keels, and Ruby Mendenhall

Introduction

The first study of the impacts of the Gautreaux residential mobility program was conducted nearly two decades ago.[1] It documented dramatic improvements in the lives of low-income African-American families placed by the program in Chicago's mostly white suburbs. Many interpreted these results as showing the power of neighbourhood context and demonstrating that families growing up in ghetto neighbourhoods can take advantage of the opportunities provided by residence in

* Authors are listed in alphabetical order. Parts of this chapter were presented at the 2009 meeting of the American Sociological Association, San Francisco. We are grateful for comments from David Harding and Larry Katz. The opinions expressed here are those of the authors and do not necessarily reflect the position of the US Department of Housing and Urban Development. This manuscript was completed while the first author was a William T. Grant Foundation Scholar and the fourth author was a Postdoctoral Fellow at the National Center for Institutional Diversity at the University of Michigan.

[1]For a comprehensive review of the history of the *Gautreaux* lawsuit, the implementation of the program, and early research results, see Rubinowitz and Rosenbaum (2000).

S. DeLuca (✉)
Johns Hopkins University, Baltimore, MD, USA
e-mail: sdeluca@jhu.edu

G.J. Duncan
Department of Education, University of California, Irvine, Irvine, CA, USA
e-mail: gduncan@uci.edu

M. Keels
Department of Comparative Human Development, University of Chicago, Chicago, IL, USA
e-mail: micere@uchicago.edu

R. Mendenhall
African American Studies and Sociology, University of Illinois, Urbana, IL, USA
e-mail: rubymen@illinois.edu

M. van Ham et al. (eds.), *Neighbourhood Effects Research: New Perspectives*,
DOI 10.1007/978-94-007-2309-2_9, © Springer Science+Business Media B.V. 2012

much better neighbourhoods. Gautreaux results also helped to inspire the Moving to Opportunity (MTO) program, an ambitious residential mobility experiment launched by the U.S. Department of Housing and Urban Development (HUD) in the mid-1990s.[2] Unfortunately, interim results from the MTO experiment, drawn 4–7 years after families entered the program, were not nearly as positive as the early Gautreaux research would have led one to believe.[3]

More recent research tries to reconcile the mixed Gautreaux and MTO results, noting important differences in the way the programs were implemented and the historical context of the programs (e.g. DeLuca and Rosenblatt 2010.; Rosenbaum and Zuberi 2010). Extensive qualitative research conducted with both Gautreaux and MTO families shows surprising agreement regarding the likely mechanisms behind little-noticed program successes and the well-documented but heretofore unexplained null findings. This chapter focuses on the contributions of MTO's mixed methods studies and on the highly productive synergies that have arisen between MTO's structured and rigorous quantitative research and a collection of qualitative studies fielded shortly after and then several years after the program first began (both the chapters by Galster 2011 and by Small and Feldman 2011 in this volume call for such a mixed methods approach). Our chapter provides a unique contribution to the literature on neighbourhood effects in several ways. First, we explore the results from a large scale federal housing experiment designed to directly test the effects of offering poor families a chance to move to better neighbourhoods. Most studies of neighbourhood effects can only examine the outcomes of families who live in different communities, but (owing to individual selection and structural constraints) rarely move between them (DeLuca and Dayton 2009). Second, we demonstrate that the data from in depth interviews reveals the mechanisms behind the quantitative experimental estimates of how the housing intervention did and did not lead to better social and economic outcomes for mothers and children. Last, we highlight how the mixed methods approaches used to examine MTO outcomes

---

[2] Starting in the mid-1990s, HUD began a massive public housing revitalization effort known as the HOPEVI program (see Popkin et al. 2004). Cities across the country have been using HOPEVI funding to demolish substandard public housing projects and replace them with a mixed income communities comprised of subsidized rental units and market rate home ownership units. In the process of redeveloping these communities, many original public housing project residents had to move from their apartments to other rental units in the private market using Section 8 vouchers. It is important to note that the moves families made through the Gautreaux and MTO programs are fundamentally different from the HOPEVI induced moves. First, Gautreaux and MTO were voluntary programs, while families in housing projects slated to be torn down were forced to relocate. Second, Gautreaux families and MTO experimental group families were given housing counseling and assistance to secure housing in higher opportunity neighborhoods, while HOPEVI residents were only given the traditional Section 8 voucher and no additional relocation assistance. Therefore, the results from the Gautreaux and MTO research cannot be generalized to the families who moved involuntarily because their project was being redeveloped under HOPEVI.

[3] Orr et al. (2003) is the official report. Other MTO papers using the interim data are listed on www.nber.org/mtopublic. See also the July, 2008 issue of the *American Journal of Sociology* for an exchange on the meaning of the interim MTO results.

provide a window into the mechanisms through which neighbourhoods are theorized to affect families.

There are few guidelines for combining evidence obtained from different research methods into a comprehensive picture of why social programs succeed and fail (Borkan 2004; Small 2011). In the case of interventions such as MTO, studies that capitalize on different 'design and analysis' approaches can help to describe how interventions interact with the individuals they are meant to help, further basic research, inform the implementation of future interventions, and make the case for or against future investments in particular policies (Borkan 2004).[4]

Our synthesis of the mixed methods work done with the MTO program begins with a review of the experiment's process model – its "theory of change" – outlining the most important ways in which program impacts were expected to come about. We then highlight several key unexpected findings and utilize the evidence from mixed methods and qualitative studies to illuminate the processes that may account for some of the program's unexpected outcomes. Throughout our account, we attempt to generalize MTO's qualitative findings with those emerging from qualitative studies of the two rounds of the Gautreaux residential mobility program.[5]

Taken together, our mixed-method evidence enables us to extend MTO's original logic model (and its assumption that neighbourhood improvement would be a sufficient condition to enhance child and family outcomes) to a broader model of how individual actions and larger social conditions reinforce or limit the power of neighbourhood interventions to bring about change. An important strength of mixed methods approaches to evaluating social interventions is their ability to show how changing opportunities operate within the realities and constraints of people's lives as well as the dynamics of social structure.

Background

The Moving to Opportunity for Fair Housing demonstration program was designed to answer the question: What are the long-term effects of moving poor families out of subsidized housing in high-poverty communities and into low-poverty neighbourhoods in five cities: Baltimore, Boston, Chicago, Low Angeles, and New York? (Orr et al. 2003) Families were randomly assigned to three groups: experimental, Section 8, and control. The experimental group received vouchers to relocate to areas with less than 10% poverty, assistance in finding a unit, and housing counselling

[4] Gibson-Davis and Duncan (2005) make a similar argument for the value of mixed methods based on the evaluation of the New Hope work support program.

[5] In addition to early qualitative work conducted by Rosenbaum and colleagues (Rosenbaum et al. 2005; Rosenbaum et al. 2002), Mendenhall and Keels conducted interviews with a stratified subset in 2000–2001. A new round of Gautreaux moves began in 2001, with qualitative interviews conducted by a team led by Kathryn Edin.

to help them prepare for the move to the private rental market. Section 8, or the Housing Choice Voucher Program, provides a subsidy so that very low income families can choose and rent a unit in the private market.[6] The Section 8 group received the conventional Section 8 counselling and assistance but could relocate to any type of community. The control group received no vouchers, only the usual project-based assistance. About 4,600 families were part of the MTO program across all five cities, and over 1,700 were randomly assigned to the group offered the low poverty vouchers. We concentrate on the contrasts between families in the experimental and control groups.

MTO's random-assignment design provides a strong basis for estimating the causal consequences on families' lives of being offered the chance to move to "better" neighbourhoods. It avoids selection bias – perhaps the most significant threat to the internal validity of neighbourhood effects research (Shadish et al. 2002; DeLuca and Dayton 2009). But experiments are not without problems (Ludwig et al. 2008), one of which is the external validity – the generalizability of results – when unique historical or policy events occur at the same time as the treatment. We argue below that historical circumstances are vital for understanding MTO's labour market effects.

Experiments provide limited information on the processes behind the experimental impacts. In the case of MTO, HUD and its advisers developed a general logic model of hypothesized pathways – at both the community and individual/family level – through which the move to low-poverty neighborhoods should have an effect on family outcomes (see Fig. 9.1, which is adapted from Orr et al. 2003). Logic models represent the underlying assumptions about how policies affect outcomes. In the case of MTO, a convincing demonstration of mediation requires causal estimation of both the effect of the MTO program on the mediator *and* the effect of the mediator on the family and individual outcomes. An experimental design is ideally suited to produce the first, but not second, of these estimates.[7]

Underlying the MTO logic model is the assumption that program participants make rational choices regarding voucher take-up and the selection of destination neighbourhoods, based on a comparison of the payoffs of a given set of possible moves against their monetary and psychic costs (Kennedy and Finkel 1994; Shroder 2002). As a result, HUD's MTO logic model envisions that families moving to low-poverty neighbourhoods will encounter and engage with a range of community, family, and individual factors that could affect – for good or ill – the six outcome domains listed on the right side of the figure. We elaborate on MTO's logic model

[6]The Housing Choice Voucher covers the difference between 30% of the household's monthly income and the locally determined payment standard for rent. (see: http://www.hud.gov/offices/pih/programs/hcv/about/fact_sheet.cfm).

[7]It should be noted that non-experimental estimates of mediation typically fail to provide arguably unbiased estimation of *either* of these two meditational components. And it is also the case that experiments such as MTO can be used to estimate instrumental variables models of the effects of mediators on outcomes (e.g., as in the Ludwig and Kling 2007 analysis of the effect of neighbourhood poverty on crime).

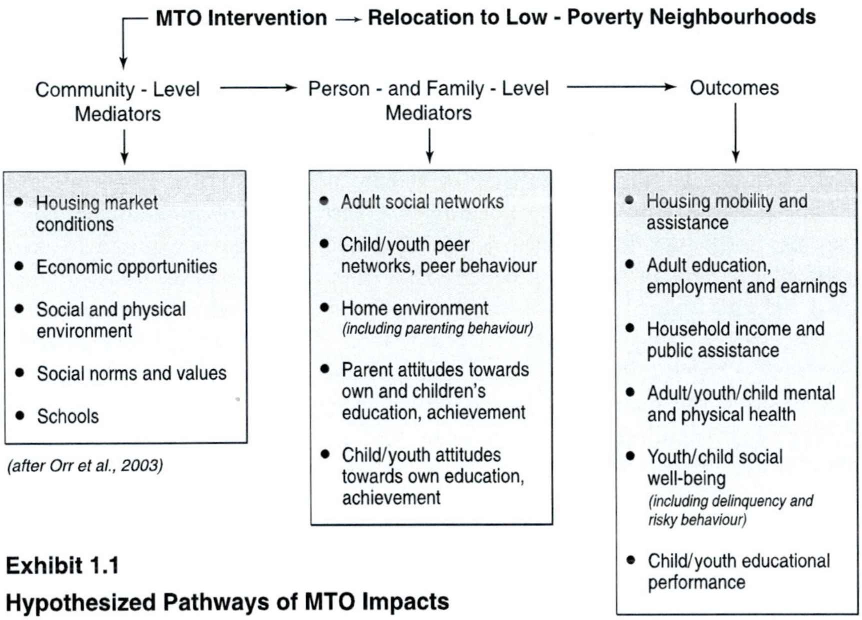

Fig. 9.1 Process model of the moving to opportunity experiment

for multiple outcomes – adult mental health, employment, youth delinquency, and children's educational performance – which provides the basis for understanding some of the contributions of MTO's mixed-method research we describe below.

Mental Health

MTO's logic model for health outcomes (Orr et al. 2003: 70, Exhibit 4.1) posits that moving to a safer community with less crime and violence should have direct, positive impacts on psychological distress, depression, and anxiety. In addition, more affluent communities may provide access to more and better quality jobs, which could also improve movers' mental health. On the other hand, the mental health of MTO movers may decline as they leave their old communities, social networks, and organizations and experience cultural and social isolation.

Employment

Orr et al. (2003, p.124) suggest pathways through which relocation to wealthier communities should improve employment prospects for parents. First, more affluent neighbourhoods, compared to poorer areas, will likely have less unemployment and faster job growth. These conditions may lead to steadier employment, higher earnings and better fringe benefits for MTO families. Second, residential proximity to jobs may reduce costs associated with looking for work and daily commuting to and from work. Third, leaving violent communities may decrease participants' stress and anxiety and thus improve their mental health and sense of self-efficacy. These improvements in personal well-being may increase efforts to find work thus leading to higher employment and earnings. Fourth, living in more affluent areas may also positively affect physical health because of lower risk of exposure to environmental hazards or receipt of better health care. Fifth, the lower transportation costs and time saving benefits of residential proximity may induce some otherwise nonemployed MTO participants to accept low-pay jobs (i.e., reduce their reservation wages) because they have fewer work related expenses. On the other hand, community norms that emphasize the importance of employment at any cost may decrease MTO families' reservation wages, leading to lower wages and fringe benefits. Moves may also rupture long-standing pre-existing informal childcare, job network, and employment relationships.

Education

MTO's logic model for children's educational pathways (see Exhibit 6.1, Orr et al. 2003, p. 102) envisions that moves to better neighbourhoods will increase children's academic achievement owing to more time spent with peers who value educational success, and increased safety and access to community institutional resources such as after-school and park district programs. Moves to more affluent neighbourhoods are expected to provide students with access to zone schools that have higher quality teachers, more rigorous courses, smaller class sizes, and higher expectations for learning and achievement. On the other hand, while an increase in school quality may foster students' commitment to education, it might also elevate grading standards which in turn might lead MTO children to become discouraged and lose self-confidence if these standards differ appreciably from their old schools.

Community-level mediators such as increased safety could also affect educational outcomes by decreasing counterproductive authoritarian parenting, as parents experience lower levels of stress and anxiety and higher levels of personal control. It was also expected that if children feel safer in these communities, the benefits will carry over into their academic and personal lives as they feel comfortable walking to school and less worried about being victimized in school. A final community hypothesis is that families will have access to greater economic resources (like better paying jobs) and this may allow the family to invest more in educational items like books, computers, etc.

On the individual and family level, the educational logic model predicts that changes in attitudes and behaviours of both students and parents serve as key mediators in improved educational outcomes. The model hypothesizes that academic achievement will be influenced by students' beliefs and attitudes (e.g., ideas about themselves as successful students, whether teachers care about them, and their peers' and parents' school beliefs). Also important are students' behaviours like taking challenging courses, committing to studying, and participating in school activities. The hypothesized parental mediators (attitudes and behaviours) include expectations about school success, the level of support for and active involvement in the school, involvement in homework completion, and parenting practices around students' actions and consequences.

Youth Delinquency and Risky Behaviour

The logic model proposes three community level mediators likely to promote changes in problem behaviours and substance abuse among the youth who moved to more affluent neighbourhoods with the MTO program: community norms and values; social and physical environment; and economic opportunities (Orr et al. 2003; Exhibit 5.1). For example, exposure to more peers and local adults in the new neighbourhood who value and encourage education and employment might decrease involvement in drugs and criminal activity. Higher quality schools with more after-school programs and a wider array of neighbourhood recreation centres could also provide an alternative to delinquent behaviour. If the local job opportunities are plentiful and appealing, youth might perceive payoffs to their school efforts and stay more engaged in school and less likely to engage in risky behaviours that could get them into trouble. Safer neighbourhoods with less drug trafficking reduces both the anxiety from exposure to violence and the pressure to get involved in the drug trade. All of these neighbourhood level benefits can directly reduce substance use and risk behaviour through socialization and the provision of local opportunities, but also indirectly through improving the mental health of the young men and women who relocated with the MTO vouchers.

On the other hand, these neighbourhood level processes might work in the opposite direction. For example, Orr et al. (2003) suggest that competition with higher performing peers in more challenging schools can provoke a sense of inadequacy and anxiety, which can lead to behaviour problems or substance use. In addition, relocation to an unfamiliar neighbourhood could lead to social isolation, or an attempt to fit into the wrong crowd – both leading to an increase in depression, anxiety, substance use or problem behaviours.

When MTO researchers tested these and other hypotheses for the Interim Impacts Evaluation, they encountered a mixed bag of program effects, including: *unexpected findings* (mental health benefits); a *weak 'treatment'* for many families (initial and subsequent moves to segregated, economically declining areas); *"null" findings* where large changes in family and child well-being were expected instead (no effects on employment and education); and a set of *conflicting findings* (low poverty moves were beneficial for girls, but harmful for boys). Using the mixed methods and qualitative studies conducted on the heels of the interim experimental impacts evaluation, we describe what was learned about the MTO program, the potential for neighbourhood interventions and the challenges of social programs meant to improve the lives of very disadvantaged families.

The "Unexpected" Finding: Mental Health Improvements

Although blindingly obvious in retrospect, MTO's potential for improving the mental health of families who move from violent, socially disorganized neighbourhoods into much safer ones would not have been investigated carefully had it not been for mixed methods research conducted by economists willing to cross disciplinary boundaries. As defined by its authorizing legislation, MTO's purpose was to reduce neighbourhood poverty, enhance the employment prospects of parents and improve the educational outcomes of children.[8] There was no mention of improving mental health, although it may have been considered as a way in which hoped-for improvements in work and schooling may have come about.

Shortly after MTO was launched, but before it conducted the 5 year interim study mandated by MTO authorizing legislation, HUD funded pilot studies in all five of the MTO sites. A mixed methods approach was taken by Larry Katz, Jeffrey R. Kling and Jeffrey B. Liebman in the Boston site, which consisted of observing orientation sessions for participants, conducting open-ended interviews with counsellors and 12 participants chosen at random from the experimental and Section 8 groups and conducting a full-scale conventional survey of all families in the Boston site.[9] The striking responses from the early qualitative interviews in Boston of the experimental respondents indicating they had moved to "tranquillity" (rather than to "opportunity") motivated the team to add some mental health questions to the short-run impacts Boston survey done in 1997–1998. Survey-based results in Boston (Katz et al. 2001) found strong improvements in indicators for mental health, which motivated an expanded mental health module implemented across all five sites in the Interim Evaluation survey (analyzed in Orr et al. 2003; Kling et al. 2007).

Abundantly clear across these various sources was the importance to participating families of getting away from neighbourhood violence and the potential for moves to improve mental health. The burden of neighbourhood violence was soon confirmed when MTO's baseline surveys were tabulated. Some 82% of MTO families across all five cities reported that "getting away from drugs and gangs" was either their first or second most important motivation for signing up for the chance of getting a program voucher (Orr et al. 2003). In distant second and third places were getting a better apartment or better schools, both of which were a first or second mention of 49% of respondents.

Victimization rates in origin neighbourhoods were high. Based on the general survey of Boston MTO families, Kling et al. (2005) report that one fourth of their respondents said that someone who lived with them had been assaulted, beaten,

[8]For example, as it geared up to launch the MTO demonstration, HUD characterized MTO's long term assessment goals as consisting of the "housing, educational, and employment outcomes of families assisted through the program."

[9]The survey results are reported in Katz et al. (2001), while the mixed-methods are described in Kling et al. (2007). HUD also funded a larger qualitative study conducted by the Urban Institute, which is reported in Popkin et al. (2001) and contributed to the design of the interim survey.

Table 9.1 MTO impacts on neighbourhood conditions and mental health

	Control group mean (%)	Impacts for movers (%)
Census neighbourhood poverty rate in 2000	37	−18%
Adult report:		
Feels safe at night	55	+30
Likely neighbours would do something if children misbehaving	54	+24
Depressed in last year	22	−8

"Control group mean" entries are mean values for control group members at the point of the interim survey. "Impacts for movers" are "treatment on the treated" estimates of differences between those who moved in conjunction with the program and the subset of control-group members who would have moved had they been offered the chance
Source: Orr et al. (2003)

stabbed, or shot within the past 6 months and an added quarter reported that someone had tried to break into their home, or that someone who lives with them had been threatened with a knife or a gun or had their purse or jewellery snatched in the past 6 months. Comparisons with data from a national survey showed that these victimization rates for MTO families were four times higher than among a national sample of public housing residents.

Kling and colleagues' (2005) qualitative interviews with a randomly chosen subset of MTO families were filled with chilling descriptions of parents' fear that their children would become the victims of violence if they remained in the high-poverty housing projects. In one case a mother reported on how the neighbourhood crime affected her family:

> (O)ne night they had a drive by shooting. The kids had to jump on the floor. Even the baby, she was under two year old. And then my son was coming home from school the next day – and because they didn't hit their target, they wanted to come back. I hear pow-pow-pow. My baby was laying on the bed sleepin'. It was like a quarter to two. And I knew my son was comin' round the corner. And I went outside and I didn't see him…(T)hen my son, instead of him comin' down the street his usual way, he came down the street where the person who was shootin' went up the street. And he like clashes between 'em. And I said, "Oh my god, I got to move out of here." (p. 252)

MTO successes in improving neighbourhood quality were striking (Table 9.1, taken from Orr et al. 2003). Four to seven years after baseline, program movers enjoyed neighbourhood poverty rates that were half those of control-group families – 19% vs. 37%. The lower poverty rates translated into many more (85% vs. 55%) of program movers who reported feeling safe at night as compared with controls. A question about collective efficacy (whether it was likely that neighbours would do something if children were "doing graffiti on local buildings") produced a 24-point differential favouring the experimental movers.

What about mental health impacts? Qualitative accounts provided by experimental families suggested transformative changes. Kling et al. (2007) quote one of their qualitative respondents as saying that, after enrolling in MTO, "the doors opened on my behalf." She went on to describe her new neighbourhood as follows: "It's so beautiful. So nice. The neighbours are very friendly.... I like the peace and quiet ... I have peace of mind. I'm closer to the stores, and the transportation, too."

A mother in the Popkin et al. (2001) qualitative study[10] provided a similar account of her new neighbourhood: "[It's] totally different. It's a totally different neighbourhood because there is no drug activity, no kids hanging on the corner, not kids fighting each other. It's totally different from the city. It's somewhere you can call home. You can just sit down and be comfortable and have no worries at all" (Popkin et al. 2001, p. 42).

Largely in response to the Boston work, considerable interviewing time was devoted to a comprehensive depression measure (the CIDI-SF Major Depressive Episode scale) in the Interim Impacts Evaluation. At the time of the follow-up interview, 22% of control-group adults reported experiencing at least one depressive episode in the 12 months prior to the interview. MTO moves lowered this to 14% – a difference that compares favourably with medical trials of best practice depression care (Orr et al. 2003). Favourable impacts were found for most other components of mental health included in the follow-up survey.

All in all, mental health improvements emerged as one of the most important personal changes wrought by MTO moves. Had it not been for the mixed method work in the Boston site, far less effort would have been devoted to assessing mental health in MTO, and this key result might have been missed altogether.

"Weak Treatment": Why Weren't MTO Families in Higher Opportunity Neighbourhoods?

Relocation to more affluent neighbourhoods was the primary lever behind why MTO was expected to improve the life chances of adults and children, and experimental movers experienced dramatic decreases in neighbourhood poverty. However, across all five cities, more than 40% of the experimental movers had used their vouchers to move to neighbourhoods where the poverty level was already increasing during the 1990s. Four to seven years after beginning the program, while experimental movers were living in neighbourhoods that were significantly better than those of controls on a number of measures (Orr et al. 2003), their neighbourhoods were above the 10% low-poverty threshold and racially segregated. Why weren't the MTO families living in higher opportunity neighbourhoods?

To answer this question, Rosenblatt and DeLuca (2010) conducted a mixed methods study in Baltimore to examine how the MTO families considered neighbourhood choices and the different barriers they encountered in the search for

[10]Popkin et al. (2001) interviewed 58 adults and 39 children, sampled across all five cities within the following strata: MTO Experimental Movers to Higher Poverty Areas, MTO Movers to Lower Poverty Areas, Section 8 Movers, and In Place Controls (families still located in their original public housing project).

housing with their experimental vouchers. Combining the Census, GIS, and interim MTO address data, Rosenblatt and DeLuca first conducted a 'choice set analysis' to examine where families *could have* used their voucher in the central Maryland area. In 1990, there were 419 tracts that were low-poverty (less than 10% poor) in central Maryland; the 146 experimental mover families leased up in only 55 of these 419 available tracts (an additional seven families leased up in tracts that were more than 10% poor according to the 1990 census). This represents only 13% of the available tracts they could have moved to. However, interviews revealed the majority of families who moved with MTO did not have a car. This meant that they had to rely on public transportation for everything from shopping to getting to work to visiting friends. As Cookie explains, she ultimately decided not to use her MTO voucher because:

> The buses only run a certain time and then they cuts off. So I don't believe nobody dictating to me that I gotta move here, and no transportation even though I have driver license but I don't have a car. If my child gets sick you can call an ambulance, but if I need to get to the store I gotta walk down the road...Like right now my job hours are 1–9 so if I'm out way in the county, and the bus stop running at 5 o'clock that's not good to me right now.

When Rosenblatt and DeLuca (2010) excluded census tracts that are not serviced by bus lines, the number of low-poverty tracts in central Maryland drops by 100, to 318. Fifty-four of the 55 experimental families who moved to low-poverty tracts moved to ones that were serviced by bus. When Rosenblatt and DeLuca further considered tracts with a rental vacancy rate above 6% (a common measure of tightness), there is even less available housing to choose from, leaving only 111 (out of 419) low-poverty, bus line accessible tracts for families to lease-up in. By the time of the interim survey, experimental movers only lived in 16 of the 111 low-poverty, public transportation accessible, slack rental market tracts, down from 26 of these tracts at first move. In other words, structural constraints like access to public transportation or rental availability explain some of the reasons families ended up where they did. Yet subsequent moves still brought families to neighbourhoods that were not as opportunity rich as those in which they could have potentially used their vouchers, even among places with bus lines and available housing. The in-depth interviews help to further explain why MTO families were not living in less poor, higher quality neighbourhoods.

Landlords

Across the interviews, families talked of frequent mobility and the desperate search for their next home when they had to move. Landlords emerged as a primary cause for some of this instability. In addition to encountering many landlords not wanting to rent to voucher holders, common complaints involved landlord neglect that resulted in structural damage, plumbing malfunctions, or rodent infestations. Other families had the houses they were living in sold out from underneath them.

In the face of landlord neglect, some families saw no other choice but to leave, while others were forced to leave after the house failed to comply with Section 8 inspection standards. Candy, who had lived almost all of her life in a high rise housing project, described this house below to which she and her four children moved with her MTO voucher as her "dream house."

> The landlord didn't have, you know, when things needed to be fixed, he didn't couldn't come and fix what needed to be fixed and then the garage would flood from the rain so it caused rats to come inside the basement through the window. So the Section 8 inspector came out here and inspected the whole house. I cried because I really didn't want to leave the house because I was so excited to have that house, you know, so after he came to inspect the house and told me: 'Ms. Jackson, I'm sorry to disappoint you but you have to find another house.

Affordability and Space Trump Neighbourhood Quality

In addition to the instability of landlords, families had to weigh a number of concerns when thinking about where to move. Over half of the sample mentioned a need for space (such as additional bedrooms or storage) as something that attracted them to a particular unit, kept them in their current unit, or as a reason to think about moving out. This is not unlike what most families look for in a place to live. However, an interesting finding across the interviews was that families often conflated their housing units with their neighbourhoods. These poor families were not selecting *neighbourhoods*, they were selecting *apartments* and *townhomes* based on amenities and space, with little attention to the quality of the neighbourhood (which is very different from middle class housing search considerations). Unfortunately, the need for space was often balanced against what was affordable, leading to serious tradeoffs in neighbourhood quality. Jane, a mother of four boys who has worked numerous part time jobs to support her family, explains how she renewed her lease in public housing because it meant more space for less money than she would pay elsewhere:

> I thought about renting at first but I knew for a fact, anywhere I would have inquired big enough to hold me and my family would have run me at least 700 dollars a month or more. And it was like, oh no, I found a place big enough to hold everybody comfortably even if they get bigger, still enough space regardless. This was perfect for my income and for you know enough room for my kids. And then I'm like well I'm getting central air, I'm getting 2 bathrooms. If I go anywhere else trying to get all this I'm really going to pay for it.

It may be surprising to think of affordability as a problem for families when that was what the voucher was designed to ameliorate. However, some experimental movers eventually forfeited their voucher during later moves when they couldn't find a place to rent within the window of Section 8 search eligibility, and were subject to regular market rents which were often far too high in low poverty neighbourhoods. For others, private market rent also meant having to pay utility bills, which were not an issue in public housing.

"We Don't Live Outside, We Live in Here"

Mothers' considerations of neighbourhood quality were also often filtered through prior experiences living in unsafe neighbourhoods. After years of living in high crime public housing developments, many MTO families had developed strategies for negotiating neighbourhoods that were unsafe. A number of women felt that knowing neighbours and having neighbours know them made spaces safer and others avoided certain blocks or made sure to be inside after a certain time of night. Mothers kept closer reins on kids by implementing curfews, keeping in regular cell phone contact, or designating areas where kids could and could not play. As one parent explained:

> Yeah, and they fight up in the next block, I mean big, huge fights where the police come and mace people...you probably would hear the ambulance or the police around here, you might hear it every night or whatever. But this is about the quietest the block is, this block right here is nice, don't get me wrong. I wouldn't change the block I live on for the world, this is a nice block. But these surrounding blocks is a mess. And my kids never wanna stay on this street, they always somewhere else...And I keeps them in the house and they be so mad. I find everything to keep them in the house.

These strategies were employed in their original public housing projects, but also in neighbourhoods to which MTO families moved. More than 40% of the MTO experimental movers broke down their neighbourhood by blocks when talking about safety. Like the focus on the housing units, this restricted focus on the block face was common; both practices meant that families were less likely to consider the larger neighbourhood to which they were moving and whether it provided the kinds of resources that could improve their employment and their children's education. "Minding one's own business", or physically avoiding unsafe areas meant that families rated neighbourhoods that might appear dangerous to outsiders as safe or manageable. Jane, who grew up in public housing, explains how her ability to negotiate space gave her confidence that she could "live anywhere":

> It's pretty much the same because it's still living in the city...If push come to shove, yeah I could live there, I could pretty much live anywhere. And I tell people all the time, as bad as people make it seem, this is not that bad living in the projects. It's really not.

Another neighbourhood outlook that emerged from the interviews was the belief that violence could be encountered in any neighbourhood. As Sharon put it, "it's trouble everywhere, it's not where you live, it's how you live. You mind your business, you don't have to worry about nothing".

Rosenblatt and DeLuca's (2010) mixed methods analysis makes it clear that MTO families made a set of constrained choices when they moved and looked for housing and that the 'treatment' of a low poverty neighbourhood rested on assumptions about how families find and select housing. A family's search for neighbourhoods or apartments takes place within a context shaped by landlords and Section 8 regulations, as well as the availability of transportation and affordable units with enough space for the family. MTO families didn't just move back to poorer areas to be with family or because they preferred to live there; there were serious structural constraints that shaped where they *could* live. The MTO families were also making trade-offs between affordability, unit space and neighbourhood quality, with the latter usually at the bottom of the list of priorities. However, the interviews also revealed that many families ended up in poorer areas because of coping mechanisms developed in response to years of living in violent neighbourhoods. By focusing on the quality of the housing unit or the condition of the block face, families were less likely to consider the larger neighbourhood context. By believing they could 'live anywhere' and that most neighbourhoods were unsafe, they settled for higher poverty and higher crime neighbourhoods that often lacked important institutional resources.

"Nuance-Ing the Null", Part 1: Why Didn't MTO Boost Employment?

Boosting employment has always been seen as a key potential benefit of offering families living in public housing the chance to move to better neighbourhoods. The earliest research on the Gautreaux housing mobility program found that participants living in the suburbs worked 25% more than those living in the city (Popkin et al. 1993). However, more recent research using administrative data from a much larger subset of Gautreaux families found *no* significant differences in either earnings or welfare receipt between families placed in the city or suburbs (Mendenhall et al. 2006).[11]

Consistent with the more pessimistic evidence on employment gains in Gautreaux, the MTO interim evaluation showed that experimental and control families differed little across a range of employment and earnings measures, some of which are listed in Table 9.2. Qualitative evidence, as well as a consideration of the policy context in which MTO was implemented, helps to illuminate mechanisms hidden from the MTO logic model.

Historical Policy Context and Employment Outcomes

For Elder (1998:3), the "life course of individuals are embedded in and shaped by the historical times and places they experience over the lifetime." In the case of MTO, four events and conditions unfolding during the early years of the experiment

[11]While the city/suburban difference was not important, neighborhood quality still appeared to matter somewhat: participants placed in Black segregated areas with the lowest level of community resources (safety, jobs, family income, and education) spent significantly less time (6–9%) employed and had lower earnings ($2,400 and $2,900 per year) when compared to participants placed in more integrated (11–60% Black) or predominantly White (0–10% Black) areas with higher levels of resources in both city and suburban neighborhoods.

Table 9.2 MTO impacts on employment

	Control group mean	Impacts for movers
Currently employed	52%	ns
Currently employed full time	39%	ns
Annual earnings	$8,899	ns
	% employed in control group	% employed in experimental group
At the point of random assignment	31	31
4–7 years after random assignment	49	48

"Control group mean" entries are mean values for control group members at the point of the interim survey. "Impacts for movers" are "treatment on the treated" estimates of differences between those who moved in conjunction with the program and the subset of control-group members who would have moved had they been offered the chance. "ns" means that the estimated impact was not statistically significant at $p<.05$
Source: Orr et al. (2003)

pose a major threat to the generalizability of its findings. The HOPE VI program tore down public housing in control neighbourhoods (1993 Pub.L. 102–389), while the 1996 welfare reforms, a major expansion of the Earned Income Tax Credit and an extremely tight labour market were pushing or enticing low-income single women – both experimental and controls in the case of MTO mothers – into paid employment.

The HOPE VI program ushered in dramatic changes in the urban landscape of many of the neighbourhoods where MTO families lived, especially with its demolition of severely distressed public housing. The MTO program targeted families in public housing who lived in extremely poor (40% or more) neighbourhoods. Nearly one-fourth (22%) of MTO families lived in some of the first public housing developments scheduled for demolition under HOPE VI. This likely affected the relocation behaviours of the control group. At the time of interim evaluation, 70% of the control group moved from their original location. On the face of it, the disruptions for control families caused by the demolitions might be expected to *increase* the employment advantages for experimental movers.

The passage of the 1996 Personal Responsibility and Work Opportunity Reconciliation Act produced "the single greatest break from past [welfare] policy" (Grogger 2003, p. 394). Welfare agencies began transitioning poor single-parent families – the kind enrolling in MTO – from welfare to work. Between 1994 and 2000, welfare caseloads plunged by 59% (Grogger 2003). At the same time, labour markets tightened up and a major expansion of the EITC provided as much as $4,400 in refundable tax credits to this same group of workers.

What was the net result of these shifting structural forces? For control-group women whatever disruptions that might have been caused by HOPE VI appeared to have been overwhelmed by the make-work-pay conditions of the late 1990s – employment rates of women in the control group jumped from roughly 30% to 50% (Table 9.2, bottom two rows; Orr et al. 2003, p. 131). The employment rates of women in the experimental group also increased dramatically and identically – from about 30% to 50%. MTO's experimental impact on employment, based as it is on the *difference* between experimental and control mothers, was essentially zero.

Undoubtedly, contextual changes contributed to the big employment gains enjoyed by both experimental and control families. How employment would have changed differentially for the two groups of women in a more "normal" policy and economic environment is impossible to say, hindering attempts to generalize MTO policy results. Results from MTO's qualitative studies suggest at best a modest impact.

Human Capital Barriers

Turney et al. (2006) examine employment patterns from the MTO Baltimore site using data from both the interim impacts survey and the 2003–2004 embedded qualitative study of families in Baltimore.[12] They discover several reasons why the experimental movers might not have benefited from their new communities, relative to the controls, and why controls might have experienced some advantages from remaining in the city.

First, they found that experimental-group women who moved in conjunction with the program often perceived their neighbours as working in jobs (office workers, police officers, lawyers, etc.) that required more education than they currently possessed. While nearly 40% had either 2- or 4-year college degrees or training certificates (e.g., as home health aides, pharmacy technicians, etc.), these perceptions about their neighbours' employment often made the experimentals reluctant to ask them about job information. This hesitation may have led these movers to miss beneficial opportunities in their new communities.

Second, for some movers, ill health interfered with their ability to obtain and/or maintain employment. All of the long-term unemployed experimentals cited debilitating health issues, often several at the same time, as the causes of their lack of labour market participation. One woman in the experimental group reported suffering from panic attacks in addition to having HIV, diabetes and depression. Another woman reported having a nervous breakdown and suffering from depression, and others complained of severe arthritis. Although MTO's positive impacts on mental health may have enabled some experimental to take jobs, the striking set of health problems undoubtedly limited the scope of MTO's potential employment impacts.

[12]The Baltimore based qualitative studies described in this chapter are all derived from the same fieldwork period and sample. Interviews were conducted in 2003 and 2004, following the interim survey which was done in 2002. A stratified random subsample of 124 heads of household from all three treatment groups (experimental, Section 8 and control) were interviewed for the Baltimore study. The heads of household are all female and African-American, and many had low incomes at the time of the interim survey. Interviews lasted between 3 and 5 h, and covered a range of topics from family history and neighborhood issues to employment, welfare use and children's schools. Refer to each individual paper for specific details on the analyses conducted.

Limited Access to Job Networks

Both control and experimental families (79% and 60%, respectively) in the qualitative study reported hearing about or getting their jobs through ties to people with similar educational levels and jobs as MTO participants (Turney et al. 2006). Although women in the control group had lower numbers of employed social ties in their communities, they were more likely to run into these individuals as they went about their daily routines (work, commuting to work or school, shopping, etc.) than experimental movers.[13]

While experimental movers were surrounded by employed families, they were less likely to encounter neighbours with similar training and employment that might help refer them to an available job. Gautreaux qualitative interviews showed similar results, especially in terms of how respondents reported getting job information from people they already knew, such as co-workers or teachers (Mendenhall 2004). One difference was that more affluent suburban neighbours did provide some information about jobs to Gautreaux mothers. However, the information was mostly about entry-level jobs and helped participants with the least education more than those with certificates (Mendenhall 2004).[14]

Reverse Spatial Mismatch and Inadequate Public Transportation

Experimental families' new neighbourhoods were also further away from the jobs they would typically apply for based on their education and skill levels. This seems to contradict the employment logic model (and spatial mismatch hypothesis) that experimental movers may be closer to jobs.[15] In addition to living further away from jobs, some experimental movers reported a lack of adequate public transportation as an employment barrier. When interviewed for this study, MTO experimental families lived, on average, 5.8 miles from their original public housing units and a similar distance from many of the local city jobs their social ties knew about. Getting to these jobs often required families to juggle several bus schedules and routes. In addition, the distance from their original neighbourhoods put families farther away from their social support networks, which were important for providing some of their transportation and child care in the past.

---

[13]For additional qualitative research on the specific nature of MTO families' social networks, as well as a consideration of the costs and benefits to these social ties, see Kissane and Clampet-Lundquist 2010.

[14]Prior Gautreaux qualitative research found that, while mothers did not mention getting direct job assistance from suburban neighbors, these acquaintances supported their efforts to find work by sharing cars, helping with child care and encouraging mothers' efforts to go back to school (Rosenbaum et al. 2005).

[15]See Turney et al. 2006 for a detailed discussion of job mapping in Baltimore County.

"Nuance-Ing the Null", Part 2: How Housing Choice Relates to Schooling Opportunity

The first results from MTO's Baltimore site suggested that moves to better neighbourhoods led to improvements in children's test scores and school behaviours (Ludwig, et al. 2001). But the interim data, collected 4–7 years later, and providing a direct measure of test scores, showed that there were no educational benefits for youth in the experimental group and surprisingly small improvements in school quality (Table 9.3; Orr et al. 2003; Sanbonmatsu et al. 2006).

Solving the mystery of the missing educational effects was an important priority for the Baltimore qualitative team when it entered the field in 2003. Fieldwork involved talking to the teachers of the experimental mover and control children, observing classrooms, and asking parents about homework and school quality. However, when researchers began to analyse these interviews, the question the qualitative data seemed more suited to answer was not so much "why test scores didn't improve?", but rather, "why didn't school quality improve?" In a mixed methods study, DeLuca and Rosenblatt (2010) combined surveys, GIS and in-depth interviews to examine one of the main assumptions of the MTO program and the primary mechanism through which educational effects was theorized to occur: that better housing opportunities would lead to access and attendance at better schools.

While a handful of the Baltimore MTO children attended high performing, affluent schools in surrounding suburban counties, the vast majority either remained in their original city schools or relocated to other low performing schools (a result more or less replicated in all five MTO cities). While there was a significant difference in the poverty rates of experimental movers' schools relative to controls (54% vs. 70% poor), school reading test score rankings were abysmal for both groups (Table 9.3; the 26th percentile for experimental movers and the 15th percentile for controls). Why didn't school quality improve *more* for experimental children after their moves?

Residential Change Does Not Mean School Change

Analyses of the Baltimore site of MTO show that not only were some children attending their original city schools or low performing suburban schools at the interim evaluation, but that this was the case for many of the children in the period *directly following* their families' moves. In other words, some parents kept their children in their original city schools, even if they moved elsewhere in the city or to another county. Additionally, other moves to low poverty neighbourhoods did not yield attendance in high quality schools. Keels (2009) found a similar pattern among parents who moved with the second round of the Gautreaux program. These results are surprising, given that the violence and poor academic records of the original

Table 9.3 MTO impacts on neighborhood conditions, school quality and achievement

	Control group mean	Impacts for movers
Census neighbourhood poverty rate in 2000	46%	–25%
Whether state rank of neighbourhood poverty rate is >20th percentile	12%	+45%
State percentile rank of school attended	15th percentile	+9 percentiles
Woodcock-Johnson test scores (mean 0, *sd* =1)	0	ns
Age 6–10		ns
Age 11–14		ns
Age 15–19		ns

"Control group mean" entries are mean values for control group members at the point of the interim survey. "Impacts for movers" are "treatment on the treated" estimates of differences between those who moved in conjunction with the program and the subset of control-group members who would have moved had they been offered the chance. "ns" means that the estimated impact was not statistically significant at $p<.05$
Source: Sanbonmatsu et al. (2006)

schools could be expected to push most mothers to transfer children out of those districts and into the higher performing county schools outside of Baltimore. Why didn't this happen?

First, DeLuca and Rosenblatt (2010) discovered that about a third of the parents were resistant to transferring their children because they thought it would be too disruptive for them and that it would be hard for them to be away from familiar faces. One MTO mother said, "I can't keep pulling them from school to school… moving them from house to house because I don't like this house or I don't like that house,…What am I going to do keep dragging them out of school letting them catch the bus? I'm not going to do it to my child."

Second, despite the fact that some families did relocate and send their children to schools in suburban counties with very low poverty rates and much higher test scores, most residential moves didn't place families in the suburban communities with the highest quality schools (following the discussion above on neighbourhood change). For example, DeLuca and Rosenblatt show that even though there were over 400 census tracts with fewer than 10% poor residents in the central Maryland and Baltimore metropolitan region, MTO families only moved to only 46.[16] Half of the 46 were experiencing increases in poverty between 1990 and 2000, and most were majority black. This meant that even though the average reading exam scores across the central Maryland counties (outside of Baltimore city) were close to the 60th state-wide percentile, the zone schools in the census tracts to which MTO families moved were at the 33rd percentile.

[16]This number is slightly smaller than the number of tracts cited in the Rosenblatt and DeLuca (2010) study described above. The difference is due to the fact that the DeLuca and Rosenblatt (2010) paper analyzed only those families who had school aged children (6 years or older) at the time of random assignment.

At the same time, the average African American concentration in the zone schools across the other counties ranged from 2 to 26%, but the zone schools in the areas where MTO families moved averaged about 70% African American students. While the Baltimore MTO families moved to more affluent communities (11% poor on average), the zone schools in these communities averaged about 45% free or reduced lunch students. Possibly the most important reason why the MTO housing opportunity did not translate into large increases in school quality is that families did not relocate to the communities with the highest performing schools. In other words, their residential choice in large part determined their school choice.

School Choice Does Not Equal School Quality

Despite structural impediments, MTO mothers could have translated their residential moves into higher increases in the quality of their children's school. DeLuca and Rosenblatt's in-depth interviews suggested that poor information, low expectations and parenting practices explained the paradox. Parents lacked critical information, such as understanding their options when a school is put on probation or how to transfer a child to a new school.[17] Many had low expectations about what schools in general were able to accomplish, or were discouraged by obdurate school bureaucracies and persistent problems and less likely to push for a higher quality school setting. Still others left the decision about where to attend school up to their children.

Most striking of all, DeLuca and Rosenblatt found that, despite the poor conditions in their children's neighbourhood schools, two-thirds of the parents in the qualitative study believed that school quality mattered much less for learning than a child's work effort and "good attitude." These attitudes undoubtedly stemmed in part from little personal experience with high performing schools. Parents like Tisha, an experimental mover and mother of two children who attended the zone school back in their city neighbourhood, explained:

> That school is crazy. I have to pray for her, it's like I send my child to hell every day and then I expect her to get good grades and learn. But like I said it's up to the individual 'cause she could separate herself from that and she could get what she needs. And she could keep going or she could fall into that crowd to which she's a follower and she'll mess herself up.

Kim, a control mother who has lived in public housing on Baltimore's West Side for 13 years, had a similarly optimistic attitude about what children can accomplish:

> I just don't care for that school much, but like I say, it all depends on how the children make it. If you go up there and you're willing to learn, then you're gonna learn. If you ain't willin to do nothin, then you're gonna do nothin.

[17] Keels's (2009) qualitative interviews with Gautreaux families found similar results.

Similarly, Tisha dismisses private schooling in light of what children contribute:

Interviewer: Did you ever think about sending him to another school?

Tisha: Mmm, not really...a lot of parents think if they, if I send my child to a private school, he would learn better. Well, you can send a hard head to a private school and it's not gonna make a bit of difference. You can send a good child to what you might think a not so good school and as long as they focus and pay attention it'll benefit them.

Parents' decisions about schooling often had little to do with academic quality. For many poor families, moving priorities began with proximity to transportation, family members, and mothers' jobs—with schools sometimes coming after that, if at all. For almost 70% of the Baltimore mothers in the qualitative study, what makes a good school had less do with academics, and more to do with proximity to work and whether teachers care about children. Even when parents did take school characteristics into account, they sought a sense of comfort and a welcoming atmosphere rather than academic rigor. Some parents simply wanted to be allowed to visit or be given "some general idea that you know, my child is in this school somewhere here." Many mothers valued such non-academic aspects of their children's schools as uniforms, security guards and disciplinary policies. These considerations make sense given that children were coming from chaotic, violent city schools. However, school decisions based on these characteristics were unlikely to result in higher academic quality.

DeLuca and Rosenblatt also found that continued poverty and the myriad challenges facing these families before they entered the program made it difficult for them to fully benefit from the initial moves to new communities, or to prioritize schooling decisions. While most parents emphasized the importance of school and wanted better things for their children, these good intentions and hopes were often thrown off course by constant instability and chaos. It is jarring how frequently severe substance use and death entered into already disrupted young lives. Parents were in and out of jail, rehab, and abusive relationships; children were shuffled between caregivers and homes. Troubles with landlords and the irregularity of part time low-wage work all contributed to frequent mobility and affected the context in which schooling decisions took place.

What about the children who did attend the higher performing suburban schools through their MTO moves? The few Baltimore interviews with the caregivers of these children suggested that they enjoyed the benefits of a richer academic environment and more attention from their teachers. One mother who used her voucher to move to Howard County makes this point clearly:

> In county schools, the classrooms are not overcrowded. The teachers are willing to work with you and help you and tutor you after school, before school. They were more concerned than the teachers in city public schools. So I would prefer any child to go to a county school... They're there to help you in the county schools.

Unfortunately, these success stories were rare among the MTO families, and children encountered other challenges after relocating to higher performing schools. For example, interviews with the Gautreaux II children showed that even when families successfully transfer their children into high performing schools, the mismatch between children's previous schooling experience and their own characteristics can interfere with academic achievement (Keels 2009).

The isolated nature of inner-city public housing communities and the high poverty urban schools that the majority of these children attended led many of those who relocated to the suburbs to be cultural outsiders. This in turn fuelled behaviour problems in both the MTO and Gautreaux II programs, especially for the boys (Keels 2009; Clampet-Lundquist et al. 2011). Children who transferred to high-achieving schools in the second round of the Gautreaux program also had substantial difficulties bridging the curriculum gap between their old and new schools (Keels 2009), which produced alarming post-relocation drops in classroom grades, and difficulties managing classroom assignments and homework. In Gautreaux I (and II), some children in predominantly white suburban schools were racial outsiders. They experienced discrimination in the form of suspensions from school or special education placement because they fought back when other children hit them or called them racial slurs (Mendenhall 2004).

"Conflicting Findings": Why Boys and Girls Respond Differently to New Neighbourhoods

Some of the most intriguing findings from the MTO research were the differences in how young men and women responded to the moves to low poverty communities. Boys between the ages of 12–19 had significant increases in self-reported behaviour problems after moving into less disadvantaged areas. In addition, boys in the experimental group demonstrated substantial increases in the proportion ever arrested and in arrests for property crime (Orr et al. 2003). These outcomes were similar to or worse than the outcomes for males in the control group. Boys in the experimental group were also more likely than control boys to report smoking cigarettes or drinking alcohol. Lastly, they demonstrated similar levels of psychological distress and generalized anxiety as their control counterparts. The story for girls in the experimental group was dramatically different. Their levels of delinquency went down and they reported significant improvements in mental health. Their psychological distress and generalized anxiety were significantly lower than control girls. These puzzling effects were part of the impetus for the qualitative studies launched in several of the MTO cities. After collecting in depth interviews with young adults and their parents, the mechanisms behind some of these gender based patterns became clearer, while others still require further research.

Clampet-Lundquist et al. (2011) analyzed the interviews conducted with a stratified random sample of 86 MTO teenagers (between 14 and 19 years of age) living in Baltimore and Chicago. In an attempt to uncover the hidden social processes in the lives of MTO teens, they examined daily routines, schooling experiences, neighbourhood norms, strategies for negotiating the community, peer relationships, friendship processes, and geographic distance from father figures.

Socializing and Daily Routines

One of the patterns to emerge from these conversations with youth was that the experimental girls were more likely than their male counterparts to spend substantial amounts of their leisure time in public places outside of their neighbourhoods such as the movies, shopping malls, or places downtown. When girls were in their neighbourhoods, they either stayed inside or spent time on porches or stoops. The way girls socialized allowed them 'exit strategies' from the negative aspects of the neighbourhood, and also put them in more contexts where there was some adult supervision. Girls also reported more selectivity when it came to who they chose to socialize with. Experimental boys on the other hand, although visitors to those same places, typically spent the majority of their leisure time in the larger neighbourhood, exposed to bullying and drug activities. They often used schools, parks, and alleys to play football or basketball, but these activities were rarely done as part of an organized school or community team, where coaches and parents would be involved. In addition, they gathered socially on street corners, at bars and at convenience stores.

Clampet-Lundquist et al. suggest that the locations where experimental boys hung out may have made them more vulnerable to negative social outcomes. For example, an experimental teenager name Jay reported that he often played basketball in the neighbourhood and that it was difficult to find a place to hangout that was free of active drug dealing. In fact, Jay identified for MTO researchers a location near the basketball court where a dealer stored his drugs.

Another finding was that experimental boys had more negative views of their neighbourhoods than experimental girls and control boys. Male teens felt the new neighbourhoods were too quiet or boring and that neighbours and police were less accepting of teen males getting together socially in public spaces. Seventy-three percent of experimental boys reported being questioned or harassed by the police compared to 58% of control males.

Police

According to the youth who moved to new neighbourhoods, both black and white neighbours called the police. Tiah, an experimental female reported that one of her white neighbours called the police on her and her friends. She believed the white neighbours were "concerned about their safety" due to the number of blacks moving into the area. Based on the youth's stories it appeared that a power struggle was taking place between the neighbours (black and white) and the youth over their rights to use public spaces. This power dynamic was often solved by neighbours engaging the police to reinforce neighbourhood norms. These social processes appeared to have particularly negative consequences for experimental males.

Alongside the calls from neighbours, police appear to engage in their own surveillance and harassment of males. Clampet-Lundquist (2011, p. 1170) described Roger as a 16 year old experimental male who moved with his family to a predominantly black suburb in Chicago. He stated, "I was in the suburbs. There wasn't nothing to do at all. Police always messing with you. Talking about you doing this, you're doing that." Another male teen, Ed, who attended church frequently, stated that the police questioned him at least once every 3 weeks. The authors quoted Ed describing an encounter with the police as he was leaving church in his current neighbourhood:

> Everywhere we go, we going to get stopped by the police because…they can always say we look suspicious and stuff…. We got stopped, me and my friend, coming out of the church gate before, by the detectives…. They talking about there was gunshots on the next block and we match the description or whatever…. They was like, "Do y'all got guns?" or something. "We heard shooting on the next block. Y'all match the description. Where y'all just come from?" We like, "We just came out the church. Y'all done seen it." You know, just, they stopping us for no reason (p. 1170).

Street Level Efficacy

The interviews also revealed some reasons why the experimental boys' behavioural outcomes might have looked worse than control boys in the interim evaluation. The narratives of the control boys demonstrated their detailed knowledge of each of the blocks in their neighbourhoods and which were 'hot' (drug heavy) and which were safer. They employed many strategies to stay safe and out of trouble in their neighbourhoods and were more selective about their friends than many of the experimental boys who later moved back to the city. One common strategy was one they learned from their parents (as noted above), which is 'staying to oneself' or 'being in your own business.' Clampet-Lundquist et al. (2011) presented Scott's account of this kind of strategy:

> I don't want to get caught up in nothing I don't got nothing to do with. I just stay in the house. I be wanting to go around there but I don't. It be pressure on me. [I: Is there any place that you can go to try and get away from that pressure?] I go to the library sometimes [and] get on the computer or read a book or something (p. 1173).

These stories were largely missing from the interviews with the experimental boys. Clampet-Lundquist suggested that the control boys had an advantage over the experimental movers when it came to living in the city: because they had always lived in high crime areas, they knew how to navigate them. When the experimental boys' families returned to city neighbourhoods, they were at a loss for some of these skills, having never had a chance to develop them because they spent a portion of their childhood in lower poverty neighbourhoods after the MTO move.[18]

[18]In another paper, Clampet-Lundquist (2011) compared the mental health differences between the experimental and control males. While the interim evaluation did not show significant benefits for the boys in the experimental group relative to controls, she found that control boys demonstrated higher levels of anger in their interviews than their experimental counterparts. She also found that the types of stressors reported by control boys were more severe than those described by the experimental males.

Female Fear and Sexual Pressure

Using data from the other three MTO cities, Popkin et al. (2010) focused more specifically on how to explain the improvement in mental health among the young women who moved with MTO. They employed a mixed methods, mostly qualitative, approach in Boston, Los Angeles, and New York. They randomly selected 122 families and performed 276 semi-structured interviews with parents, youth, and young adults all three treatment groups (Section 8, those who used vouchers in low-poverty areas for 1 year, and those who received a voucher but did not move). They identified several major themes: safety, female fears, escaping risk, pressure for early sexual activity and its costs, and overall well-being.

While Clampet-Lundquist et al. (2011) described how the social activities of young men seemed to make them vulnerable to involvement in the juvenile justice system because of their exposure in public places, young experimental women were vulnerable in public for different reasons. Popkin and colleagues' (2010) found that experimental girls reported that they were less afraid of harassment from men and boys and experienced less pressure to engage in sexual activity. One respondent noted that in the lower poverty areas guys "know not to touch them" (p. 729). This is in stark contrast to how Javon, a control group boy in Los Angeles, described attitudes and behaviours towards women in his neighbourhood:

> They pretty much treat them like animals. . . . They be calling them bitches and all that. . . . They just say bitches, whores, that's it...People, they treat women . . . like they was just objects, as if they owned them or whatnot. . . . You can either mistreat a woman or the woman is going to mistreat you. (Popkin et al., p. 732)

Some of the experimental girls talked about how leaving their old neighbourhoods, and the sexual pressures associated with them, may have allowed them to avoid early pregnancy and temptations to engage in drug use. Popkin et al. (2010: 720) quoted Antoinette, now 20, but who moved with her family as a teenager in New York:

> Because a lot of kids in my [old] neighbourhood, like the girls, wound up not finishing junior high or just starting high school like one of my best friends. I mean, we were in every single class since we started school together. We even went to the same high school. And then like ninth grade she had a kid and that was it.

The relief from general safety issues and sexual pressure appeared to have positively affected the overall well-being of both parents and daughters in the experimental group. The final report promises to shed additional light on the structural and social processes that gender the experience of neighbourhood contexts for boys and girls.[19]

[19] In the summer of 2010, Susan Clampet-Lundquist, Stefanie DeLuca and Kathryn Edin launched a 10 year qualitative follow up with young men and women In Baltimore (ages 15–24) whose families participated in MTO when they were children. This in depth study of over 150 young adults focuses on the long term consequences of neighbourhoods for young men and women, with an emphasis on the gender differences discovered in the Interim Evaluation. The study also explores the transition to adulthood for disadvantaged youth across important domains such as risk behaviour, family formation, transition to work and college, substance use and family relationships.

Conclusions

MTO's interim report has prompted critics to question the relevance of housing policies and neighbourhood context for improving life outcomes for poor families and children (e.g., Mathews 2007). MTO's randomized design lends heft to the arguments of these critics, although it also reinforces the credibility of the program's large positive effects on mental health so often ignored in media and policy discussions. In this chapter, we showcase how we and other scholars have used mixed methods to expand the scope of the policy discussion surrounding MTO and other residential mobility programs. Such techniques lead us to question whether MTO-based policy approaches are simply ineffective or might instead be an *insufficient* part of the solution for solving the entrenched problems some poor families face.

The use of combined data sources and methodologies also allows us to show how the potential of these housing programs is affected by space-based structural factors and the dynamics of poor families' beliefs, backgrounds and constraints. In the case of MTO, mixed methods approaches show us that we need to consider how program effects (whether unforeseen or expected) are conditioned by existing structural inequalities (such as housing markets and urban school quality) and how the conditions of life for poor families facilitate or impede their ability to engage new structural opportunities. For example, the in-depth interviews with Baltimore MTO participants put the null employment results in context by showing that the assumptions behind how social networks function and the availability of jobs in more affluent areas were off base. The interviews also demonstrated that residential choices and barriers, mothers' decisions and beliefs around schooling opportunity, and major life challenges all made it difficult for the MTO children to experience the benefits of improved school environments.

Previous research led policymakers to assume that the opportunity for neighbourhood change provided by MTO could sufficiently promote families' escape from poverty and improve children's educational opportunities. While neighbourhood change could be a necessary condition to protect children and improve their schooling, it is not always enough in light of the deep tangle of issues that characterize the lives of most very poor families. The families participating in social programs like MTO have often been living in poverty for generations and have needs that exist beyond those that the vouchers are meant to remedy. The constraints of continued financial hardship and the instabilities that accompany that hardship make it difficult for them to not only relocate to higher opportunity areas, but to utilize some of the higher quality services in those communities (such as the readily available but higher cost extracurricular activities in more affluent areas, as Keels 2009 documents).

Improving outcomes in residential mobility programs might well require intensive and sustained housing counselling to ensure relocation to areas that provide access to rigorous and high performing schools if we are to expect these programs to increase access to educational resources. Efforts must also be made to address mechanisms that foster residential segregation and school inequities such as zoning laws and race and class discrimination. We also learned that parenting practices, beliefs and information matter significantly when it comes to determining housing and school choice. Housing counselling with a component that educates parents specifically about the implications of residential choice for school opportunity is one step in that direction.

As an example of how this might work, a partnership between fair housing lawyers, a local foundation and the agency administering the Thompson mobility program in Baltimore has trained the housing counsellors so that they can help parents make a connection between schools and new neighbourhoods (such as looking up school characteristics online). They have also hired a liaison to work with the parents to help transition students into new schools, which is critical for special needs children as well as for parents who feel intimidated by the unfamiliar settings. Finally, the match between children's developmental skills and the demands of the new school is an important factor to consider in interventions that involve changes school quality. Assistance like that being introduced in Baltimore could also ameliorate some of the adjustment problems that come from the curriculum gaps between new and old schools.

References

Borkan, J. M. (2004). Mixed methods studies: A foundation for primary care research. *Annals of Family Medicine, 2*(1), 4–6.

Clampet-Lundquist, S. (2011). Teens, mental health, and Moving to Opportunity. In E. Birch, H. Newburger, & S. Wachter (Eds.), *Neighborhood and life chances: How place matters in modern America* (pp. 204–220). Philadelphia: University of Pennsylvania Press.

Clampet-Lundquist, S., Edin, K., Kling, J., & Duncan, G. J. (2011). Moving teenagers out of high-risk neighborhoods: How girls fare better than boys. *American Journal of Sociology.*

DeLuca, S., & Dayton, E. (2009). Switching social contexts: The effects of housing mobility and school choice programs on youth outcomes. *Annual Review of Sociology, 35*, 457–491.

DeLuca, S., & Rosenblatt, P. (2010). Does moving to better neighborhoods lead to better schooling opportunities? Parental school choice in an experimental housing voucher program. *Teachers College Record, 112*(5), 1443–1491.

Elder, G. H. (1998). The life course as developmental theory. *Child Development, 69*(1), 1–12.

Galster, G. (2011). The mechanism(s) of neighbourhood effects: Theory, evidence, and policy implications. In M. van Ham, D. Manley, N. Bailey, L. Simpson, & D. Maclennan (Eds.), *Neighbourhood effects research: New perspectives.* Dordrecht: Springer.

Gibson-Davis, C. M., & Duncan, G. J. (2005). Qualitative/quantitative synergies in a random assignment program. In T. Weisner (Ed.), *Discovering successful pathways in children's development: New methods in the study of childhood and family life* (pp. 283–303). Chicago, IL: University of Chicago Press.

Grogger, J. (2003). The effects of time limits, the EITC, and other policy changes on welfare use, work, and income among female-headed families. *The Review of Economics and Statistics, 85*(2), 394–408.

Katz, L., Kling, J., & Liebman, J. (2001). Moving to opportunity in Boston: Early results of a randomized mobility experiment. *Quarterly Journal of Economics, 116*(2), 607–654.

Keels, M. (2009). Unpacking the null effects of relocating poor children into non-poor neighborhoods: Focus on school experiences. Working paper under review.

Kennedy, S. D., & Finkel, M. (1994). *Section 8 rental voucher and rental certificate utilization study: Final report*. Washington, DC: U.S. Department of Housing and Urban Development, Office of Policy Development and Research.

Kissane, R., & Clampet-Lundquist, S. (2010). Exploring the social ties and social capital of families after moving to opportunity. Working paper under review.

Kling, J., Liebman, J., & Katz, L. (2005). Bullets Don't Got No Name: Consequences of fear in the Ghetto. In T. S. Weisner (Ed.), *Discovering successful pathways in children's development: Mixed methods in the study of childhood and family life* (pp. 243–281). Chicago: University of Chicago Press.

Kling, J., Liebman, J., & Katz, L. (2007). Experimental analysis of neighborhood effects. *Econometrica, 75*(1), 83–119.

Ludwig, J., Ladd, H. F., & Duncan, G. J. (2001). Urban poverty and educational outcomes. *Brookings Wharton Papers on Urban Affairs, 2*, 147–201.

Ludwig, J., Liebman, J. B., Kling, J. R., Duncan, G. J., Katz, L. F., & Kessler, R. C. (2008). Can we learn about neighborhood effects from the moving to opportunity experiment? *American Journal of Sociology, 114*(1), 107–143.

Mathews, J. (2007). Neighborhoods' effect on grades challenged. *Washington Post*, August 14, 2007.

Mendenhall, R. (2004). *Black women in Gautreaux's housing desegregation program: The role of neighborhoods and networks in economic independence*. Unpublished doctoral dissertation, Northwestern University.

Mendenhall, R., DeLuca, S., & Duncan, G. (2006). Neighborhood resources, racial segregation, and economic mobility: Results from the Gautreaux program. *Social Science Research, 35*(4), 892–923.

Orr, L., Feins, J., Jacob, R., Beecroft, E., Sanbonmatsu, L., Katz, L., et al. (2003). *Moving to opportunity interim impacts evaluation*. Washington, DC: U.S. Department of Housing & Urban Development.

Popkin, S. J., Harris, L. E., & Cunningham, M. K. (2001). *Families in transition: A qualitative analysis of the MTO experience*. Washington, DC: The Urban Institute.

Popkin, S. J., Katz, B., Cunningham, M. K., Brown, K. D., Gustafson, J., & Turner, M. A. (2004). A Decade of HOPE VI: Research findings and policy challenges. Washington, DC: The Urban Institute & The Brookings Institution. http://www.urban.org/UploadedPDF/411002_HOPEVI.pdf.

Popkin, S. J., Leventhal, T., & Weismann, G. (2010). Girls in the hood: How safety affects the life chances of low-income girls. *Urban Affairs Review, 45*(6), 715–744.

Popkin, S. J., Rosenbaum, J. E., & Meaden, P. M. (1993). Labor market experiences of low-income black women in middle-class suburbs: Evidence from a survey of Gautreaux program participants. *Journal of Policy Analysis and Management, 12*(3), 556–573.

Rosenbaum, J. E., DeLuca, S., & Tuck, T. (2005). Crossing borders and adapting: Low-income Black families in Suburbia. In X. de Souza Briggs (Ed.), *The geography of opportunity: Race and housing choice in metropolitan America* (pp. 150–175). Washington, DC: Brookings Institution.

Rosenbaum, J. E., Reynolds, L., & DeLuca, S. (2002). How do places matter? The geography of opportunity, self-efficacy, and a look inside the black box of residential mobility. *Housing Studies, 17*(1), 71–82.

Rosenbaum, J. E., & Zuberi, A. (2010). Comparing residential mobility programs: Design elements, neighborhood placements and outcomes in MTO and Gautreaux. *Housing Policy Debate, 20*(1), 27–42.

Rosenblatt, P., & DeLuca, S. (2010). *We don't live outside, we live in here: Residential mobility decisions of low-income families*. Paper presented at the Annual Meeting of the American Sociological Association, Atlanta.

Rubinowitz, L. S., & Rosenbaum, J. E. (2000). *Crossing the class and color lines: From public housing to white suburbia*. Chicago: University of Chicago Press.

Sanbonmatsu, L., et al. (2006). Neighborhoods and academic achievement: Results from the moving to opportunity experiment. *The Journal of Human Resources, 41*(4), 649.

Shadish, W. R., Cook, T. D., & Campbell, D. T. (2002). *Experimental and quasi-experimental designs for generalized causal inference*. Boston: Houghton-Mifflin.

Shroder, M. (2002). Locational constraint, housing counselling, and successful lease-up in a randomized housing voucher experiment. *Journal of Urban Economics, 51*, 315–38.

Small, M. L. (2011). How to conduct a mixed methods study: Recent trends in a rapidly growing literature. *Annual Review of Sociology*.

Small, M. L., & Feldman, J. (2011). Ethnographic evidence, heterogeneity, and neighbourhood effects after moving to opportunity. In M. van Ham, D. Manley, N. Bailey, L. Simpson, & D. Maclennan (Eds.), *Neighbourhood effects research: New perspectives*. Dordrecht: Springer.

Turney, K., Clampet-Lundquist, S., Edin, K., Kling, J. R., & Duncan, G. (2006). Neighborhood effects on barriers to employment: Results from a randomized housing mobility experiment in Baltimore. *Brookings-Wharton Papers on Urban Affairs*, 137–187.

Chapter 10
School Outcomes and Neighbourhood Effects: A New Approach Using Data from Finland

Venla Bernelius and Timo M. Kauppinen

Introduction

Much of the neighbourhood effects literature focuses on the effects that social environments of neighbourhoods might have on individual outcomes. The social composition of neighbourhood populations and the social organization at the neighbourhood level are seen as the most relevant factors in the formation of neighbourhood effects (Kauppinen 2007). Broader conceptualisations of neighbourhood effects consider the neighbourhood as a more complex combination of physical environment, social structure, local institutions and opportunity structures (e.g. type of employment offered locally). Such a broad conceptualisation does not only expand the list of neighbourhood characteristics potentially affecting various individual outcomes but it may also help in analyzing the pathways through which the effects of, for example, the social environment may occur (see Galster 2011 in this volume for a list of potential mechanisms).

When investigating neighbourhood effects on young people's educational outcomes, schools are an important part of the social, institutional and physical environments that young people experience in their daily lives. Most young people go to local schools, sometimes even in the same neighbourhood in which they live (depending on the availability of schools, their catchment areas and the degree of choice available in the schooling system). Therefore, school effects can in part be viewed within the framework of neighbourhood effects (see Bramley and Karley 2007). Because of this relationship between residential locations and school locations,

V. Bernelius (✉)
Department of Geosciences and Geography, University of Helsinki,
Helsinki, Finland
e-mail: venla.bernelius@helsinki.fi

T.M. Kauppinen
Department of Social Research, University of Turku, Turku, Finland
e-mail: timo.kauppinen@utu.fi

M. van Ham et al. (eds.), *Neighbourhood Effects Research: New Perspectives*,
DOI 10.1007/978-94-007-2309-2_10, © Springer Science+Business Media B.V. 2012

studies of school effects on educational outcomes, that do not take into account where the pupils live, may overestimate the importance of the school. Correspondingly, studies of neighbourhood effects on educational outcomes that do not take into account the school environment, may overestimate the importance of the neighbourhood. Sometimes both contexts are acknowledged but the link between them is often not made explicitly.

If the aim is to provide a full picture of neighbourhood effects, then both the neighbourhood and the school context should be included in the explanation, and the link between them should be explicitly assessed. In this chapter, neighbourhood effects on school outcomes are understood in accordance with broader definitions that include the effects of neighbourhood institutions, including schools. School effects are therefore included in the explanation of neighbourhood effects, including, for instance, the effects of the school's social environment (especially peer groups), quality of teaching and the school's resources.

Tackling Neighbourhood Effects on Educational Outcomes with a New Study Design

The aim of this chapter is to explore some of the central methodological problems in neighbourhood and school effects research, and some of the potential solutions. The principal contribution of this chapter is the introduction of a bespoke research project. The project is designed to overcome some of the recognised methodological challenges, and to allow for the analysis of both neighbourhood and school effects and their mediating mechanisms. The project focuses on neighbourhood and school effects on educational outcomes in the Helsinki Metropolitan Area, Finland, which we hope will expand the European knowledge base on these effects in a country context which is particularly interesting for research.

Finland is an egalitarian welfare state with relatively low levels of urban socioeconomic and ethnic segregation. If neighbourhood effects are observed in this context, where social differences are small and strong public policies exist with aims to enhance equality between individuals and neighbourhoods, then they can be argued to exist across a variety of other social contexts. Low levels of social and spatial differentiation and strong welfare policies are among the key elements assumed to diminish the probability of observable neighbourhood effects (Musterd and Ostendorf 2005). The Helsinki Metropolitan Area is also interesting for research purposes because it is undergoing a process of increasing segregation. Recent research suggests that differences between neighbourhoods and schools are growing, which makes the region an attractive "urban laboratory" for research within a dynamic setting. The different parts of the large region also have varying levels of segregation, which makes it possible to compare effects in different types of municipalities and neighbourhoods.

Finland is also a particularly attractive setting for educational research because of its egalitarian educational system and its continuous success in the OECD comprehensive education PISA tests. The average high scores for the country as a whole are likely to hide local variations in school outcomes, which can be expected to grow with increasing levels of school segregation. Education is also an attractive topic for neighbourhood effect research as it is one of the most powerful factors in defining the life chances and welfare of an individual in modern society. Children are of particular interest, as they are arguably the most vulnerable to these effects due to their on-going socialization process and their localised everyday lives (e.g. Rankin and Quane 2002).

The structure of the chapter is as follows. Firstly, we present the empirical context with a brief description of Finnish society and the Helsinki Metropolitan Area. The aim is to give the reader some contextual understanding of the research project presented later, and to explore the ways in which the Finnish context is of general theoretical interest. After this, we present a literature review on educational neighbourhood effects and previous Finnish research evidence. Our central aim is to outline some of the key methodological challenges in neighbourhood effect research. We then introduce the project design, which aims to tackle some of the methodological and empirical questions arising from the former Finnish and European neighbourhood effect research on education.

Social Spatial Segregation and Educational Outcomes in Finland

Throughout the last decade Finland has been continuously ranked very high in the OECD PISA educational comparisons. The ranking is not only based on a high overall educational achievement, but also on exceptional equality (small variations) in outcomes for individual pupils and schools. The point of departure for this chapter is that if the PISA comparisons had been carried out in the Helsinki Metropolitan Area (HMA) only, the results would have showed much more variation in educational outcomes. The differences between student composition in the schools of the HMA is greater than in the rest of the country put together (Jakku–Sihvonen and Kuusela 2002). As the pupils' socio-economic background is strongly connected to the educational outcomes of schools, the "urban PISA" would show a more polarised pattern resembling trends observed in most of the other Western countries (Bernelius 2010; Kupari 2005). Particularly the schools in the central municipalities of the capital region show a segregated pattern, while the suburban edges are more socio-economically balanced (Bernelius 2010; Kuusela 2010). There is a strong link between the composition of pupils in schools and the population composition of the surrounding urban area.

Neighbourhood effects are generally assumed to intensify as socio-spatial segregation increases (Musterd and Ostendorf 2005). For instance, the heavily segregated North American cities tend to be associated with more pronounced effects when compared with most of their European counterparts with lower levels of urban differentiation. Generalisation of findings from cities with high levels of segregation to less segregated contexts is viewed as problematic. Neighbourhood effects are also assumed to be stronger if there is a lack of social political interventions: strong welfare policies are thought to counteract possible neighbourhood effects (Musterd and Ostendorf 2005). As discussed in the introduction, Finland is an interesting setting for neighbourhood effect research because the social and spatial differentiation is still modest by European standards and the welfare policies and services are strong and universal which would be expected to maintain low levels of neighbourhood effects.

The Nordic welfare state has been efficient at narrowing societal differences in income and welfare, and the low levels of spatial segregation can mostly be attributed to low levels of socio-economic and ethnic differentiation together with a policy of social mix (Vaattovaara and Kortteinen 2003). Although the HMA is still socio-economically and spatially a relatively unsegregated city region with a high level of welfare and a highly educated population when compared to other European capitals, the differences between neighbourhoods are significant. For instance, in the metropolitan core the share of working age residents with tertiary education ranges from approximately 10–80% (Kortteinen and Vaattovaara 2007). The most recent change in these neighbourhoods has been the rapid increase of the immigrant population; the share of immigrants has grown to approximately 10% in 2009 from practically zero at the beginning of the 1990s. The immigrant population is predicted to make up over a fifth of the population of the HMA in 15 years. There are already indications of an uneven spatial distribution of especially non-Western immigrants (Vilkama 2008; Kauppinen 2002). Recent metropolitan developments in growing income differences, a rapid increase in immigration, and cuts in social and housing policies have also contributed to the trend of growing differences between neighbourhoods. Together with the educational level of the adult population, the ethnic element is also highly significant for the challenges met by schools, the educational attainment of pupils and the possible neighbourhood effects on education (Bernelius 2008; Kuusela et al. 2008; Andersson and Molina 2003).

The growing socio-economic gaps in Finnish society and the growth of urban segregation within the HMA resonate in schools. A simple means of observing the development of differences between schools is to look at the *intra*-class correlations of pupils' attainment in standardised tests (Fig. 10.1). Intra-class correlation refers to the share of total variance in the test scores that can be attributed to differences between schools, instead of differences between individual pupils. In other words, it describes the share of between-school variance as a proportion of the total variance. When intra-class correlations are higher, the differences between schools are greater. As can be seen from Fig. 10.1, the intra-class correlation in the schools in the HMA is more pronounced than in other large Finnish cities and far greater than in rural municipalities and small towns. This observation shows that the differentiation between schools is greater in the larger cities, than in smaller towns and rural areas where pupils are dispersed more equally between the schools. The most significant trend, however, is the overall growth of the differences between schools in cities; increasingly, the majority of variation occurs between, rather than within, schools.

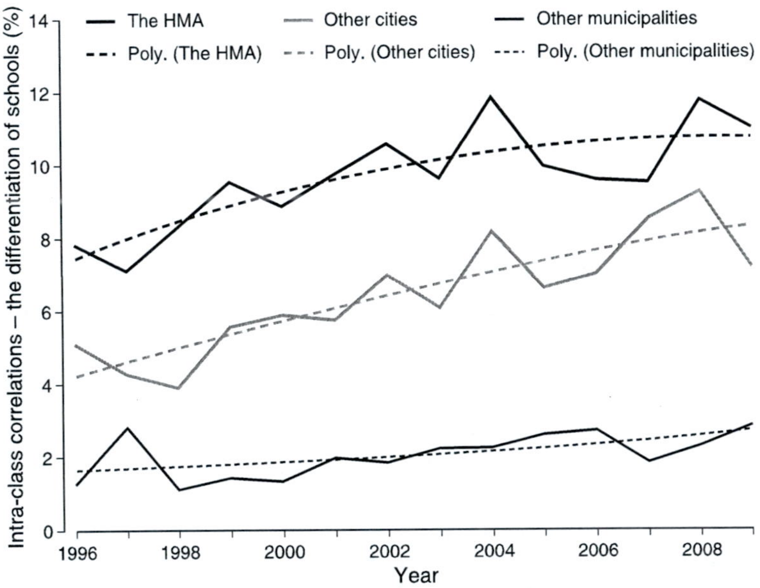

Fig. 10.1 Intra-class correlations (differences between schools) in educational attainment 1996–2009 for several regions (Kuusela 2010: 41). Actual developments and smoothed curves

Neighbourhood Effects in the Finnish Context

The main body of neighbourhood effects research on education is of North American origin and has been produced during the last two or three decades (Friedrichs et al. 2003). Although some North American and especially European studies have questioned the existence of neighbourhood effects (for education, see e.g., Burgess et al. 2008; Musterd and Ostendorf 2007; Plotnick and Hoffman 1999), many studies claim to have found at least some evidence supporting the hypothesis of neighbourhood effects (see for instance: Ammermueller and Pischke 2006; Friedrichs et al. 2003; Robertson and Symons 2003; Zimmer and Toma 2000; Kauppinen 2007). The argument in favour of the existence of neighbourhood effects is particularly strong in the case of children and adolescents.

Previous European research implies that school context is an important link between neighbourhoods and educational outcomes. Neighbourhoods seem to have little additional effect, when school effects are accounted for (Kauppinen 2008; Brännström 2008; Sykes and Musterd 2010). The composition of the pupils in schools has been found to have effects on cognitive outcomes (Thrupp et al. 2002; Butler and Hamnett 2007) and on educational careers (see Kauppinen 2008). Similar results have been also found vis-à-vis health outcomes (Sellström and Bremberg 2006; see also West et al. 2004). However, there seems to be less agreement regarding effects of school resources and organizational characteristics. For example Rutter and Maughan (2002) and Nash (2003) point out some contradictory results.

The presence of contradictory results suggests that the North American or even Central European research cannot be directly generalized to Nordic countries and especially not to the Helsinki region, where the social and urban contexts differ noticeably with regard to the social policies and the level of socio-spatial segregation (see Kauppinen 2007; Friedrichs et al. 2003). Crucially, if neighbourhood effects only occur when certain 'threshold' levels in neighbourhood segregation are exceeded (Crane 1991), then they might not be observed in less segregated societies and cities.

The body of Finnish research on neighbourhood effects is relatively small, and only a few studies are concerned with adolescents and education. The main works have been carried out by Karisto and Montén (1996), Karvonen and Rahkonen (2002), Kauppinen (2004, 2007, 2008) and Bernelius (2005, forthcoming) in Helsinki. The first two studies have investigated educational attitudes or 'pedagogical ethos', while the work by Kauppinen concerned educational careers. Bernelius's work has dealt with the relationship between school outcomes and neighbourhood characteristics and school effects on educational outcomes and attitudes.

All the above studies have been plagued to some extent by difficulties of getting appropriate data and in particular the lack of information about the adolescents' individual social background, educational outcomes or postal addresses. Finnish legislation and administrative institutions have been particularly restrictive with regard to releasing information on test scores linked to pupils' background data, even for research purposes. However, both Karisto and Montén (1996) and Karvonen and Rahkonen (2002) were able to establish some indications of possible neighbourhood effects on adolescents' educational attitudes, but the evidence remains questionable due to deficiencies in the data used.

Kauppinen's (2004, 2007) data were more extensive and detailed, but his findings do not support the hypothesis of neighbourhood effects to the same extent as the other two studies (Karisto and Montén 1996; Karvonen and Rahkonen 2002). In Kauppinen's analysis of the type of secondary education chosen by adolescents across various neighbourhoods, the only finding supporting neighbourhood effect theories was the overrepresentation of individuals selecting academic options when they lived in neighbourhoods with the a high proportion of well-educated adults. Further analyses showed that this effect was mostly mediated by schools (Kauppinen 2008). The pupil composition of the school, in terms of the socioeconomic status of the parents, was found to have a linear effect on the educational choices of the pupils. However, this study had only limited amount of school-related information, which restricted the analysis.

Bernelius's research has focused on the comprehensive schools in the HMA. In Finland, compulsory education is delivered through comprehensive schools, which are funded publically. Although there are a few private schools they are also publicly financed, bound to the same curricula as the comprehensive schools and tuition fees are not charged. The studies which focus on comprehensive schools (see e.g. Bernelius 2005, 2008, 2010, forthcoming) indicate that the association between urban segregation and educational outcomes appears to be clear even in the relatively egalitarian Finnish context. The socio-economic background of the pupils is a strong determinant

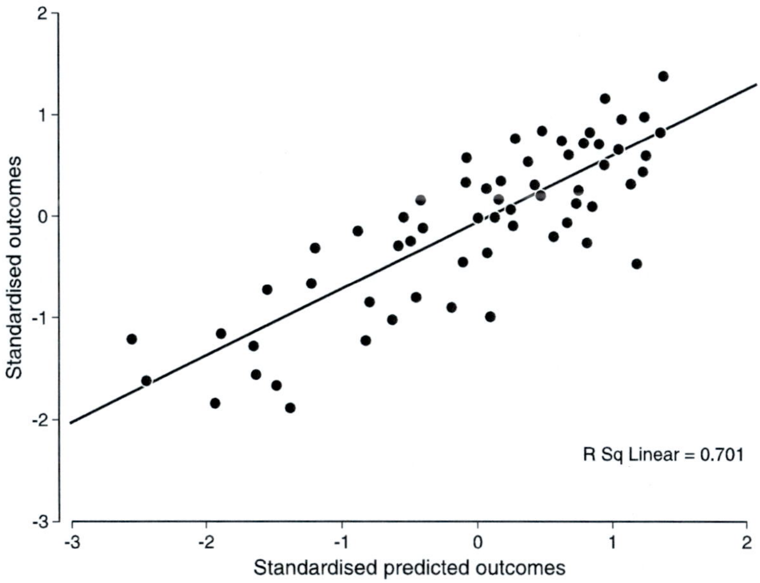

Fig. 10.2 Standardised regression model outcomes against observed outcomes for Helsinki's primary schools

of educational success both when observing individual pupils or school averages (see also Jakku–Sihvonen and Kuusela 2002; Kupari 2005; Kuusela 2006).

The school-level educational outcomes, measured using scores from standardised tests, are strongly correlated to the socio-economic structure of the catchment area population. Approximately 60–70% of the variation in the schools' educational outcomes can be accounted for by the school catchment area characteristics, at least when applying linear regression analysis (see Fig. 10.2). In this analysis, the socio-economic status of a school's catchment area was described using three key characteristics which were strongly correlated to educational outcomes: the percentages of social housing, immigrants and adults with a low educational status. The sorting of pupils with similar backgrounds into the same schools, influenced by the socio-economic and ethnic composition of neighbourhoods, is evidently the main mechanism differentiating the schools' educational outcomes. The school-level research of the HMA did not give indications of strong, "non-linear" neighbourhood effects, which would require a certain threshold level of urban poverty or particularly high welfare (see Bernelius 2005, 2010). The relationship between the school catchment area characteristics and school outcomes is linear throughout all types of neighbourhoods (see Fig. 10.2). In the case of pronounced neighbourhood effects, targeting especially deprived or privileged neighbourhoods, one should expect to see the privileged neighbourhoods and schools attaining higher results than predicted by the linear model, or deprived neighbourhoods producing a significantly lower attainment level than expected.

However, Bernelius (forthcoming) found some indication of possible neighbourhood or school effects in an analysis of individual pupils and pupil groups from different schools. These effects appear to exist in all types of schools instead of only the very disadvantaged or privileged, and thus the effects are not observed in the school-level analysis described above. Multi-level modelling reveals that a small portion of the variance in educational outcomes is related to the level of parental education across the school rather than to the educational level of a pupil's parents. Based on these results, it appears possible that the overall pupil composition of a school may exert a small, independent effect on the outcomes of individual pupils. Observing the variation within schools shows an "averaging out" of educational outcomes within schools. That is to say, educational outcomes in schools do not to cover the full range of possibilities from low to high, but are concentrated so that outcomes tend to be consistently higher or lower. Importantly, this is not controlled away within the multi-level model when parental education achievement is included. Thus, the different educational outcomes observed at the school level appear not to be merely a product of an uneven distribution of pupils with differently educated parents *between* schools, but are also driven by pupils with similarly educated parents achieving different outcomes relative to the school context. It is also interesting to note that the strongest disparity between family background and educational outcomes can be observed in the best performing schools. In these schools, pupils from all backgrounds tend to do well, whereas in the low-performing schools the effect of the school appears to be smaller, and the individual background is a stronger predictor of pupils' outcomes.

The results of this study are similar to the few previous Finnish studies on educational neighbourhood effects. The indication of school-level characteristics influencing individual pupils resonates with Karvonen and Rahkonen's (2002) results, and the observation of a slightly more notable positive effect in the schools with highest attainment is in line with Kauppinen's (2004, 2007) research on the shift to secondary education in Helsinki.

Shortcomings and Challenges of the Previous Neighbourhood Effects Studies

The current evidence base on the effects of neighbourhood characteristics on individual level educational outcomes has some shortcomings. This is largely a consequence of the lack of suitable data availability. In this section, we will first discuss these limitations and then the challenge of demonstrating causality in studies of contextual effects on educational outcomes.

In Finland, none of the larger national or metropolitan datasets on cognitive outcomes and educational attitudes have linked socio-economic background characteristics with individual level information (see Jakku–Sihvonen and Kuusela 2002), mainly because of strict legislation limiting the collection and availability of data. As a result, it has been impossible to implement a full multi-level approach, where the neighbourhood and school are treated as separate contexts, and the pupil's individual background and personal characteristics are taken into account. The lack of multi-level datasets may lead to a misidentification of the explanatory factors for educational outcomes (see Diez–Roux 1998). The available data is also cross-sectional, which prevents the study of causal processes. Longitudinal data would allow researchers to study the temporal ordering of events to establish causal relationships, and would allow the analysis of the development of educational achievement and attitudes over time. As with many national contexts, it will only be possible to address these problems if new data are collected.

Perhaps the most fundamental methodological problem in studies of contextual effects is that *associations* are observed while conclusions are presented as if they were about causal *effects* (see also Small and Feldman 2011 this volume). One of the main challenges to overcome is the problem of selection bias. If non-random selection of families into neighbourhoods and pupils into schools could be taken into account, the remaining associations between contextual characteristics and individual-level outcomes could more convincingly be reported as causal effects. The most common way to attempt to eliminate selection bias is to control within a model for observed characteristics of families and pupils that may affect both the selection into neighbourhoods and schools and the individual-level outcomes under analysis. Unfortunately, not all relevant characteristics can be measured, which leaves the outcomes vulnerable to selection bias.

The most convincing way to eliminate selection bias is a random assignment experiment, whereby families are randomly allocated to neighbourhoods, or pupils are randomly allocated to schools (or a naturally occurring quasi-experiment, see Sampson 2008, for a discussion of the complexities related to these experiments). Because (quasi-) experimental data is rarely available, many studies use statistical methods to try to deal with selection bias. Such methods include fixed-effects and difference models, and instrumental variable approaches (see Galster 2008). The use of fixed-effects and difference models is not without problems, for example because they cannot control for unmeasured characteristics which change over time and they typically do not control for changing or lagged effects of the constant characteristics. Instrumental variable approaches are also not without problems because it is often difficult to find good instruments.

Although these advanced statistical methods aim to establish causal relationships between contextual characteristics and individual-level outcomes, the underlying causal mechanisms remain a black box (cf. Jencks and Mayer 1990). The use of increasingly sophisticated statistical methods do not advance our understanding of why particular neighbourhood characteristics cause (or do not cause) particular outcomes (see also the chapters by Galster 2011; Small and Feldman 2011 in this volume). It is important to show what kind of mediating factors or mechanisms cause certain outcomes and therefore it is important to *observe the processes* linking the contexts to the individual-level outcomes.

The ideal way to show these processes would be to observe 'micro-level' sequences of events or states over an individual's life course, and to analyse what kind of sequences generally link particular contexts or contextual characteristics to particular outcomes. Especially social interaction and the individual's own perceptions of her social environment would be important objects of observation, in addition to the simultaneous measurement of the development of educational attitudes. At least it would be important to measure the educational norms prevailing among the pupils and teachers in the school (see Donnelly 2000) and factors like bullying at school or engagement in unhealthy behaviours. An ideal research design would also collect characteristics of several contexts which are likely to influence pupils. These contexts include not only schools, but also the neighbourhood in which they live and the family in which they grow up. Combining contexts may show that the school environment is an important link between neighbourhood characteristics and individual educational outcomes (see Kauppinen 2008). The school effect itself might be decomposed into school-level and class-room-level components, the first presumably being more neighbourhood-related. The remaining sections of this chapter will describe a new research project which will attempt to tackle some of the above mentioned challenges.

A New Design to Study Neighbourhood Effects

The national research council of Finland, the Academy of Finland, is currently funding a 4 year research project to address some of the recognised challenges in neighbourhood effects research on educational outcomes. The new "MetrOP"-project aims further our understanding of neighbourhood and school effects in Finland by using a new data collection approach which is designed to tackle some of the current deficiencies in data and some of the more general methodological challenges in neighbourhood effects research (Rimpelä and Bernelius 2010).[1] The project focuses on the educational and health outcomes of comprehensive school pupils, observing their changes as a result of the process of metropolitan segregation. The central research question is whether increased segregation of schools and neighbourhoods in the Helsinki Metropolitan Area lead to poorer health and learning outcomes among children.

The main contribution of the project is to produce a more comprehensive analysis through the collection of a large dataset that is multi-level, longitudinal and rich in background factors and personal characteristics of individual pupils. The data combine information on pupils' educational outcomes and attitudes, as well as their health, welfare and health behaviour, together with individual background

[1] This research project is designed with contributions from the whole MetrOP team, which includes in addition to the authors: Project leaders Jarkko Hautamäki (University of Helsinki) and Matti Rimpelä (University of Tampere), and Riittakerttu Kaltiala-Heino, Hannu Oja, Arja Rimpelä, researcher Lotta Alho (University of Tampere); Airi Hautamäki, Mari Vaattovaara, researcher Sirkku Kupiainen (University of Helsinki); Sakari Karvonen, Timo Ståhl, researcher Vesa Saaristo (National Institute of Health and Welfare); Jorma Kuusela (Finnish National Board of Education).

characteristics and information about the school and neighbourhood. The dataset tracks the same cohort of students over 3 years in lower secondary school (grades 7–9). This type of a data enables the use of multi-level models in the analysis of both neighbourhood and school effects. The dynamic setting makes it possible to observe changes in students as they happen, and it is hoped that together with the richness of individual characteristics measured, this will help to overcome some of the selection bias problems and to achieve causal explanations, as well as to identify possible mechanisms of the effects.

The limitations of data discussed above, including the legislative data problems, access to educational outcome data and the ability to combine it with socio-economic indicators at an individual level can be overcome by this research setting. Firstly, the research is designed to be carried out in close collaboration with the institution authorised to collect sensitive educational data, the Finnish National Board of Education. Secondly, the individual research subjects are further protected by not building a register of pupil identification numbers for the research team, but by storing the identification information in the schools and municipalities, who are authorised to keep short term student registers for the collection of longitudinal data.

Research Aims and Questions

The MetrOP project has been designed to lead to a comprehensive analysis of to determine if there are contextual effects in the HMA and what mechanisms lie behind them. A key scientific goal of the research is to increase theoretical and empirical understanding of socio-spatial segregation and the consequences of this on educational outcomes and attitudes, as well as a range of social well-being outcomes. The project seeks to accomplish this aim by using bespoke data and through the combination of methodological and theoretical contributions from various fields of social research. Ultimately, this will strengthen the understanding about the early development of factors in children which contribute to welfare issues later in their life.

The main research questions to be addressed are:

- Has the increase in the social and spatial segregation of the HMA lead to a growing differentiation of the educational outcomes of comprehensive school pupils? Do health and wellbeing outcomes and health behaviours exhibit similar trends?
- Is it possible to find evidence of an independent neighbourhood or school effect on the learning and wellbeing of pupils?
- How are the main welfare factors interlinked between individual and contexts (classroom, school, neighbourhood, municipality)? Can the interaction explain the differentiation in educational outcomes, wellbeing, health and the social exclusion of children?
- What role do schools play in the process of segregation within the HMA?

Research Design

The research project combines the topics of socio-spatial segregation, learning and health, which have thus far been studied separately. The work is carried out by a group of 15 researchers in three universities and two public research and administration organisations. The research design is *quantitative, multi-level and longitudinal*, and allows for the effects of the developmental contexts of children and their individual characteristics to be analysed separately. Besides supporting the analysis of neighbourhood effects and the possible mechanisms behind them, the research design and the database created will enable continued monitoring of future metropolitan developments, after the project has ended. The wish to achieve as comprehensive and robust an analysis as possible also means that the data will contain the *entire population*; all the pupils studying at a certain grade in one year will participate in the surveys, making up a dataset of approximately 16,000 individuals. The research focus is on school grades 7–9 when pupils are aged between 13 and 16 years old as the importance of peer groups (as context effects) are thought to be especially strong during the teenage years.

In the project, spatial, socioeconomic and ethnic segregation is understood as a complex system consisting of several levels. These levels are set out as the individual child, their family, their peers, their class at school, the school, the neighbourhood and the wider region in which they reside. These levels are mutually interconnected: the composition of pupils in a school is largely influenced by the population and social structure of the local neighbourhood(s) and region(s) from which they school catchment is comprised. The inclusion of both school and neighbourhood structures as independent entities makes it possible to differentiate between the social contexts in the neighbourhood and school and analyse their effects independently. In the metropolitan region, children are assigned to their neighbourhood school based on catchment areas, but the municipalities also offer a possibility to choose another school. In the HMA just under half of lower secondary school pupils choose a school outside their own catchment area. Recognising that neighbourhood effects can operate at different as well as at multiple spatial scales (see for instance van Ham and Manley 2010) a range of different spatial scales from postal code areas to larger functional and administrative units are included in the geocoding for the data. The MetrOP research design treats the different levels as developmental contexts and will analyse the effects that each level has and seek for links between them (Fig. 10.3).

A number of outcome variables have been included in the analysis in recognition that outcomes rarely exist in a vacuum. Instead, educational outcomes are correlated with health outcomes and behaviours, which in turn are correlated with well-being. Analysing a fuller range of outcomes will help to outline the individual-level connections and interactions between the different aspects of welfare. In the part of the project focusing on education, educational outcomes are treated as the primary outcome variables, and factors related to health and well-being can be

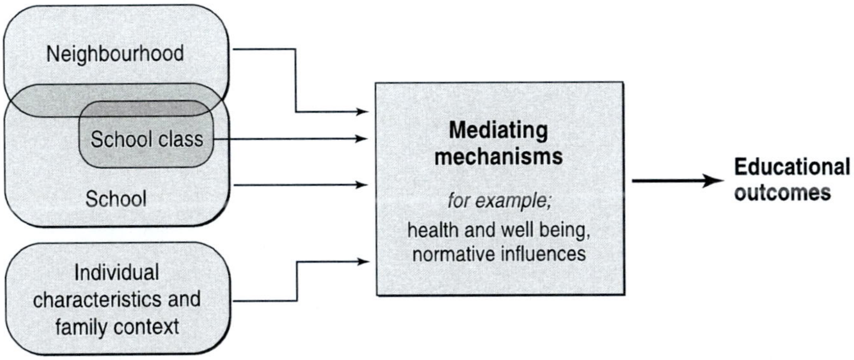

Fig. 10.3 Research design: the developmental contexts, individual characteristics and mediating mechanisms contributing to the formation of educational outcomes

used both as mediating mechanisms and control variables. Ultimately, it is hoped that this approach, will also shed more light on the mechanisms of neighbourhood effects. For example, health-related selection into educational tracks as shown by Koivusilta (2000) may provide a pathway to understand a neighbourhood's contextual characteristics and how they impact on educational outcomes. Similarly, an analysis of pupils' views on the attitudes of peers and family members may provide evidence of normative mechanisms. This can be explored by linking together pupils' own attitudes, pupils' assessment of their peer group's attitudes, and the pupils' background characteristics and observing the changes over time. In short, it is hoped that the richness of background data and individual characteristics combined with a longitudinal analysis of individual change will help the identification, and controlling, of some of the selection issues and lead to the convincing identification of causal processes.

Data and Methods

MetrOP combines a number of existing administrative datasets with new data to be collected through surveys (see Tables 10.1 and 10.2). The existing datasets allow for both individual-level and aggregate-level analyses. These data have been collected by various public institutions. All of the following datasets are separate, but they can be linked together by the common schools and geographical areas. The new data will be produced at the individual level and aggregated to other levels of analysis. Two repeated surveys produce a longitudinal dataset tracking the same cohort of students through lower secondary school undertaken in 2011 and 2013. An anonymous classroom questionnaire lasting around 90min is given to 7th grade pupils in the beginning of the school year over 168 schools containing approximately 16,000 pupils.

Table 10.1 Administrative data used in MetrOP

Dataset	Level
Pupils' health and health behaviour: School Health Promotion Survey (SHP). Family background and use of school health services, chronic diseases, obesity, health complaints, depressive symptoms, school climate, school-related burnout, school grades, working conditions, learning difficulties, health behaviours, bullying, delinquency. Data are self-reported by the pupils, collected yearly since 2000. (Kaltiala–Heino et al. 2000; Konu et al. 2002; Karvonen et al. 2005; Ritakallio et al. 2005)	Child, school, municipality
School resources: Health Promotion Capacity at School (HPCS): School characteristics, health-promotion capacity, resources of welfare and health services at school, working conditions, and school-based statistics on truancy, harassment, bullying and injuries reported by headmasters of comprehensive schools in 2007 and 2010 (Rimpelä et al. 2008)	School, municipality
Urban context: Socio-spatial data and school data. Statistics Finland's geo-coordinate based statistical data; population structure (education, employment, income, ethnic status, single parent households, children from immigrant backgrounds and the use of child welfare services, etc.), and school statistics from Statistics Finland and The Finnish National Board of Education	Neighbourhood, school catchment area, municipality

Table 10.2 New data to be collected for MetrOP through a single survey

Survey elements	Level
Background information: former school and family: 6th grade school, postcode of home address, family structure, ethnic background/mother tongue; education, occupation and unemployment of parents, smoking parents; parental monitoring (Fröjd et al. 2006)	Individual
Health and health behaviours: HBSC symptom checklist (Hetland et al. 2002), Strengths & Difficulties Questionnaire (Koskelainen et al. 2001), chronic diseases, self-reported weight and height, perceived learning difficulties, self-rated general health and health behaviours, bullying (Kärnä et al. 2007), School Burnout Inventory (Salmela–Aro et al. 2009) and the School Well-Being Model (Konu et al. 2002).	Individual
Educational outcomes and attitudes: 1. School marks in key academic subjects 2. Educational outcomes: PISA-type criterion-based tests for assessing the academic achievement in mathematics (Mattila 2005; Niemi 2008) and Finnish/Swedish languages (Lappalainen 2003, 2008) 3. FILLS-scales: competence and beliefs: cognitive competence and educational attitudes, including formal thinking, ability to use a rule where arithmetical operations are used, deductive reasoning, learning motivation, self-efficacy, academic self-concept and self-esteem, assessment of peer group attitudes, educational aspirations and expectations, assessment of parental attitudes (Hautamäki et al. 2002, 2005, 2006)	Individual
Mechanisms of neighbourhood and school effects: The other sections of the survey contain questions that can be used in the analysis of mechanisms (such as assessment of peer group attitudes), but there are also specifically designed questions targeted for this end. These include assessment of own and parents' social networks in the neighbourhood, time spent with friends from school/neighbourhood, trust and fears in the neighbourhood, adult supervision in the neighbourhood etc.	Individual

The cohort will be followed up by repeating the survey in 2013, when they have reached the 9th grade, producing a longitudinal dataset. Low level administrative and geocoding will be included to enable contextual characteristics to be linked. To ensure the validity of the data, all of the major parts in the new survey are internationally established and have been used before in the HMA. Most of the datasets, including the new one, will contain the entire population of the target group (children age 13–16 years old) within the HMA.

The analysis of data is built on interdisciplinary collaboration, and the research group brings together a wide array of tools for data analysis, consisting of teams of geographers, welfare researchers, educational experts and statisticians. The data will be analysed using statistical methods including multi-level modelling and through the use of geographical analysis tools (GIS). The most appropriate spatial level (municipalities, schools, class-room contexts) which captures the strongest variation in health, health behaviours and educational outcomes will also be captured by the use of multi-level modelling. Furthermore, a multi-level modelling tool will be used for identifying the statistical connections between different phenomena in the multiple spatial scales available in the data. The extent to which differences in these components locate where they do e.g. in the municipal and school levels, will be further analysed (Goldstein 2003; Karvonen et al. 2005).

Insights into the causal nature of associations between neighbourhood and school characteristics and individual-level educational outcomes are sought in several ways: firstly, controlling for observed characteristics of the pupils and their families that may be related both to neighbourhood and school selection and educational outcomes; secondly by including questions measuring potential pathways of contextual effects, and; thirdly, utilising the longitudinal design, and applying panel-data methods to control for unobserved and potentially confounding characteristics. The emphasis will be placed on the first two ways. The information regarding the health and well-being of the pupils is used both to control for individual characteristics in the educational models and in order to gain insight into the mechanisms of contextual effects.

Implementation and Expected Results

Within MetrOP, there are four sub-studies that are designed to yield information regarding the processes of urban segregation, school segregation and the potential connections between individual outcomes and neighbourhood context. The first sub-studies are carried out before the longitudinal dataset is complete. From the viewpoint of analysing contextual effects only when the longitudinal data are available will the major contributions of the MetrOP project be realised. The first sub-studies will set the context and help to test the statistical tools that are used. The first sub-studies will also provide information about the significant individual-level connections between the various factors, such as health behaviour and educational attitudes, and help formulate the analysis in the final sub-study.

The Development of the Socio-Spatial Differentiation from 1990 to 2009 (2010–2011)

The first sub-study sets the research context and provides a benchmark for the outcomes of the subsequent sub-studies. The objective is to examine changes in the patterns of socio-economic and health indicators, patterns of health behaviours and patterns of educational outcomes for neighbourhoods and schools. The analyses will answer the questions of whether there is a trend towards increasing segregation, and if there is a correlation between the trends in different indicators list above. These analyses will significantly add to the literature as longitudinal analysis of differentiation in these factors at the school catchment area level have not been carried out previously. The analyses will also help uncover the processes of differentiation and show how they relate to changes in resources and policies. As described above, this sub-study is based on existing administrative data combining school surveys measuring education and health.

Interaction between Health and Educational Outcomes and the Effect of School Choice on School Segregation in a Cross-Sectional Setting (2011 Onwards)

The second sub-study analyses the interaction of education and health at an individual level using the new survey data. The objective is to study the relationship between indicators of health and educational outcomes and to examine, whether this relationship is modified by individual and neighbourhood socio-economic and school characteristics. The study will also seek to determine how health, health behaviours and educational outcomes are related to each other and how they relate to the individual background. To counter selection effects, the effect of the choice of comprehensive school will be included and the analysis. This analysis will determine if the choice of school increases differentiation in educational outcomes or health and health behaviours, compared to a hypothetical setting where all children would attend the nearest school in their own catchment area. To enable this, all pupils in the 7th grade are reallocated to their nearest school and the student base and outcomes are compared to the actual, observed student base and outcomes of the schools.

Neighbourhood and School Effects on Health and Education Outcomes in a Cross-Sectional Setting (2011 Onwards)

The objective of this sub-study is to produce preliminary findings on neighbourhood and school effects by using the multilevel aspect of the data. The findings will provide input for the next final sub-study and for developing interventions and support mechanisms to promote children's learning, health, and wellbeing outcomes. The possibility of neighbourhood and school effects on all the outcomes will be analysed through the combination of updated data from the first sub-study with the new survey data. The analysis will also focus on finding the mechanisms behind the neighbourhood and school effects. Specific questions to be addressed include: How are the neighbourhood and school characteristics related to the differences in individual-level health and educational outcomes? What are the connections between factors constant throughout the region and between genders and social backgrounds? Can different tendencies be observed in more segregated neighbourhoods or school catchments? Are some groups of pupils more vulnerable to certain effects than others?

In-Depth Study of Neighbourhood and School Effects and Changes in School Differentiation During the Transition Through the Lower Secondary School – A Longitudinal Analysis (2013/2014)

The final sub-study is the most important in terms of analysing contextual effects and combines the all the previous sub-study data including the 2011 school survey with a repeated individual-level survey, when the pupils have passed through secondary school. The objective is to observe individual development during the transition through secondary school and to gain deeper insights into whether neighbourhood and school effects exist and if they do what mechanisms may be operating behind them. Separating the contexts of school, class, and neighbourhood will be a key challenge. The sub-study is also designed to yield information on the development of school-level segregation during secondary school.

The main analysis in this sub-study will be based upon refining the neighbourhood and school effect analyses using the longitudinal data. The results from the first survey are combined with the results from the second survey, using student codes given to each student by the school. Low level administrative and geocoding will enable the linking of school and neighbourhood data. The wealth of individual-level data on background and personal characteristics helps to assess selection bias in the study, as well as allowing for the search for possible explanatory factors or mechanisms behind observed effects in students. Tracking the students will also allow for observing actual changes, at the level of the individual and help to recognise possible causal processes during secondary school and reduce the selection bias, compared to a cross-sectional setting. A secondary analysis examines the changes in socio-spatial segregation and in the differentiation in children's health and educational outcomes between individuals and schools in the transition of the cohort of the pupils between the 7th and 9th grade. Further research questions relate to whether the observed changes are related to the pupils' distribution by gender, ethnicity or family background, characteristics of the school or of the school catchment area. All data will be analysed using statistical methods suitable for repeated measurements, along with multilevel modelling.

Discussion: Scientific and Social Contributions

In this chapter, we have described some of the main challenges in (Finnish) neighbourhood effects research. These are mainly related to the availability of suitable data, and the limitations this places on the analytical tools, as well as on the possibility of demonstrating genuine causal processes and mechanisms. We have also presented a research project designed specifically as an attempt to tackle some of the analytical problems in this field of research.

In their analysis of the European research base, Friedrichs et al. (2003) outlined two basic methodological approaches to the study of neighbourhood effects: the comparative neighbourhood case study and the analysis of non-experimental, longitudinal databases. In many ways, the research project presented in this chapter combines these approaches by collecting longitudinal data for different types of neighbourhoods; allowing for detailed analysis of individuals over time, as well as a comparative study of the populations in different neighbourhoods. The wealth of individual data collected in this study exceeds the depth and breadth of any standard register data available for statistical research, and the large research area and the inclusion of the whole metropolitan age cohort exceeds the size and scope of small groups researched in typical case studies.

We have discussed several ways in which the new research project presented in this chapter tries to improve upon existing Finnish research and also neighbourhood effects research on educational outcomes more generally. Firstly, we have emphasised that when studying educational outcomes of young people, the effects of both neighbourhoods and schools should be analysed. Studying one without the other may lead to biased conclusions. To achieve this it is of primary importance to be able to differentiate the pupils who attend their own neighbourhood school from the ones choosing a school in another part of the city (and therefore another neighbourhood), and assign each individual to a specific neighbourhood and school. Similar, it is crucial to include identifiers of school class membership, as they are the primary context in the everyday functioning of schools, and they have been shown to be strongly differentiated within and between schools.

Secondly, certain basic data requirements should be met in order to make causal claims about contextual effects. These include (1) an individual level longitudinal design, (2) identification of several spatial contexts: at least one level of neighbourhood and schools but preferably also classes within schools, and (3) measurement of factors that may influence both the selection into the contexts and also influence the educational outcomes. The last of these requirements can be defined in a more general way as a requirement that a serious effort should be made to deal with the problem of selection bias. This may include both better measurement of relevant factors and the application of certain statistical methods that attempt to overcome the selection problem. However, designing ways to overcome the selection problem is not the only approach that should be taken when studying contextual effects. Another is to study more closely the associations between contexts and individual-level outcomes, which leads to our third point.

The third point relates to attempts to open the 'black box' of neighbourhood effects (see Jencks and Mayer 1990), and to show what happens between an individual's exposure to a context and the measurement of that individual's outcomes. In the research project presented, this means attempts to measure educational norms in the school context, for example, and a combined analysis of several contexts in order to show their interconnectedness. Friedrichs et al. (2003) summed up the assumed mechanisms behind neighbourhood effects in four basic categories: neighbourhood resources (e.g. services), model learning (e.g. nature of peer groups), socialisation and collective efficacy (e.g. sense of control in public space) and resident perceptions of deviance (e.g. crime) (see also the chapter by Galster 2011 in this volume). The design of the MetrOP-project should, theoretically allow study of neighbourhood effects in all four of these categories. Neighbourhood resources can be analysed through detailed information on neighbourhoods, their services and resources and educational culture in schools. Model learning can be explored through the information collected from the pupils themselves and could be derived from questions dealing with their assessment of peer group attitudes, and how they use time in school and in their neighbourhood. As the last point, collective efficacy is studied through a set of questions targeted to measure the pupils' views on, amongst other things, social control in the neighbourhood.

Richer data also allows the crossing of disciplinary boundaries. Separate traditions of research on contextual effects on young people have co-existed in educational research, health research and urban research. Combining these different traditions in a single research project may bring out some insights that would not have been reached in a more narrowly focused setting, and much can be learned from literatures and research traditions from other disciplines. In practical terms, data related to domains such as health and well-being may help in controlling for the individual backgrounds of the young people or in showing what kind of mechanisms are at work.

As a more unique point, the project presented provides an opportunity to explore both the existence and mechanisms of neighbourhood effects in the context of a Nordic welfare state with growing urban segregation. The Finnish context contrasts dramatically with the American context from which much of the neighbourhood effects literature originates. Even in European terms, and although the spatial differences and school segregation are growing, the metropolitan region is still relatively unsegregated. Ultimately, the MetrOP project will bring new insights for the understanding of neighbourhood effects, and to the international research base on which factors contribute to the PISA success that Finland has, so far, been able to boast. The project will shed some light on the interplay between neighbourhood and school segregation contextual effects and children's educational outcomes. There is already evidence of growing gaps in educational outcomes between students and schools, and this project provides a chance to quantify not only these differences as well as the contributing factors and the change in time. The project provides a longitudinal analysis of the last 15 years in the aggregate level of neighbourhoods and schools, as well as following one cohort of individual students through lower secondary school.

Beyond adding to the academic knowledge-base, this research aims at producing tools for monitoring and developing the metropolitan school system in Finland. The goal is to provide simple welfare indicators and data collection and management tools that can be used for observing the developments within the HMA after the project has ended. The inter-disciplinary research setting combining a wealth of research traditions and tools is designed to contribute to the creation of social, analytical tools through the perspectives and traditions in different scientific fields and public institutions. These indicators are particularly valuable for administrative purposes within the region, and they may also be used for comparative research in the future.

The current research setting is not designed to be internationally compatible, although many of the scales used in the survey are internationally developed and tested. However, international comparison would provide a potentially valuable theoretical and administrative input. Repeating the surveys within different national contexts with different welfare structures might yield valuable insights into understanding if and how a neighbourhood and (or) school context influences the educational and health outcomes of children.

References

Ammermueller, A., & Pischke, J. (2006). *Peer effects in European primary schools: Evidence from PIRLS*. London: Centre for the Economics of Education, London School of Economics.

Andersson, R., & Molina, I. (2003). Racialization and migration in urban segregation processes – Key issues for critical geographers. In K. Simonsen & J. Öhman (Eds.), *Voices from the North* (pp. 261–282). Aldershot: Ashgate.

Bernelius, V. (2005). *Onko oppimistulokset valettu betoniin? Tutkimus Helsingin kaupunkirakenteen ja peruskoulujen oppimistulosten välisestä yhteydestä ja kouluvalintojen vaikutuksista* (Are educational outcomes real estate? A study of the connection between urban segregation and educational outcomes in public schools, and the effects of the public school market in Helsinki). Masters Thesis, University of Helsinki.

Bernelius, V. (2008). *Lähi(ö)koulu – Helsingin koulut ja kaupunginosat EU:n Urban II –ohjelman kouluhankkeissa.* (The school in the suburbs – The schools and neighbourhoods of Helsinki in the EU Urban II school initiatives). *Tutkimuksia 3*. Helsinki: City of Helsinki Urban Facts.

Bernelius, V. (2010). *Alueellinen eriytyminen heijastuu kouluihin* (Urban segregation affects schools). Peruskoulujen oppimistulokset ja oppilaiden hyvinvointi eriytyvällä Helsingin seudulla: MetrOP–tutkimus 2010–2013 (pp. 19–23). *Tutkimuksia B1*. Helsinki: Helsinki University Press.

Bernelius, V. (forthcoming) *Osoitteenmukaisia oppimistuloksia? Kaupunkikoulujen eriytymisen vaikutus peruskoululaisten oppimistuloksiin Helsingissä* (Street numbers, school grades? The effects of school segregation in Helsinki). Research article.

Bramley, G., & Karley, N. K. (2007). Homeownership, poverty and educational achievement: School effects as neighbourhood effects. *Housing Studies, 22*(5), 693–721.

Brännström, L. (2008). Making their mark: The effects of neighbourhood and upper secondary school on educational achievement. *European Sociological Review, 24*(4), 463–478.

Burgess, S., Wilson, D., Briggs, A., & Piebalga, A. (2008). Segregation and the attainment of minority ethnic pupils in England. CMPO working paper 08/204.

Butler, T., & Hamnett, C. (2007). The geography of education: Introduction. *Urban Studies, 44*(7), 1161–1174.

Crane, J. (1991). The epidemic theory of Ghettos and neighborhood effects on dropping out and teenage childbearing. *American Journal of Sociology, 96*(5), 1226–1259.

Diez–Roux, A. V. (1998). Bringing context back into epidemiology: Variables and fallacies in multilevel analysis. *American Journal of Public Health, 88*(2), 216–222.

Donnelly, C. (2000). In pursuit of school ethos. *British Journal of Educational Studies, 48*(2), 134–154.

Friedrichs, J., Galster, G., & Musterd, S. (2003). Neighbourhood effects on social opportunities: The European and American research and policy context. *Housing Studies, 18*(6), 797–805.

Fröjd, S., Kaltiala–Heino, R., & Rimpelä, M. (2006). The association of parental monitoring and family structure with diverse maladjustment outcomes in middle adolescent boys and girls. *Nordic Journal of Psychiatry, 61*(4), 296–303.

Galster, G. (2008). Quantifying the effect of neighbourhood on individuals: Challenges, alternative approaches, and promising directions. *Schmollers Jahrbuch, 128*(1), 1–42.

Galster, G. (2011). The mechanism(s) of neighbourhood effects: Theory, evidence, and policy implications. In M. van Ham, D. Manley, N. Bailey, L. Simpson, & D. Maclennan (Eds.), *Neighbourhood effects research: New perspectives*. Dordrecht: Springer.

Goldstein, H. (2003). Multilevel modelling of educational data. In D. Courgeau (Ed.), *Methodology and epistomology of multilevel analysis* (pp. 25–42). London: Kluwer.

Hautamäki, J., Arinen, P., Eronen, S., & Hautamäki, A. (2002). *Assessing learning–to–learn. A framework. Evaluation Reports 4*. Helsinki: Finnish National Board of Education.

Hautamäki, J., Kupiainen, S., Arinen, P., & Hautamäki, A. (2005). *Oppimaan oppiminen ala–asteella 2. Tilanne vuonna 2003 ja muutokset vuodesta 1996*. (Learning to learn in the primary schools. The situation in 2003 and the changes from the year 1996). *Oppimistulosten arviointi 1*. Helsinki: Finnish National Board of Education.

Hautamäki, J., Kupiainen, S., Arinen, P., & Hautamäki, A. (2006). Learning–to–learn assessment in Finland – Versatile tools to monitor and improve effectiveness and equity of the educational system. In R. Jakku–Sihvonen & H. Niemi (Eds.), *Research–based teacher education in Finland. Research in Educational Science 25*. Turku: Finnish Educational Research Association.

Hetland, J., Torsheim, T., & Aaro, L. E. (2002). Subjective health complaints in adolescence: Dimensional structure and variation across gender and age. *Scandinavian Journal of Public Healthh, 30*(3), 223–230.

Jakku–Sihvonen, R., & Kuusela, J. (2002). Mahdollisuuksien koulutuspolitiikan tasa–arvo (Equality of opportunities in the educational policies). *Oppimistulosten arvointeja 7*. 2nd edn. Helsinki: Finnish National Board of Education.

Jencks, C., & Mayer, S. E. (1990). The social consequences of growing up in a poor neighborhood. In L. E. Lynn & M. G. McGeary (Eds.), *Inner–city poverty in the United States* (pp. 111–186). Washington, DC: National Academy Press.

Kaltiala–Heino, R., Rimpelä, M., Rantanen, P., & Rimpelä, A. (2000). Bullying at school – An indicator of adolescent at risk for mental disorders. *Journal of Adolescence, 23*(6), 661–674.

Karisto, A., & Montén, S. (1996). Lukioon vai ei: Tutkimus alueellisista eroista helsinkiläisten lukionkäynnissä ja lukiolakkautusten vaikutuksista (College or not: A study of neighborhood differences in the college education and effects of college dropouts). *Tutkimuksia 6*. Helsinki: City of Helsinki Urban Facts.

Kärnä, A., Salmivalli, C., Poskiparta, E., & Voeten, M. (2007). *Bullying and victimization: variation and risk factors at individual, classroom and school levels*. The 13th European Conference on Developmental Psychology, Germany, August 21–25, 2007.

Karvonen, S., & Rahkonen, O. (2002). Kuka vastustaa koulutusta: Kouluvastaisuuden alueelliset erot Helsingissä (Who's against education: Local differences in school attitudes in Helsinki). *Yhteiskuntapolitiikka, 67*(4), 324–332.

Karvonen, S., Vikat, A., & Rimpela, M. (2005). The role of school context in the increase in young people's health complaints in Finland. *Journal of Adolescence, 28*(1), 1–16.

Kauppinen, T. M. (2002). The beginning of immigrant settlement in the Helsinki metropolitan area and the role of social housing. *Journal of Housing and the Built Environment, 17*(2), 173–197.

Kauppinen, T. M. (2004). Asuinalueen ja perhetaustan vaikutukset helsinkiläisnuorten keskiasteen tutkintojen suorittamiseen (The effects of neighborhood and family background in completing secondary degrees among youth in Helsinki). *Tutkimuksia 6.* Helsinki: City of Helsinki Urban Facts.

Kauppinen, T. M. (2007). Neighborhood effects in a European city: Secondary education of young people in Helsinki. *Social Science Research, 36*(1), 421–444.

Kauppinen, T. M. (2008). Schools as mediators of neighbourhood effects on choice between vocational and academic tracks of secondary education in Helsinki. *European Sociological Review, 24*(3), 379–391.

Koivusilta, L. (2000). Health–related selection into educational tracks. A mechanism producing socio–economic health differences. *Tutkimusjulkaisuja D 394.* Turku: Turku University Press.

Konu, A. I., Alanen, E., Lintonen, T., & Rimpelä, M. (2002). Factor structure of the school well–being model. *Health Education Research, 17*(6), 732–742.

Kortteinen, M., & Vaattovaara, M. (2007). Miten Helsingin käykään? *Yhteiskuntapolitiikka, 72*(2), 137–145.

Koskelainen, M., Sourander, A., & Vauras, M. (2001). Self–reported strengths and difficulties in a community sample of Finnish adolescents. *European Child & Adolescent Psychiatry, 10*(3), 180–185.

Kupari, P. (2005). Kotitausta näkyy matematiikan oppimistuloksissa (Social bacground is reflected to PISA outcomes in maths). In P. Kupari & J. Välijärvi (Eds.), *Osaaminen kestävällä pohjalla – PISA 2003 Suomessa* (pp. 115–127). Jyväskylä: Jyväskylä University Press.

Kuusela, J. (2006). Temaattisia näkökulmia perusopetuksen tasa-arvoon (Thematic insights to equality in primary education). *Oppimistulosten arviointi 6/2006.* Finnish National Board of Education. 144 s.

Kuusela, J. (2010). MetrOP–alueen kouluista toisen asteen yhteishakuaineiston perusteella (Schools in the MetrOP–area, analysis based on secondary education choice register). In M. Rimpelä & V. Bernelius (Eds.), *Peruskoulujen oppimistulokset ja oppilaiden hyvinvointi eriytyvällä* Helsingin seudulla: MetrOP–tutkimus 2010–2013 (pp. 38–43). *Tutkimuksia B1.* Helsinki: Helsinki University Press.

Kuusela, J., Etelälahti, A., Hagman, Å., & Hievanen, R. et al. (2008). *Maahanmuuttajaoppilaat ja koulutus – tutkimus oppimistuloksista, koulutusvalinnoista ja työllistämisestä* (Immigrant pupils and education – A study of educational outcomes, school choice and employment). Helsinki: National Board of Education.

Lappalainen, H. (2003). Osaat lukea – miten osaat kirjoittaa? (You can read – how well can you write? – Evaluation of educational outcomes). *Oppimistulosten arviointi 4.* Helsinki: Finnish National Board of Education.

Lappalainen, H. (2008). On annettu hyviä numeroita. (They have given good grades – Evaluation of educational outcomes). *Oppimistulosten arviointi 3.* Oppimistulosten arviointi Helsinki: Finnish National Board of Education.

Mattila, L. (2005). Perusopetuksen matematiikan kansalliset oppimistulokset 9. vuosiluokalla 2004 (National evaluation of Maths outcomes in the 9th grade.) *Oppimistulosten arviointi 2.* Helsinki: Finnish National Board of Education.

Musterd, S., & Ostendorf, W. (2005). Social exclusion, segregation and neighborhood effects. In Y. Kazepov (Ed.), *Cities of Europe: Changing contexts, local arrangements, and the challenge to urban cohesion* (pp. 170–189). Oxford: Blackwell Publishing.

Musterd, S., & Ostendorf, W. (2007). Spatial segregation and integration in the Netherlands. In K. Schönwälder (Ed.), *Residential segregation and theintegration of immigrants: Britain, the Netherlands and Sweden* (pp. 7–40). Berlin: Social Science Research Center Berlin.

Nash, R. (2003). Is the school composition effect real? – A discussion with evidence from the UK PISA data. *School Effectiveness and School Improvement, 14*(4), 441–457.

Niemi, E. (2008). Matematiikan oppimistulosten kansallinen arviointi 6. vuosiluokalla vuonna 2007 (National evaluation of Maths outcomes in the 6th grade in 2007). *Oppimistulosten arviointi 1.* Helsinki: Finnish National Board of Education

Plotnick, R. D., & Hoffman, S. D. (1999). The effect of neighborhood characteristics on young adult outcomes: Alternative estimates. *Social Science Quarterly, 80*(1), 1–18.

Rankin, B., & Quane, J. M. (2002). Social context and urban adolescent outcomes: The interrelated effects of neighborhoods, families, and peers on African–American youth. *Social Problems, 49*(1), 79–100.

Rimpelä, M., & Bernelius, V. (Eds.) (2010). Peruskoulujen oppimistulokset ja oppilaiden hyvinvointi eriytyvällä Helsingin seudulla: MetrOP–tutkimus 2010–2013 (Educational outcomes and the welfare of pupils in the segregating Helsinki Region: MetrOP–research 2010–2013). *Tutkimuksia B1*. Helsinki: Helsinki University Press.

Rimpelä, M., Rigoff, A., Kuusela, J., & Peltonen, H. (Eds.) (2008). *Hyvinvoinnin ja terveyden edistäminen peruskouluissa 1–2* (Promotion of wellbeing and health in comprehensive schools based on surveys to comprehensive schools 1–2). Helsinki: Finnish National Board of Education.

Ritakallio, M., Kaltiala–Heino, R., Kivivuori, J., & Rimpelä, M. (2005). Delinquent behaviour and depression in middle adolescence: A Finnish community sample. *Journal of Adolescence, 28*(1), 155–159.

Robertson, D., & Symons, J. (2003). Do peer groups matter? Peer group versus schooling effects on academic attainment. *Economica, 70*(277), 31–53.

Rutter, M., & Maughan, B. (2002). School effectiveness findings 1979–2002. *Journal of School Psychology, 40*(6), 451–475.

Salmela–Aro, K., Kiuru, N., Leskinen, E., & Nurmi, J. (2009). School burnout inventory (SBI): Reliability and validity. *European Journal of Psychological Assessment, 25*(1), 48–57.

Sampson, R. J. (2008). Moving to inequality: Neighborhood effects and experiments meet social structure. *American Journal of Sociology, 114*(1), 189–231.

Sellström, E., & Bremberg, S. (2006). The significance of neighbourhood context to child and adolescent health and well–being: A systematic review of multilevel studies. *Scandinavian Journal of Public Health, 34*(5), 544–554.

Small, M. L., & Feldman, J. (2011). Ethnographic evidence, heterogeneity, and neighbourhood effects after moving to opportunity. In M. van Ham, D. Manley, N. Bailey, L. Simpson, & D. Maclennan (Eds.), *Neighbourhood effects research: New perspectives*. Dordrecht: Springer.

Sykes, B., & Musterd, S. (2010). Examining neighbourhood and school effects simultaneously: What does the Dutch evidence show? *Urban Studies*. doi:10.1177/0042098010371393.

Thrupp, M., Lauder, H., & Robinson, T. (2002). School composition and peer effects. *International Journal of Educational Research, 37*(5), 483–504.

Vaattovaara, M., & Kortteinen, M. (2003). Beyond polarisation versus professionalisation? A case study of the development of the Helsinki Region, Finland. *Urban Studies, 40*(11), 2127–2145.

van Ham, M., & Manley, D. (2010). The effect of neighbourhood housing tenure mix on labour market outcomes: A longitudinal investigation of neighbourhood effects. *Journal of Economic Geography, 10*(2), 257–282.

Vilkama, K. (2008). Maahanmuuttajaväestön kasvu alueittain 1992–2007 (The growth of immigrant population in the neighbourhoods of Helsinki 1992–2007.) In Bernelius, Venla. Lähi(ö) koulu – Helsingin koulut ja kaupunginosat EU:n Urban II –ohjelman kouluhankkeissa. Helsingin kaupungin tietokeskus, *Tutkimuksia 3* (pp. 20–21).

West, P., Sweeting, H., & Leyland, A. (2004). School effects on pupils' health behaviours: Evidence in support of the health promoting school. *Research Papers in Education, 19*(3), 261–291.

Zimmer, R. W., & Toma, E. F. (2000). Peer effects in private and public schools across countries. *Journal of Policy Analysis & Management, 19*(1), 75–92.

Chapter 11
Recasting Research on 'Neighbourhood effects': A Collaborative, Participatory, Trans-National Approach

Michael Darcy and Gabrielle Gwyther

Introduction

This chapter discusses an alternative approach to examining links between individual disadvantage and the neighbourhood context. Although 'neighbourhood effects' have long been of interest to policy makers and academics, research into the issue has tended to draw on conventional empirical and case study methods to elicit insights and understandings. Most empirical studies purport to be spatial investigations, yet this often means modelling the effects of neighbourhood level measures of poverty on individual outcomes, rather than investigating the effects of genuine spatial factors. We argue that such studies are vulnerable to three types of flaws – omitted variable bias, implied causality and ecological fallacy – and require the incorporation of situated local knowledge to understand the dynamics between spatial factors such as comparative location and spatial scales, social and physical infrastructure, development history and contemporary culture, local economy and governance, levels of public and private investment, and the implications of government policy on the place. Some more qualitatively oriented studies – in their quest to present different perspectives in the production of place-based knowledge – have attempted to incorporate residents' experiences and insights. This approach, however, remains premised on an academic convention of knowledge production which inherently partitions community and academic knowledge bases. Moreover, insofar as they are required to respond to questions and propositions framed and formulated in the language and context of academic research or policy making, the approach has the tendency to represent research participants as an (often invisible)

M. Darcy (✉)
School of Social Sciences, University of Western Sydney, Penrith, NSW, Australia
e-mail: m.darcy@uws.edu.au

G. Gwyther
School of Social Sciences, University of Western Sydney, Penrith, NSW, Australia
e-mail: g.gwyther@bigpond.com

M. van Ham et al. (eds.), *Neighbourhood Effects Research: New Perspectives*,
DOI 10.1007/978-94-007-2309-2_11, © Springer Science+Business Media B.V. 2012

'other' (Subedi and Rhee 2008). The lived experience of residents is thus objectified, examined, interpreted and reported through the lens of what middle class intellectuals consider to be 'disadvantage'.

Place and Disadvantage – Academic and Policy Discourses

Academic and policy interest in the social consequences of concentrated disadvantage is often said to have been sparked by the publication of William Julius Wilson's *The Truly Disadvantaged* (1987), but actually has a much longer history encompassing the 'social hygiene', slum clearance and 'garden city' movements of the early twentieth century (see Pugh 1976), through Orwell's highly reflexive and influential sociology in *The Road to Wigan Pier* (1937), and importantly Oscar Lewis's (1961, 1998) description of the 'culture of poverty' (of which more later). Wilson's work came in response to the specific consequences of deindustrialisation and the flight of jobs and workers from many U.S. inner cities in the 1980s. While he is deeply concerned about 'concentration effects' in the residual populations leading to isolation of the urban poor from the institutions and opportunities of the metropolis, Wilson explicitly rejects the idea of cultural causation of poverty and insists that local conditions and the social practices of residents of poor areas cannot be understood independently of the macro social and economic forces which shape them (Arthurson 2002).

In what could be described as a 'spatial turn', the impact of 'place' – or more specifically, residential location – on the social and economic situation of households has emerged as a fundamental concern of social science and social policy over recent decades. Illustrative constructs including 'social exclusion' and 'social capital' have been developed by social scientists to describe, and attempt to explain, the apparent persistence of social disadvantage amongst poor households residing together in defined 'poor areas' of major cities. However, measurement of these constructs, and demonstration of causal links between place and disadvantage or opportunity, has proved elusive, largely because of their capacity to be colonised by normative or prescriptive discourses.

Social exclusion and inclusion are 'dualising' discourses which support the concept of a 'mainstream' or 'normal' society to which all should aspire to belong. As Levitas (1998) points out, the construct of social exclusion imagines a boundary, and focuses attention on those outside it rather than on the features of society which systematically generate widespread poverty and disadvantage. The metaphor of 'exclusion' has contributed to the construction of disadvantage as a largely spatial phenomenon, drawing attention to the location and concentration of people in poverty, usually within small and defined areas, as evidence of a causal link between place and poverty which is frequently understood to be mediated by local cultures.

To the extent that exclusion, and hence poverty, is perceived as (at least partly) a function of residential propinquity, then its persistence is understood as primarily a process of cultural reproduction strongly reminiscent of Oscar Lewis's (1961)

'Culture of Poverty' theory referred to above – social exclusion/poverty persists in certain geographically defined areas because people are exposed to dominant local values and locally acceptable behaviours which inhibit personal economic advancement and independence. That is, unemployment is not due to structural problems but due to "a cultural commitment to dysfunctional and irrational values resulting in the wish of the 'underclass' to follow alternative values counter to the norms of society" (van Ham and Manley 2010, p. 259). The insidious nature of the Culture of Poverty discourse itself is particularly difficult for stigmatised residents to surmount. As outlined above, Wilson (1987) was a key critic of the culture of poverty, and argued that it was structural economic factors that brought about poverty. However, it is his identification of an urban 'underclass', whose values and behavioural norms reinforced spatial and social separation, which has done the most to stimulate this field of research.

The paradigm of neighbourhood effects rests on the notion that the geographic propinquity of large numbers of disadvantaged households creates a social or cultural dynamic at the local level which compounds and perpetuates their disadvantage, while conversely 'social capital' is understood as the form of social interaction/relationships which leads to or sustains economic well-being, through role modelling, networking and civic participation. Social capital is thus commonly associated with 'face-to-face' relationships and interaction in the 'non-economic' sphere of local neighbourhoods or communities, and so also has a spatial character (Defilippis 2007). Regardless of what we know about the operation of affinity and economic networks, especially in the cyber-age, physical space and the built form are deeply embedded in ideas concerning the production of social (civic) behaviour and anti-social behaviour. Aspects of residential areas, including questions of density, distinctive building design and specific combinations of public and private space are all assumed to be implicated in the production of social capital, as evidenced for example by the New Urbanist principles guiding redevelopment of public housing and new town development in the U.S. and Australia.

The fact that the places where prosperous households reside appear to manifest better commercial facilities and public infrastructure – and house richer people – is assumed to be evidence of local social capital at work. Conversely, when applied to the analysis of public housing space this logic has deemed such spaces to lack social capital, or to have the wrong type of social capital ('bonding' or even 'negative' social capital) which cannot lead to economic capital because at best it is unproductive and at worst produces anti-social behaviour.

A lively international debate has emerged in academic and policy circles concerning what to do about 'disadvantaged places'. New books have appeared (Allen 2008; Bennett et al. 2006; Goetz 2003); two international journals have produced special issues in 2008 , a number of national and international conferences have been convened, and a welter of scholarly and research articles have appeared in leading journals. In the policy domain, state policies which historically contributed to, or even actively pursued, the geographical separation and containment of poor households have more recently been concerned with the emergence of an urban economic and social 'underclass' and now emphasize dispersal or penetration of low income communities as a way of generating social order in disadvantaged neighbourhoods (Uitermark et al. 2007).

De-concentration of poverty, and the development of mixed income communities, has become a central theme of (public or social) housing policy across the English speaking world even though there is little convincing evidence that neighbourhood effects exist. Nevertheless, a key premise of de-concentration is that poor communities lack the 'social capital' required to build wealth, that they are 'socially excluded' and that by living in more economically mixed areas, poor people will learn skills and build social networks necessary to engage in cultural practices which will improve their lives (deFilippis 2007). The problem is that if neighbourhood effects do not exist, introducing social mix into neighbourhoods is not going to 'improve' the lives of people living in them. In the meantime, the process of realising social mix requires removing residents from their neighbourhood, taking away their local knowledge and disrupting their social networks.

Whether based in Wilson's underclass or Lewis's culture of poverty, critics of so-called 'neighbourhood effects' have identified serious conceptual and methodological difficulties (Lupton 2003; Manley and van Ham 2011 in this volume) and, at worst, see its underlying premise as culturally bound and ideologically driven (Bauder 2002). Despite the contentious nature of the concept and the equivocal evidence produced to date, housing managers and policy makers on three continents have enthusiastically adopted the theoretical notion of neighbourhood effects as a rationale for radical dispersal and redevelopment projects, particularly focussed on public housing neighbourhoods. The fact that many of these projects are structured around the privatisation of public assets, and the effective gentrification of valuable and well located urban areas, appears to lend weight to the view that social scientists have been captured by an ideological policy agenda. Researchers who are genuinely concerned with understanding the connections between place and socio-economic outcomes for poor households need to take these criticisms seriously and to embrace approaches which do not replicate and reinforce the phenomenon of 'exclusion' which they purport to study.

Causes and Effects

Inherent in research that seeks to establish 'effects' are principles of logical positivism. That is to say, it relies on the notion that observed phenomena have identifiable causes which can be verified by rigorous observation and measurement. Debates over the appropriateness of positivist science to understanding social phenomena have raged for decades (see Phillips 1987 for a useful review) and this is not the place to rehearse them. However, even within the frame of positivism, neighbourhood effects research, as commonly practised, has serious flaws.

Firstly, the majority of studies are based on cross-sectional individual or household data extracted from census, survey, and administrative data sets. Sometimes this data is aggregated to indicate the level or relative concentration of disadvantage within a chosen boundary and to establish correlations between pre-selected variables. Studies using aggregated data can be said to represent the *compositional* attributes of a given population within the boundary, but cannot predict or explain outcomes for individuals and are thus vulnerable to the ecological fallacy. This type of data often includes limited contextual and spatial characteristics of place. When used as individual level data, researchers are at best able to identify correlations between individual level outcomes and area characteristics. To be able to properly sequence supposed causes and effects it is necessary to use longitudinal data (see for some recent examples: van Ham and Manley 2010; Musterd, et al. 2003). Research of this kind is at best able to offer proof of an unsurprising association between population characteristics and social outcomes – it cannot speak to the issue of causality, but takes it as implied. Longitudinal studies which take account of the temporal sequencing of their observations are far better placed to speculate about causal relationships, but still suffer from the problem of implied causality insofar as they rely on statistical data which cannot possibly take account of all the variables which may lead to a family moving house, or improving its economic circumstances – or make conclusions about the importance or direction of dependency between them. Moreover, few studies attempt to account for how and why individual households came to be located in particular areas in the first place and to the extent that this was an outcome of the housing market or a public housing allocation process, questions can be raised as to whether persistent disadvantage is an effect or a cause (see also Hedman and van Ham 2011 in this volume). These issues are described as the 'omitted variable' problem (Galster 2003).

Secondly, the mechanisms by which effects follow from neighbourhood characteristics are assumed rather than discovered by this type of social science whether quantitative or qualitative. Because most studies consider specific settings, usually pre-determined through secondary data analysis to be 'disadvantaged', they have no way of identifying these processes or the particular local factors and conditions which might mediate or modify them.

Nonetheless, a number of possible mechanisms by which the disadvantage of individual households might be compounded by residential propinquity have been suggested in the literature, and are summarised by Jencks and Mayer (1990) and reviewed by George Galster (2011, in this volume). These fall into two categories: those under which concentration of disadvantage is thought to produce negative effects for those households; and those whereby mixing of advantaged and disadvantaged households in an area could produce negative outcomes, either for poor or better off households, or both.. The first group includes 'epidemic' or 'contagion' models and 'collective socialisation' theories (referred to above), both of which emphasise the spread of 'negative' values and behaviours through peer influences on individuals, and the ways in which this might be interrupted by the presence of more affluent neighbours, although the latter is less committed to the mathematical predictability of contagion and takes more account of the importance of local conditions and wider social relationships. Also in this category are 'institutional' explanations of neighbourhood effects which rely on the economic and political resources brought to an area by more affluent neighbours and which are said to improve the quality of local services such as schools.

Importantly, as mentioned above, Jencks and Mayer's second category identifies mechanisms under which *de*-concentration might actually compound households' disadvantage by increasing the sense of relative difference that poor households experience, forcing them to compete with more advantaged households for scarce resources, and possibly leading to a heightened feeling of cultural division and even conflict. The bulk of policy driven research in this field appears to ignore these possible negative neighbourhood effects of mixing, and to assume that 'mixed' neighbourhoods are inherently good for poor households.

Despite a growing body of research questioning the extent to which mixed income housing developments lead to new mixed income peer networks (Arthurson 2002; Chaskin and Joseph 2010), the bulk of research in the field does not seek to confirm or dispense with explicit hypotheses, but goes to great lengths to attempt to demonstrate a measureable 'effect'. The explanations offered vary in the degree to which the causes of persistent disadvantage are perceived to be endogenous in local neighbourhoods, and even in the direction of the effect of particular social conditions, but frequently treats the behaviours and relationships of disadvantaged people as evidence of a 'problem'.

Most interest in neighbourhood effects arises from such a problem orientation articulated by 'experts', the 'policy community' and other outsiders, and, as outlined by Bauder (2002), reflects an ideological or normative view that certain social or behavioural traits which are frequently associated with poor neighbourhoods are inherently pathological and represent social dysfunction. Yet many studies have shown that, when asked, residents of disadvantaged neighbourhoods, and especially public tenants, often express a desire to stay in the area, or are at best ambivalent about leaving (Vale 1997; Stubbs et al. 2005) and that for them 'neighbourhood effects' are as often as not experienced as positive. Rather than seeing this as cause for questioning the normative nature of their own research and the policies it supports, some have seen this as simply more evidence of the negative cultural norms in place (Imbroscio 2008). Clearly a more reflexive and engaged approach is needed to avoid such self-fulfilling accounts.

Validating Residents' Experience

In response to the identified limitations and flaws in existing research approaches discussed above, this chapter argues a need for more in-depth qualitative research focussing on people, their context and how people and context interact, and how people experience their environment. The approach proposed in this chapter draws on the alternative social science traditions of phenomenology and participatory action research. Phenomenology is "*the study, in-depth, of how things appear in human experience . . . [*and*] involves the 'bracketing' or laying aside of preconceptions (including ones derived from science), in order to be able to inspect (one's own) conscious intellectual processes more purely*" (Phillips 1987, p. 205). Such an approach will more thoroughly reveal the complexity of experiences of living in areas of poverty concentration. Phenomenology is not simply a different research method, but requires a different theory of knowledge in which the researcher steps down from the position of detached expert to enter and validate the viewpoints and the experience of people who are most closely embedded in the phenomena under study – allowing them to define what is a 'problem', what questions need to be asked, and what associations and dependencies it is most important to explore. Given the characterisation of neighbourhood effects research by some critics as intrinsically ideological, the demands of such a radical approach represent a major challenge for the field. However, if social science is to do more than simply reproduce the power-knowledge systems which underpin existing dominant discourses and explanations of social structure, then that challenge must be accepted.

In the following sections we describe an evolving research project that is methodologically driven by the need to explore the issues of entrenched disadvantage and community and neighbourhood effects, through techniques which attempt to produce alternate ways of knowing, based in the lived experience of residents. This approach does not start from pre-assumed mechanisms or hypotheses, but centres the investigation of neighbourhood effects around the lived experience of those most affected. Underpinned by the paradigm of engaged scholarship and the notion that community contributors are competent agents and experts in their lives, this project is designed using a 'collaborative university – community research' approach whereby community collaborators, as situated knowledge producers, inform each phase of the research, from developing specific research questions, to data collection, analyses and dissemination. This approach enables a foregrounding of residents' own constructions of community, advantage and disadvantage within social housing estates and to realise the perspectives of other residents in comparable situations. Such an approach has the potential to seriously challenge conventional notions of place-based disadvantage and dispersal and redevelopment policies which flow from them. The approach is also an attempt to respond to the fundamental problem for Housing Studies identified by Allen (2009, p. 55): *"Challenges to Housing Studies' knowledge and understanding from other epistemic universes ('lived experience' etc.) are often dismissed as inferior. I argue that this is a form of fraudulent scientific imperialism because it fails to recognize the epistemological value of 'lived experience' and the 'local knowledge' that is constituted through it."*

The research approach and strategy discussed in this chapter is intended to support residents of publicly subsidised housing – as situated knowledge producers – to examine the so called 'problem' of public housing/subsidised housing concentration, and to scrutinize the strategy of de-concentration and the *mixed income* neighbourhood model underpinning housing renewal programs in Australia, Europe and the United States. While disadvantaged communities have often been the focus of university and government sponsored research, this project is intended to go well beyond gaining residents' insights or responses to the predetermined questions of professional researchers. It sets out to provide an opportunity for residents of so-called 'excluded' or disadvantage neighbourhoods to be involved in setting the research and policy agenda. By synthesising situated local knowledge with other sites of local knowledge this approach will produce new ways of understanding the issues surrounding *social exclusion* and proposed policy solutions such as redevelopment and *mixed income* housing projects. Findings produced by the cross-site and transnational linking of situated local knowledges can be used to provide credible alternative perspectives to views that de-concentration is the only way of increasing opportunity in low-income communities. To reiterate, the current evidence base for neighbourhood effects is at best shaky, yet many government projects are based on the notion of neighbourhood effects and view mixed neighbourhoods are seen as the solution. There remains little evidence, however, that mixed neighbourhoods created by such policies deliver the promised benefits.

This chapter is not primarily a theoretical contribution but a discussion on an emerging research approach. The remainder of the chapter is divided into two parts: the first suggests the need to reassess the epistemology which underlies the contemporary discourse on place and disadvantage; while the second outlines our attempt to conduct research which reflects a new epistemology of housing studies.

Whose Problem, Whose Knowledge? A Question of Epistemology

In the research and policy development activity described above, with a few notable exceptions, and certainly in the policy directions which have emerged, the voice and perspective of those most affected is absent. This is despite Allen's (2009, p. 62) concession that: "*Theoretical housing researchers are very well aware that dominant definitions of housing issues (whether in social science or society at large) are simply those whose protagonists have successfully transformed them from concepts into 'established facts' that are widely agreed upon.*" The 'exclusion' of public tenants can thus be seen to extend to their exclusion from the discursive practices through which disadvantaged places have become problematised, and in which policy solutions are framed. These discursive practices notably include academic research. Conventional positivist epistemology, which dominates both research and policy-making, systematically excludes important aspects of community life as experienced by those most affected (Darcy 2007, 2010). And where attempts have been made to incorporate residents' experiences and insights (see for instance Warr 2005; Bryson and Thompson 1972; Arthurson 2002; Randolph and Wood 2004; Peel 1995), this has generally been undertaken through conventional empirical and case study approaches premised on the academic convention of knowledge production and control. One of the problems with this approach, according to Allen (2009, p. 66), is that: "*The social sciences ask questions about housing phenomena that are fundamentally different to the types of questions posed (if they are posed at all) by people as they dwell in everyday life.*" Several of the research centres involved in this project have previously conducted research and community engagement work which involves developing alternative resident-centred approaches at the local level. This work demonstrates a strength of place attachment and positive social engagement by residents in identified disadvantaged areas which we believe justifies a move to far more strongly resident-driven research and collaborative theorising.

The aim of this research is to assist resident researchers to analyse deductively their links to the theoretical discourses of exclusion, inclusion and social mix, and to open a new space for debate concerning the relationship between social inclusion, geography and housing policy. It will do this by drawing on 'other' world views and bringing together the situated and experiential knowledge of residents in disadvantaged communities in contrasting urban settings, and across national and cultural differences, the aim is to assist resident researchers to analyse deductively their links to the theoretical discourses of exclusion, inclusion and social mix, and to open a new space for debate concerning the relationship between social inclusion, geography and housing policy. Nyden (2006, p. 21) notes that the ability of local research groups to share research questions, experiences and knowledge "*represents an underdeveloped source of new knowledge*". A fundamental element is the transnational linking of local knowledges to inspire and create a dynamic, cross-cultural production of knowledge. Comparative findings produced by the transnational linking of local knowledges can be used to provide credible alternative perspectives to views that deconcentration is the only way of increasing opportunity in low-income communities. The cross cultural sharing of knowledge involves the placing of one's own experience in the context of others, and drawing on the similarities and differences to develop more comprehensive explanations and understandings.

Such an approach is particularly important given the 'travelling policy' (Ozga and Jones 2006) of social mix as the orthodox solution to the 'problem' of concentrated public housing. With regard to the liberating potential of globalisation Sassen (2008) refers to: "*the multiplication of partial, often highly specialized, cross-border assemblages of bits of national territory, authority, and rights that are getting dislodged from national settings ... some of which are emergent spaces for political action, notably spaces where those confined to the nation-state (citizens) or those who are immobile (because of poverty or political vulnerability) can actually engage in global politics.*" Following on from Sassen then, this situation warrants participatory comparative research which is capable of recasting international debate concerning the relationship between poverty, place and housing. Involving local communities as integral members of this trans-national research is expected to produce new insights, understandings and knowledge.

A Methodology of Emergence

The remainder of this chapter will discuss the alternative methodology and research design which emerges from this approach. It is a methodology that is necessarily fluid, and is driven by the need to explore the issues of place, community and advantage and disadvantage through techniques which produce alternate ways of knowing and 'new *theoretical* possibilities' (Cahill 2007). Consequently, drawing on the paradigm of engaged scholarship (Holland 2006), the research is designed using a 'collaborative university – community research' approach (Nyden 2006). Methodology involves community collaborators informing each phase of the research, from developing specific research questions, to data collection, analyses and dissemination (Cahill 2004). It is an approach which recognises that community contributors are competent agents and experts in their own lives (Cahill 2007), and thus emphasises the integration of academic and community knowledge in the making of new knowledge. According to Nyden (2006), the real value of collaborative research is the emphasis on *what could be* rather than the traditional research emphasize on *what is*.

The research design is intended to support residents of 'disadvantaged' places, such as public housing estates – as situated knowledge producers – to investigate the 'problem' of public housing/subsidised housing concentration, and to scrutinize the strategies of public policy agencies which seek to address it. As discussed above, tenants have rarely had opportunities to frame their own research, and universities are generally less supportive of research that commences without well-developed plans and proposal, so this approach presents many initial challenges. For instance, university based researchers need to develop relationships and networks amongst tenants, and to carefully avoid imposing pre-emptive theoretical frameworks, while at the same time continually reinforcing the potential for residents' lived experience to be valorised as 'knowledge', disseminated, discussed and responded to by other contributors.

The following section describes an international project being undertaken by the current authors which attempts to operationalise the principles of participation and collaboration we have outlined. The study will focus primarily on sites in the metropolitan regions of Sydney and Adelaide in Australia, and Chicago in the United States. The sites have been chosen largely on the basis that the researchers have strong existing contacts with tenants groups, and also have access to technological resource and communication nodes in these cities. Nonetheless, the similarities and differences displayed in these cities offer an opportunity to divine from practice the principles driving these developments which are independent of the particular urban context. All the sites are large metropolitan centres with a small minority of public tenant households, although in each case a large proportion of public housing dwellings were 'mass produced' in relatively large scale concentrated projects developed between the late fifties and the late seventies (as part of the Keynesian/modernist welfare state, and to provide affordable housing for the low income urban industrial workforce).

In each case, eligibility criteria, underinvestment and tenancy management issues have seen this housing occupied by progressively more disadvantaged populations, fewer and fewer of whom are in the paid workforce. Policies of deinstitutionalisation of people with psychiatric, physical and intellectual disabilities over the past 20 years have also had an effect. These tenants and their housing have become ever more stigmatised and associated in policy discourse and the public mind as undesirable, crime ridden, dangerous and degraded places which are seen as encouraging welfare dependency and intergenerational poverty. Programs of deconcentration and social mix are viewed as the prescription to this problem. Where prior to the current housing market crisis Chicago was the most active U.S. city in transforming subsidised housing communities through poverty de-concentration and social mix, Sydney is emerging as the Australian leader. The juxtaposition of the similarities and differences between housing renewal project sites presents a dynamic canvas upon which new ways of understanding social mix policies can be developed.

Collaborative, Comparative Research Model

The collaborative, comparative research model informed by this methodology of emergence involves three primary components – city based research steering groups, multiple local research teams and academic research teams (see Fig. 11.1) – and the transnational linking of these knowledge bases (see Fig. 11.2). It is proposed that each city will have its own local steering group comprising of people engaged in public and subsidised housing issues including residents, tenant support networks and tenant advocates. The role of the steering groups is to guide the research and research teams, refocus the project as required and assist with recruitment of residents to situated local research teams. The local steering groups, in conjunction with the support of the academic research teams, will also be able to develop and support the transnational communication between local research teams and residents, which is fundamental to the success of the project. In the first instance this structure may appear to be organised around a hierarchy of power relationships. However, it is hoped that a solid framework will satisfy public funding bodies, ethics committees and university reporting requirements, whilst still allowing for an anarchic form of research at the grass roots level.

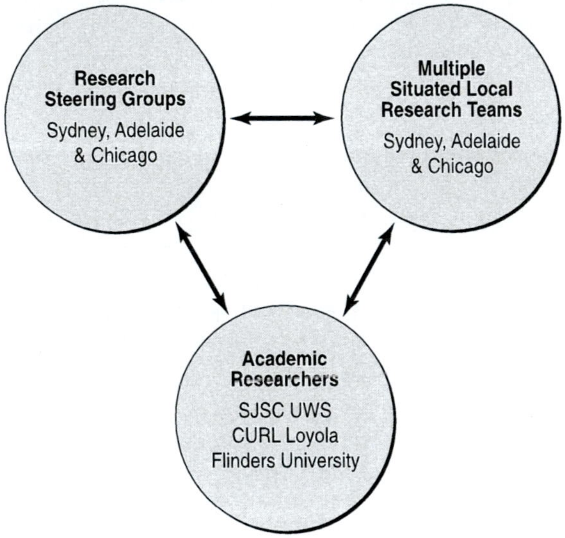

Fig. 11.1 Collaborative, comparative research model (*Source*: Resident Voices Prospectus – G. Gwyther and M. Darcy)

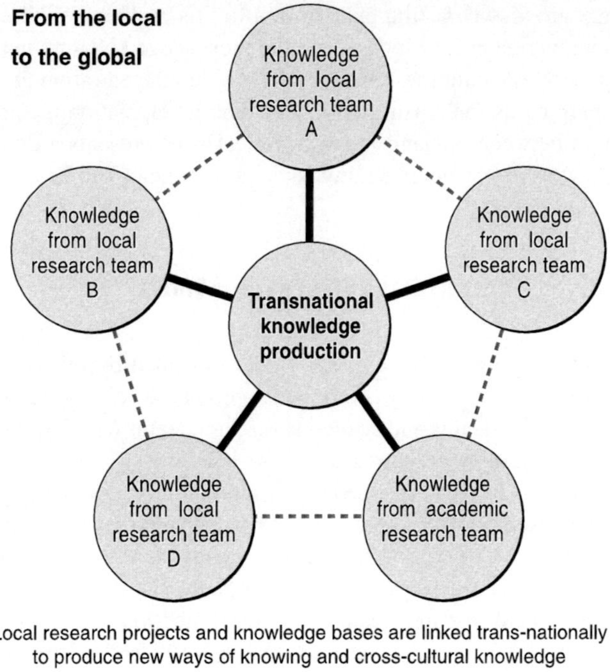

Fig. 11.2 Linking knowledge bases (*Source*: Resident Voices Prospectus – G. Gwyther and M. Darcy)

Local research teams will essentially comprise residents of social housing. The principal role of the local research teams is, with support from their local steering group and academic researchers, to determine the issues, frame research questions relevant to their context, and collect data through techniques appropriate to the groups' questions, skills and interests. Groups will also be involved in the analysis of their data and in the inter-city linking of local findings (see Fig. 11.2 – Linking Knowledge Bases).

The role of the academic researchers is to work *with* local steering groups, local research teams and individual participant researchers in conceptualising, designing, analysing and disseminating the research. The role is based on the recognition that we all have knowledge bases and capacities that are important in completing change-oriented, collaborative research. The academic knowledge base is just one of these. The research involves four developmental phases. The first is the formation of city-based Research Steering Groups in Sydney, Adelaide and Chicago. These groups will primarily comprise tenants but also housing advocates and non-government service providers where appropriate. In the second phase local and peer research groups will be established. These groups will comprise of local residents with issues and interests in common. Kesby (2000, p. 425) explains that "*these exclusive groupings enable participants to share experiences and develop ideas independently of those with different and/or competing*

Fig. 11.3 Generating new knowledge (*Source*: Resident Voices Prospectus – G. Gwyther and M. Darcy)

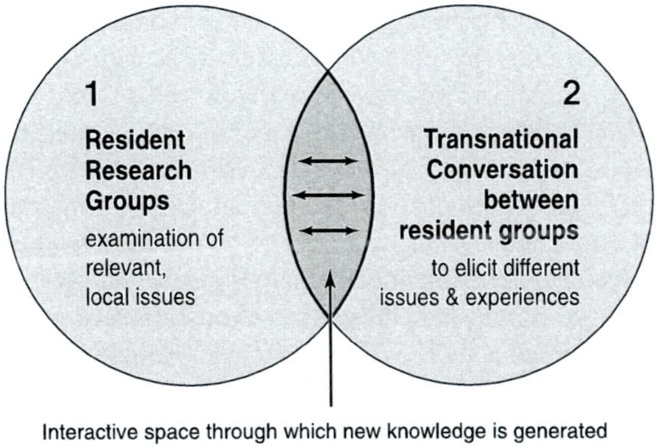

positionalities." Groups will meet locally and with the support of a member of the academic research team.

Importantly, these groups will start from the point of examining from their experience, perspective and reflection the advantages and disadvantages of living in their 'place'. This is in contrast to more standard research methods where academics develop models of residents' behaviour and outcomes, disregarding residents' perspectives in the process. Notably in this research, each resident group is differentially located in regard to the experience of concentrated disadvantage, housing form and housing renewal models. This is important for informing the third phase of the research process.

In the third phase local and peer interest groups will be linked with other local and international groups ('inclusive plenary sessions'). The purpose of establishing communicative links between research teams is to expose participants to different perspectives, experiences, issues and solutions. As Kesby (2000, p. 425) explains, "*plenaries can then become arenas in which the social re-negotiation of the phenomena under discussion can begin to take place*". 'Buzz' sessions to identify differences between co-researchers' knowledge (P.L.A. Notes 1998), context based story telling (autoethnography) and visual contributions (photographic, video) will be important methods of cross-cultural knowledge development, exchange and interpretation of findings. In collaborative emergence research, data analysis is an integral part of the research process and is not necessarily separate from the processes of data collection and production techniques. Regular "moments of analysis" (Cahill 2007, p. 306) emerge at various points of the research processes as part of the collaborators' reflective praxis.

Establishing formal and informal local and transnational communication opportunities is a particularly important methodological aspect of this project because it is within and between the sharing of experiences, ideas, arguments and rationales that new knowledge emerges (see Fig. 11.3: Generating New Knowledge). This process draws on C. Wright Mills' *Sociological Imagination* which seeks to understand issues through the relationship between private troubles and public issues (Mills 1959). Further, it is through these communication opportunities that the analysis of data takes place.

Clearly, supporting communication opportunities is a crucial aspect of the project. One way of doing this will be to establish an interactive website for supporting communication between research teams – transnational, local and institutional – and individuals involved in the project. The site will also have the potential to facilitate international blogs on pertinent aspects of the research, data collection, live feed-back loops, and the dissemination of findings and outputs. Although the general content of the website will be managed by the project's steering groups and researchers, the intention is that all manner of material emerging from the project could be posted and online, open discussion encouraged. Other methods of supporting communication exchange include virtual, transnational face-to-face meetings through teleconferencing as well as physical face-to-face meetings between members of the steering committees and local research team.

The fourth phase of the research will involve the collation of findings and dissemination of various outputs. Dissemination of insights into place-based disadvantage and social exclusion amongst research members is as important as disseminating findings to the academy and policy community. Academic outputs will be one form of information dissemination, however the emphasis is on disseminating information through more accessible mediums and publications that all research team members will have input into and the potential to access. The project's Third Sector partners are also particularly well placed to disseminate findings through the housing and welfare sectors. The data collection and analysis techniques which are applicable to 'emergence methodology' include:

- Participatory Diagramming: this is a flexible participatory visual technique which draws on the knowledge and experience of participants, reflects their own priorities and interests, and promotes information sharing and education. Participants work initially in small peer groups or teams over a period of time. Question raising, question answering and issues are worked through by the group using large sheets of paper, coloured pens, sticky dots, post-it notes and whatever other visual materials are necessary or handy. Although the visual output is important, the most valuable part of the method is the discussion which develops. Exclusive groupings are then encouraged to participate in inclusive plenary sessions to examine the perspectives of others (Kesby 2000). The photograph at Fig. 11.4 shows a diagramming exercise which was used by the authors to introduce the research concept to residents of Woolloomooloo in Sydney in 2008.
- Storytelling and collaborative autoethnography (in written and visual forms): this involves grounding research data in the concrete details of community researchers' lives (Cahill 2007).
- Other possibilities include creative techniques such as photo voice, video making, performance ethnography and other arts based inquiry (Gwyther and Possamai-Inesedy 2010).
- Social media will play a multi-function role as a communication tool, a mechanism for data collection and collation, and as a means for disseminating findings. As discussed above, a dedicated, interactive website will underpin social media practice.

Fig. 11.4 'On Being a Houso! Perceptions, Reality, Hopes and Fears' – Diagramming exercise with public housing residents of Woolloomooloo in 2008 (*Source*: *Residents' Voices* newsletter, 19 December 2008. G. Gwyther and M. Darcy)

In mid 2010 the project received funding from the Australian Research Council. Prior to this, a Sydney-based steering group comprising residents from inner city and outer suburban public housing estates was established in 2008. The group has met a number of times to discuss the nature and potential of the project, possible research questions and issues around communication including the cost and access to communication technology. These meetings also provided an opportunity for members to discuss and compare their experiences of living in public housing. This was particularly important in demonstrating the knowledge which can develop through such a process. Members of the group also participated in a video-conference with community organisers and advocates working in disadvantaged neighbourhoods in Chicago. Again the experience enabled a sharing of experiences and knowledge.

Conclusion

Whether by contagion or collective socialisation, the notion that place in and of itself undermines the life chances of residents of disadvantaged communities is the main foundation upon which neighbourhood effect theories and contemporary urban renewal policies of social mix are constructed. In this model, place is responsible for drawing residents into a distinctive local process that distorts their values, undermines their aspirations and kindles "a strong feeling of marginality, of helplessness, of dependency, of not belonging" (Lewis 1998, p. 7). Limited compositional and cross-sectional data is employed to support the model, while alternative theories and evidence suggesting that forced socio-economic mixing may itself compound disadvantage and disempowerment are frequently ignored. It is upon such specious grounds that residents of disadvantaged neighbourhoods are moved around like expendable pawns on a policy chessboard. In reality, little is known about the complexities of social life of residents of such neighbourhoods, nor about the intervening mechanisms between place and poverty. Even qualitative studies which endeavour to examine the experiences of disadvantaged people from their perspective are conducted through the lens of the academic as expert knowledge producer, such that normative questions are posed and answers duly extracted. Perhaps most problematic, the contagion model neglects the responsibility of outsiders whose 'cultural labelling' of residents of disadvantaged communities establishes processes of cultural exclusion which inevitably results in decreased opportunities and consequently increased marginalisation (Bauder 2002).

In this chapter we posit that a new knowledge is required with which to scrutinise the dominant but incomplete knowledge upon which ideas about poverty and place-disadvantage are drawn. We argue that a new knowledge requires a new epistemological and methodological framework; one that requires residents as situated knowledge producers to examine the so called 'problem' of disadvantaged neighbourhoods; one that enables situated knowledge producers to pose the questions and to seek answers through a rigorous exchange of local knowledges; and one where academics take on a supporting role that facilitates, inspires and nurtures the emergence of this knowledge – a knowledge that is truly imbued with the epistemological value of 'lived experience'. Given the nature of the methodology and the early stage in the research at which time this chapter is being written, it is not possible to provide details on what this new knowledge looks like. What we can say, however, is that the knowledge will be of a form that will be utilizable by residents as they challenge the dominant discourse on what counts as advantage and disadvantage. Further, there will be no pretence to objectivity, as the knowledge will be made explicit through the crucible of research questions and alternate research methods that are framed by residents' own world views.

References

Allen, C. (2008). *Housing market renewal and social class*. London: Routledge.

Allen, C. (2009). The fallacy of "Housing Studies': Philosophical problems of knowledge and understanding in housing research. *Housing, Theory and Society, 26*(1), 53–79.

Arthurson, K. (2002). Creating inclusive communities through balancing social mix: A critical relationship or tenuous link? *Urban Policy & Research, 20*(3), 245–261.

Bauder, H. (2002). Neighbourhood effects and cultural exclusion. *Urban Studies, 39*(1), 85–93.

Bennett, L., Smith, J., & Wright, P. (2006). *Where are poor people to live? Transforming public housing communities*. New York: M.E. Sharpe.

Bryson, L., & Thompson, F. (1972). *An Australian Newtown: Life and leadership in a new housing suburb*. Harmondsworth: Penguin.

Cahill, C. (2004). Defying gravity? Raising consciousness through collective research. *Children's Geographies, 2*(2), 273–286.

Cahill, C. (2007). Doing research with young people: Participatory research and the rituals of collective work. *Children's Geographies, 5*(3), 297–312.

Chaskin, R., & Joseph, M. (2010). Building "community" in mixed-income developments: Assumptions, approaches, and early experiences. *Urban Affairs Review, 45*(3), 299–335.

Darcy, M. (2007). Place and disadvantage: The need for reflexive epistemology in spatial social science. *Urban Policy & Research, 25*(3), 347–361.

Darcy, M. (2010). De-concentration of disadvantage and mixed income housing: A critical discourse approach. *Housing, Theory and Society, 27*(1), 1–22.

Defilippis, J. (2007). Erasing the community in order to save it? Reconstructing community and property in community development. In H. Beider (Ed.), *Neighbourhood renewal and housing markets: Community engagement in the US and UK* (pp. 271–289). Oxford: Blackwell Publishing.

Galster, G. (2003). Investigating behavioural impacts of poor neighbourhoods: Towards new data and analytic strategies. *Housing Studies, 18*(6), 893–914.

Galster, G. (2011). The mechanism(s) of neighbourhood effects: Theory, evidence, and policy implications. In M. van Ham, D. Manley, N. Bailey, L. Simpson, & D. Maclennan (Eds.), *Neighbourhood effects research: New perspectives*. Dordrecht: Springer.

Goetz, E. G. (2003). *Clearing the way*. Washington, DC: Urban Institute Press.

Gwyther, G., & Possamai-Inesedy, A. (2010). Methodologies a la carte: An examination of emerging qualitative methodologies in social research. In A. Possamai-Inesedy & G. Gwyther (Eds.), *New methods in social justice research for the 21st century* (pp. 1–17). London: Routledge.

Hedman, L., & van Ham, M. (2011). Understanding neighbourhood effects: Selection bias and residential mobility. In M. van Ham, D. Manley, N. Bailey, L. Simpson, & D. Maclennan (Eds.), *Neighbourhood effects research: New perspectives*. Dordrecht: Springer.

Holland, B. (2006). New views of research for the 21st century: The role of engaged scholarship. In L. Silka (Ed.), *Scholarship in action: Applied research and community change* (pp. 1–10). Washington, DC: Department of Housing and Urban Development.

Imbroscio, D. (2008). [U]nited and actuated by some common impulse of passion: Challenging the dispersal consensus in American housing policy research. *Journal of Urban Affairs, 30*(2), 111–130.

Jencks, C., & Mayer, S. E. (1990). The social consequences of growing up in a poor neighborhood. In L. E. Lynn & M. G. H. McGeary (Eds.), *Inner-city poverty in the United States* (pp. 111–186). Washington, DC: National Academy Press.

Kesby, M. (2000). Participatory diagramming: Deploying qualitative methods through an action research epistemology. *Area, 32*(4), 423–435.

Levitas, R. (1998). *The inclusive society? Social exclusion and new labour*. Houndmills: Palgrave Macmillan Ltd.

Lewis, O. (1961). *The children of Sanchez*. Harmondsworth: Penguin.

Lewis, O. (1998). The culture of poverty. *Society, 35*(1), 7–9.

Lupton, R. (2003). *Neighbourhood effects: Can we measure them and does it matter?* (CASE paper 73). London: Centre for Analysis of Social Exclusion, London School of Economics.

Manley, D., & van Ham, M. (2011). Neighbourhood effects, housing tenure and individual employment outcomes. In M. van Ham, D. Manley, N. Bailey, L. Simpson, & D. Maclennan (Eds.), *Neighbourhood effects research: New perspectives*. Dordrecht: Springer.

Mills, C. W. (1959). *The sociological imagination*. New York: Oxford University Press.

Musterd, S., Ostendorf, W., & de Vos, S. (2003). Neighbourhood effects and social mobility: A longitudinal analysis. *Housing Studies, 18*(6), 877–892.

Nyden, P. (2006). The challenges and opportunities of engaged research. In L. Silka (Ed.), *Scholarship in action: Applied research and community change*. Washington, DC: US Department of Housing and Urban Development.

Ozga, J., & Jones, R. (2006). Travelling and embedded policy: The case of knowledge transfer. *Journal of Education Policy, 21*(1), 1–17.

P. L. A. Notes (1998). Participatory learning and action 33: Understanding market opportunities. PLA Notes. London: International Institute for Environment and Development.

Peel, M. (1995). The rise and fall of social mix in an Australian new town. *Journal of Urban History, 22*(1), 108–139.

Phillips, D. C. (1987). *Philosophy, science and social enquiry: Contemporary methodological controversies in social science and related applied fields of research*. Oxford: Pergamon Press.

Pugh, C. (1976). *Intergovernmental relations and the development of Australian housing policies* (Research monograph, Vol. 15). Canberra: Canberra Centre for Research on Federal Financial Relations, ANU.

Randolph, B., & Wood, M. (2004). *The benefits of tenure diversification: Final report*. Melbourne: Australian Housing and Urban Research Institute.

Sassen, S. (2008). *Deciphering political possibilities for national states and citizens in a global economy – Plenary address*. 1st ISA Forum of Sociology, Barcelona (September).

Stubbs, J., Randolph, B., & Judd, B. (2005). *The Bonnyrigg living communities baseline study*. Sydney: University of New South Wales – City Futures.

Subedi, B., & Rhee, J. E. (2008). Negotiating collaboration across differences. *Qualitative Inquiry, 14*, 1070–1091.

Uitermark, J., Duyvendak, J. W., & Kleinhans, R. (2007). Gentrification as a government strategy: Social control and social cohesion in Hoogvliet, Rotterdam. *Environment and Planning A, 39*(1), 125–141.

Vale, L. J. (1997). Empathological places: Residents' ambivalence toward remaining in public housing. *Journal of Planning Education and Research, 16*(1), 159–175.

van Ham, M., & Manley, D. (2010). The effect of neighbourhood housing tenure mix in labour market outcomes: A longitudinal investigation of neighbourhood effects. *Journal of Economic Geography, 10*(2), 257–282.

Warr, D. (2005). *There goes the neighbourhood: The malign effects of stigma. State of Australian Cities 2005*. Griffith University, Brisbane 30 Nov – 2 Dec.

Wilson, W. J. (1987). *The truly disadvantaged*. Chicago: University of Chicago Press.

Chapter 12
Are Mixed Community Policies Evidence Based? A Review of the Research on Neighbourhood Effects

Paul Cheshire

Mixed Communities Policy and Income Inequality

In this chapter I review and assess the evidence which one should expect to find supporting policies designed to create 'mixed communities' in cities. Many of these policies are based on the assumption that living in poor and segregated neighbourhoods (whether segregated on the basis of income, ethnicity, or other characteristics) can have a negative effect on individual outcomes: so-called neighbourhood effects. The idea is that creating neighbourhoods in which populations are mixed will take away these negative effects. However, the evidence supporting the significance, even the existence of neighbourhood effects is remarkably thin when subjected to rigorous evaluation. This calls into question current policies designed to produce 'mixed communities'. As a result it is not clear what social gains could be derived from forcing neighbourhoods to be more mixed on the basis of any specific characteristic, whether housing tenure, income, age, marital status, educational attainments or ethnicity.

The fundamental issue in neighbourhood effects research is causation: do poor neighbourhoods make residents poorer, or do poor people simply live in poor neighbourhoods because living in affluent ones costs too much? Although empirically income mixing, even in very small neighbourhoods, is considerable (see Hardman and Ioannides 2004) it is still true that poor people tend to be concentrated in poor neighbourhoods and richer people in more affluent ones. We know that living in nicer neighbourhoods costs more and there are very obvious reasons for this. All choices, in even the most liberal or progressive of market economies, are constrained by, amongst other factors, income: we can only buy what we can afford. We know that nice neighbourhoods cost more because the value of all those characteristics that make them nice – access to good jobs, pleasant parks, good views, low crime,

P. Cheshire (✉)
Department of Geography and Environment, London School of Economics, London, UK
e-mail: p.cheshire@lse.ac.uk

M. van Ham et al, (eds.), *Neighbourhood Effects Research: New Perspectives*,
DOI 10.1007/978-94-007-2309-2_12, © Springer Science+Business Media B.V. 2012

peace and quiet, better local public goods – are fully reflected (capitalised) in the prices or rents of houses. So because they have more limited resources, poorer people cannot afford to live in such attractive neighbourhoods and have to concentrate into poor areas in which housing is cheap.

The issue is whether living in a poor neighbourhood is a separate, significant additional cause of poverty. What the evidence in fact primarily suggests is that the income gap between ‘rich’ and ‘poor’ neighbourhoods is the spatial manifestation of wider economic and social processes that cause individuals to have different incomes, which in turn constrains them to live in different places. To tackle problems of ‘disadvantaged neighbourhoods’ we have to understand how and why such neighbourhoods arise: therefore why income differences between households and individuals exist. This involves understanding better how cities function. And then we have to understand how far, and in what ways, disadvantage (as experienced by residents) may actually be caused or exacerbated by the characteristics of the neighbourhood (particularly the mix of other residents) in which they live rather than simply reflecting sources of disadvantage operating at the individual or family level.

The desire of social and urban planners for neighbourhoods to be more ‘mixed’ or ‘balanced’ is of long standing. More recently – perhaps influenced by Wilson (1987) – the aim of mixed communities as a way of addressing poverty and social exclusion has been adopted by many governments and international bodies and is a central tenet of ‘new urbanism’. Thus in 2005 the British government argued (ODPM 2005, page 6):

> People living in deprived neighbourhoods are less likely to work, more likely to be poor and have lower life expectancy, more likely to live in poorer housing in unattractive local environments with high levels of antisocial behaviour and lawlessness and more likely to receive poorer education and health services. Living in a deprived area adversely affects individuals’ life chances over and above what would be predicted by their personal circumstances and characteristics.

The report documented in great detail differences in outcomes for people living in deprived areas (defined as the 10% most deprived wards identified by the Index of Multiple Deprivation) compared to the average for England as a whole. ‘Worklessness rates’ were some 25% compared to less than 10%; a third more of the adult population of such areas had no qualifications; and life expectancy was 2 years less. To conclude from this evidence that mixed communities provided a recipe for reducing inequality and tackling social exclusion, however, was either naive or deliberately misleading. None of the figures illustrating the problems experienced by those living in deprived areas standardised for the characteristics of individuals. Similar types of people – in terms of skill, health or education – not living in deprived neighbourhoods might have had identical probabilities of being workless and similar life expectancy. By definition they would have had similar qualifications. ODPM (2005) offered as ‘evidence’ just measures of the average levels of income, health, education or worklessness of the inhabitants of the deprived areas. Such ‘evidence’ simply does not address the issue of causation.

If the policy for neighbourhood mixing had no costs – even though it was not effective in reducing income inequality – it would be harmless. Attempting to implement it, however, costs significant resources. There are direct government expenditures involved in subsidising the provision of lower income housing in higher income developments but in Britain most of the resources devoted to generating mixed communities do not register in public expenditure. Because of the extreme scarcity of developable land – the result of 60 years of 'urban containment' policies – obtaining 'planning permission' generates extraordinary increases in land values. As a condition of being permitted to build, developers negotiate so-called Section 106 Agreements[1] with the local planning authority, obliging them to provide additional community benefits. The most common form these benefits take is an obligation to build 'affordable' housing or 'social' housing within their commercial development. Until a change in political control in 2008 the Greater London Authority,[2] for example, demanded that 50% of all housing constructed by the private sector should be such 'affordable' housing built within each substantial private development.

It is, of course, possible – superficially even plausible that geographically concentrated poverty is a greater social evil than dispersed poverty. There is a strong correlation between living in a deprived neighbourhood and being poor: or living in a neighbourhood dominated by immigrants or ethnic minorities and being an immigrant or a member of a minority ethnic group. So it might seem obvious that living in poor and deprived neighbourhoods must impose costs on the inhabitants of those neighbourhoods beyond the disadvantages of low incomes, poor health, migrant status or lack of labour market skills. The families living in such neighbourhoods experience poor services, frequently have a worse environmental quality (atmospheric pollution or noise), suffer greater ill health and are much more likely to be the victims of crime (widely documented but see, for example, Wilson 1987). The schools which serve such neighbourhoods score less well measured by the exam results their pupils achieve or by truancy rates (see, for example, Gordon and Monastiriotis 2006, 2007). Children growing up in such neighbourhoods do not have the chances in life that children raised in advantaged neighbourhoods have. That seems obvious and it is what we appear to believe.

The problem with this conclusion is that it, again, ignores the issue of causation. In all cities for which there is evidence, neighbourhoods have been segregated. In ancient Rome there were neighbourhoods segregated by artisanal trade as well as income.

[1] Section 106 (S106) of the Town and Country Planning Act 1990 allows a local planning authority (LPA) to negotiate a legally-binding agreement or 'planning obligation' with a developer as a condition of granting permission to develop. The obligation is termed a Section 106 Agreement. These agreements are a way of extracting so-called 'planning gain'. Since planning policies in Britain tightly restrict the supply of land for development, obtaining permission to develop on attractive sites creates a windfall gain (see Cheshire and Sheppard 2005). S106 Agreements are used to extract a part of this windfall gain by forcing some contribution from the developer in kind, usually in the form of infrastructure, such as highways or recreational facilities or as 'affordable' housing: that is housing available at below market price for social tenants or 'key' workers such as teachers or police.

[2] In 2008 the Conservative Party candidate Boris Johnson won the GLA Mayoral election replacing Labour Party mayor Ken Livingstone.

The particular patterns of segregation seem to be remarkably stable over time. Many of the London neighbourhoods amongst the poorest in 1881 were still amongst the poorest in 2001 (Meen et al. 2007); there is substantial stability in the pattern of the local authority areas which were most segregated in 1971 and in 2001 (Meen 2006). Moreover, when policy has deliberately constructed 'mixed' neighbourhoods – as with some of the early utopian planned communities such as Hampstead Garden Suburb, developed in north west London around 1910 – over time they have tended to become segregated again. If segregation is an integral element of urbanisation then at best policy for mixed communities is trying to push water up hill; at worst it is attempting to undermine one of the advantages that make cities such a great human innovation in welfare terms.

The larger an urban area is, the larger the areas dominated by particular types of household, rich or poor, tend to be (Gordon and Monastiriotis 2006). As Krupka (2007) points out, using neighbourhoods of roughly constant size (such as Census Tracts or Local Authority areas) the larger the city is, the greater the degree of measured segregation there will appear to be, other things being equal. This is because the larger a city is, the more likely it is that there are enough households in particular income groups to fill up more completely any fixed size of 'neighbourhood'. Useful definitions of neighbourhood, therefore, are likely to vary with city size and be largest in the largest cities.

Underlying the longevity and pervasiveness of residential segregation in human settlements is one of the fundamental functions and advantages cities provide us with. Specialisation underlies the agglomeration economies cities generate, both in production and in terms of the additional choices provided for consumption and lifestyles; segregated neighbourhoods are simply the flip side of specialised neighbourhoods. A larger number of specialised neighbourhoods provide a wider choice of urban community types and social settings in which to live. As was stressed above, however, like all other choices about what to consume, choosing where to live is strictly subject to the constraint imposed by one's income.

This underlies the issue of the direction of causation. If neighbourhood choice is conditioned by income (see for instance Hedman et al. 2011), poor neighbourhoods exist because there are poor people and we live in an unequal society; as is explained below, given that degree of inequality, we may be collectively and individually better off, living in neighbourhoods with other similar households, whether we are rich or poor. For any given distribution of household incomes that is an argument for allowing specialised neighbourhoods to develop of their own accord; but not for policies promoting specialised neighbourhoods. Questioning the case for policies for promoting mixed neighbourhoods is certainly not advocating having greater inequality or poverty within a rich society.

Not only do policies for generating mixed neighbourhoods ignore inconvenient facts, they are also, as Krupka (2008) points out, at odds with three of the best established theoretical models in urban economics. Tiebout (1956) has a model of mobile urban residents voting with their feet to concentrate in communities providing the best mix of taxes and local public goods given their incomes and preferences. Alonso (1964) – still perhaps the most important single theoretical contribution to urban economics – concludes that with centrally located employment, annular rings of exclusive land use will tend to be established as business users and residents trade-off the value to them of accessibility against the costs of space. This has traditionally been interpreted as implying residential segregation by income group, with zones occupied at high densities close to the centre by poor residents and richer households living at lower densities further from the city-centre. Although this particular conclusion has been challenged (by, for example, Brueckner et al. 1999), income segregation does seem the likely outcome of monocentric models. Finally there is the model of Schelling (1969) which on the basis of a simple preference for not being the minority in ones immediate neighbourhood predicts social segregation as the equilibrium outcome.

This chapter reviews the evidence relating to why social segregation develops and generates specialised neighbourhoods in cities and why this pattern is more obvious the larger a city is and the more unequal a country's society is. It then reviews the empirical evidence on the extent to which neighbourhood characteristics influence individuals in terms of their wellbeing or their life chances (evidence for or against the significance of neighbourhood effects), evidence on which any logically or evidence based policy for mixed communities must rest; but evidence which turns out to be unconvincing. Following that, in the Section entitled "Some social advantages of 'specialised' neighbourhoods" I review the evidence about the gains that we derive from 'specialised' or homogeneous neighbourhoods." The final sentence of that paragraph can be deleted since it refers only to the conclusion.

What Causes Residential Segregation? Nicer Neighbourhoods Cost More

As we learn more about how housing markets work, so we can understand better how they may interact with labour markets to sort households and individuals into more and less expensive and desirable neighbourhoods on the basis of their incomes. Cheshire and Sheppard (2004) focus on the case of good schools. Better schools (when access to a school is determined by where a household lives) are an example of a whole class of 'goods' one might call truly 'positional'. That is goods which can only be consumed by living in the appropriate place; and for which the ability to buy houses giving access to them is chiefly determined not by absolute income but income relative to others who are competing for the same 'goods'. The most important and obvious of these in Britain is access to the best State schools.

One of the most fundamental things that the location of a house provides is access to jobs. This is at the heart of Alonso's (1964) model of urban land markets. The idea is that households will trade off the purchase of space for lower transport costs. If jobs are concentrated in the centre of a city then the price of housing space will fall with distance from the centre: but the consumption of space will tend to increase. There will be lower densities in suburban areas than in the central city.

Empirical testing by incorporating distance from the centre and actual transport routes into hedonic estimation of house prices provides good confirmation that real cities reflect transport costs and that house prices incorporate a premium for cheaper access to better paid jobs. This is shown in Cheshire and Sheppard (1995). More recently studies have shown how access to public transport networks is also reflected in house prices (Gibbons and Machin 2005). Fundamental to cities then is the way in which housing markets price access to jobs. People with poor prospects of decently paid jobs will not be able to afford the premium necessary to buy access to them. But it is not the lack of access which causes their poor job prospects: it is their lack of skills, health or education which makes it unlikely they will get well paid jobs in the first place.

There is a wealth of evidence showing that housing is a complex good, composed of many attributes or characteristics, each of which commands a price. Since Rosen (1974) 'hedonic' analysis has become the standard framework within which these prices are analysed and estimated. The price of any given house can be thought of as the sum of the prices being paid for all its individual and particular attributes given their quantities. Although the idea is simple, it has proved fruitful as a way of analysing housing markets. Hedonic studies of housing markets have mushroomed and it seems to be an area in which genuine progress of a scientific kind has been made. Studies have incrementally improved the methodology and refined the estimation process. In the process they have shown how more finely gradated and more seemingly intangible attributes of neighbourhoods (such as neighbourhood social capital – see Hilber 2010) are reflected in house prices.

Evidence that people buy local public goods through their choices in the housing market goes back at least to Oates (1969) and estimates of the price paid for school quality have improved over time. Recent studies in the US have included Haurin and Brasington (1996) and Black (1999). One of the first studies in the UK was Cheshire and Sheppard (1995) but more recent estimates by the same authors (Cheshire and Sheppard 2004) reveal much more about the process. Indeed, it has become increasingly clear how complicated housing markets are and how sophisticated are the ways in which housing attributes – and so ultimately housing itself – are priced.

What people appear to buy as they engage in house hunting is not the current set of attributes but something corresponding to the expected long run set of attributes. Cheshire and Sheppard (2004) found that it was not just the current quality of primary schools (measured by their students' performance on standardised tests), which determined the price paid for access to 'primary school quality'. The price paid also incorporated a discount for current school quality dependent on the variance in measured quality over the previous five years. There was also a discount if the house was located in an area in which new construction was concentrated. Interestingly the effect of more new construction in the local area in depressing house prices was specifically related to school quality. There was no evidence of just a general negative 'area of new construction' effect. It was only when local new construction was expressed as a discounting factor on local school quality that a statistically significant effect on house prices was found. More local construction increased the likelihood that an address could be re-assigned to another school as the Local Education Authority implemented its explicit policy of filling its available school capacity. It may also have increased uncertainty about the composition of the intake to the local school in the future as new households moved into the neighbourhood. So both more variation in performance in the past and more local new construction reduced the price buyers would pay for the current performance of the school a house gave access to.

A comparable finding relates to the value commanded in housing markets for access to open countryside. A study by Irwin (2002) analysed residential transactions in an exurban region in central Maryland, US, and found that within 400 m of a house, the conversion of one acre of developable pastureland to conservation land raised the average house price by 1.9%, while converting it to public land yielded a premium of 0.6%. That is, the more certainly the agricultural land was protected from development, the greater its 'value'. People were paying for the expected future amenity value of the land as well as for its current value.

Moreover, studies are finding increasingly complex interactions with other variables. For example, the price paid through the housing market for access to parks or open space of a given character appears to vary with the density of the neighbourhood, household incomes and local crime levels (Anderson and West 2006). The way in which the value of local parks interacted with local crime rates was shown even more explicitly by Troy and Grove (2008). They found that in relatively low crime areas the value of open space was substantially positive but as crime rates rose relative to the mean so the value of open space declined until, in high crime rate neighbourhoods, open space was a disamenity: close proximity reduced house prices. In effect being close to a nice park in a nice low crime neighbourhood is a valued amenity: being close to a park in a very high crime neighbourhood, instead of giving access to fresh air and greenery, gives an opportunity of being mugged. All these studies showed that the value of open space varied with the distance the space was from the house. The impact of open space as either an amenity or a disamenity disappears at about a kilometre.

What hedonic studies of housing markets show is that access to amenities, whether open space, natural amenities like views or proximity to water; or produced amenities such as greater security from crime, access to good transport or better quality state provided education, costs a substantial amount. The value of all such amenities and local public goods is capitalised into house prices. As an example, moving an otherwise average house from the catchment area of the worst to that of the best primary school in Reading in 1999/2000 was associated with an increase of one third in its price.[3]

[3] In the models discussed here all prices are estimated to vary not just as the quantity of the attribute in question changes (for example the first bathroom is worth a lot more than the fifth) but as the quantity of other attributes varies (for example, the price paid for more space inside a house of given size also varies with the size of the garden, or the value paid for primary school quality varies with the suitability of the house to accommodate children). As a result, this calculation of the impact on price of moving a house from the worst to best primary school catchment area can only be done by assuming some particular levels for all other attributes. Here it is assumed that all other attribute levels are equal to the sample mean, the sample being a random sample of houses sold in the local housing market in 1999/2000 used to estimate the model.

Thus, the ability to benefit from, or consume such localised goods is dependent on the ability of a household to buy or rent a house in those particular neighbourhoods which give access to them. Since the supply of such goods is relatively inelastic and varies significantly from neighbourhood to neighbourhood and demand appears to be income elastic (Cheshire and Sheppard 1998), the price rises sharply with increasing quality and rising income. But their more or less fixed supply also means that the ability to buy such goods is more determined by how rich a household is relative to other households competing for the same local amenities than it is by the household's absolute income.

More fundamental aspects of how people live and how real welfare is distributed appear to follow from this observation. As was analysed in Brueckner et al. (1999), cities have a natural geographically and topographically determined endowment of some amenities – where the best views are to be had, where the natural amenities such as river frontage are available or where, as determined by prevailing winds, air quality is better. In the context of most Old World cities, of course, there was also a fixed neighbourhood within which a particularly valuable local public good – security – was available: within the city walls. These locationally fixed amenities or public goods generated a clustering of those households who had a taste for and could afford them. In turn, this meant that there were higher local incomes, supporting better local cultural and commercial amenities and classier neighbourhoods with better schools and lower crime rates. This further re-enforced the attractions of the more attractive neighbourhoods.

While it is common to think of the prices of composite goods, such as housing, as being the sum of a set of prices for the individual attributes of which they are made up, it is less common to consider a market for each relevant attribute with its own demand and supply characteristics. Yet, that is clearly important and the supply characteristics of individual housing attributes will vary significantly. Some, such as central heating or the number of rooms in a given space, can easily be reproduced industrially and so will be elastic in supply. Others would, in the absence of regulation, normally be elastic in supply. More urban space in aggregate (except in exceptional places such as Singapore) can always be made available by the construction of additional transport infrastructure. In Britain, and increasingly in other countries, the supply of urban space is constrained by land use regulations such as height and density controls or urban containment policies. Other attributes, such as access to particular natural amenities, open spaces or the 'best' local, state funded school, may be intrinsically in very inelastic supply. As noted above, the demand for most characteristics – including housing space, classier neighbourhoods and local amenities – appears to be income elastic. Estimates in Cheshire and Sheppard (1998) were that for many attributes of this type, evaluated at mean household income, income elasticities of demand were in the range 1.5–2.

It follows from this that competition for access to better quality, locationally fixed 'goods', will price poorer households out of access to more sought after and better quality local public goods and amenities – and so generate systematic patterns of residential segregation between richer and poorer neighbourhoods. Given that poverty is correlated with other characteristics, such as lower educational attainment, poorer health, higher unemployment and membership of disadvantaged groups, this suggests that residential segregation is largely the spatial articulation of income inequality in society (though of course there may be residential segregation between households of similar mean incomes but different tastes or characteristics).

Residential segregation is associated with lower welfare for poorer groups since households derive significant welfare from access to the better quality local public goods, including better security, and amenities. However, this is just another manifestation of the price mechanism interacting with the distribution of income to allocate goods according to ability to pay and preferences. Indeed, it may be an important part of the explanation why access to public services provided out of taxation is closely correlated with the distribution of income (Goodin and Le Grand 1987). Although they appear to be distributed according to need, in many cases you have to 'buy' access to them through the housing market.

A further implication is that if the distribution of household incomes changes, this will be reflected in a changing intensity of residential segregation. If, for example, incomes become more unequally distributed – as happened in the UK, the US and several other OECD countries from the mid 1970s to the mid 1990s – under certain circumstances described below, we should expect an intensification of residential segregation with the richest and poorest households becoming relatively more concentrated in richer and poorer neighbourhoods.[4] An increase in residential segregation has certainly been documented in the US context as discussed by Massey and Fischer (2003). They show that inequality across regions has decreased while at the same time between neighbourhoods in US urban areas it has increased. This also explains why in countries with more income equality, residential segregation is less in evidence than in countries with more unequal distributions of income. Helsinki and Stockholm do not have the gated communities and desperately deprived neighbourhoods visible in Chicago or London.

This is to be expected if the supply of at least some of the sought after localised goods is inelastic (the 'best' local state school or a house overlooking Hampstead Heath or the River Thames, for example) while the demand is income elastic. If these conditions hold then the relative price of the supply inelastic but income elastic goods will increase if (where) the rich get richer relative to the poor. More expensive houses in more expensive neighbourhoods will become relatively more expensive still, pricing the poor out to less desirable neighbourhoods even more completely. For example, if only 0.05% of houses in London can overlook a feature as attractive as Hampstead Heath (a large and beautiful park in London) then the ability to 'buy' that feature does not depend so much on your absolute income as on your income relative to the incomes of other households who have a taste for overlooking Hampstead Heath.

Analysis of the structure of house prices as the quantities of particular attributes increase produces results that are consistent with this analysis. Returning to the

[4] And, of course, if property owners already living in more/less desirable neighbourhoods, experiencing rising/falling relative asset values too – see below.

findings of Cheshire and Sheppard (2004) with respect to the price paid for school quality they report a highly non-linear price function with very little change in price if the local school goes from being the worst to middling. The price change associated with moving from the 75th percentile point in the quality distribution to the best of all – the 100th percentile point – however, was very large indeed. Their estimates imply moving an otherwise average house from the catchment area of the worst primary school to that at the 10th percentile in the quality distribution made no discernable difference to its price; moving it from that of the 10th to the 90th percentile in the quality distribution increased the price by 10.4%; but moving it from the catchment area of the primary school at the 90th decile point in the quality distribution to that of the very best was associated with an additional 16.9% increase in price.

Apart from access to the River Thames, where all the price increase was associated with having frontage to the river itself, other attributes for which the premium paid for the 'best' observed was particularly large, were closeness to the town centre, and space – both inside houses and for garden space. Equally, there were some attributes for which the estimates showed a substantial proportion of the price variation was associated with going from having the very 'worst' observed to something just a little better: neighbourhood deprivation was such an attribute. Moving from the most deprived ward to the 10th percentile place in the distribution of deprivation increased the price proportionately more than going from the 90th percentile place to the most affluent ward of all.

What this suggests is that some attributes of houses, or amenities to which particular houses give privileged access, which are in fixed or limited supply take on the status of truly "positional goods" that are effectively auctioned off via the market to the highest bidders. The ability to buy is determined not by absolute income but by income relative to other households competing for the same goods. That school quality and private land and space consumption should exhibit this character is consistent with the argument presented above. Local governments act to constrain the supply of land for housing and – no doubt unintentionally – the availability of the highest quality public goods.[5] In this way many of the "non-market" interactions that are an essential component of cities (as persuasively argued by Glaeser 2001) are actually brought into a form of market allocation via the housing market.

The house and neighbourhood characteristics allocated in the housing market include not only the public goods themselves, but also risk and uncertainty concerning their future levels. As noted above, the measure of the past variation in the quality of a local school was also reflected in the price paid for a given current level of measured quality. Gibbons (2004) showed that neighbourhood crime – an indicator of real risk – was similarly reflected in house prices in London.

[5] That is not because they want there to be worse schools or public parks but because there can only be one 'best' school or park. This is not the case with land supply which is intentionally restricted in order to implement urban containment policies.

The actual price paid for any attribute will depend on the characteristics of the local housing market[6] and economy since both these influence the supply and demand characteristics of individual attributes. For a given measure of income inequality the best local school will cost substantially less in housing markets where average incomes are lower than will be the case in higher income housing markets because demand is income elastic. In the higher income housing market people will be spending a higher proportion of their incomes trying to buy educational quality. If incomes become more unequal over time (or in housing markets in which incomes are more unequally distributed) then the price of attributes in fixed supply will rise and we should expect an even stricter sorting of households between nicer and more disadvantaged neighbourhoods. The best State schools become even more strictly reserved for the richest local households (ignoring private education, access to which is explicitly determined by income not place of residence).

Thus, house prices are about much more fundamental economic and social issues than dinner party conversations might credit. The way in which the housing market works explains an important part of the underlying differences in real welfare in society both vertically between households and across space; that is the patterns of spatial segregation one observes in all cities. Many local public goods, funded from taxation and which we think of as naturally being provided on an equal basis to all households, are really much better thought of as being allocated through the housing market. Consumption of them is thus conditioned on household income in just the same way as consumption of foreign holidays, private education or personal security services is conditioned on income. But because the supply of many of them is more or less fixed within a particular urban area or housing market, income relative to others competing for access to the same goods is the real determinant rather than the absolute level of income itself.

Evidence on the Significance and Size of 'Neighbourhood Effects'

In trying to figure out the direction of causation between poverty and place, there are two major problems. The first is how to be sure when we compare the outcomes for individuals living in different types of areas that we have adjusted for all the

[6] Including, of course, local policy. If access to schools is not determined by place of residence but by some other mechanism, for example, by lot, or by selective examination, then there would be no price paid via the housing market. Consistent with both this and demand for school quality being income elastic is the estimated hedonic price of school quality in Reading compared to Darlington in 1993 and 1997 respectively. Again looking at the 'average' house, the price per GCSE point improvement in Reading, where mean sampled incomes were £28,610 pa, was £243.9 while in 1997, in Darlington, where mean incomes were £23,422, it was £30.80. However, since not only were incomes lower in Darlington but allocation to schools was less tightly tied to home address, we cannot be sure what the contribution of each factor separately was to the estimated difference in the price of 'school quality'.

relevant characteristics of individuals. This is critical because the second issue is that people select the neighbourhoods – subject of course to varying constraints – in which they live. As Goering and colleagues (2003, p.4) point out: "Since people typically select their neighbourhoods to match their needs and resources, researchers restricted to cross-sectional, nonexperiemental evidence must try to separate the impact of personal factors affecting choice of neighbourhood from effects of neighbourhood. But it is difficult if not impossible to measure all those socioeconomic, personal and local characteristics well enough to distinguish their effects."

Separating the impact of personal factors affecting choice of neighbourhood from the effects of neighbourhood requires great ingenuity and work on the part of the researcher. Comparing mean outcomes for households living in neighbourhoods of different types tells us nothing at all about the impact of neighbourhood type in individual outcomes.

Experimental Evidence from Moving Poor People to Affluent Neighbourhoods

There are two basic approaches to resolving this direction of causation problem. The first is to study the (changes) in life outcomes of people who move neighbourhoods quasi-randomly, without any significant personal control as to their new location. The second is to track individuals over time and – after controlling for as many of their characteristics as possible – see whether their neighbourhood of origin influences their life chances. Such studies are generally referred to as cohort studies.

The US Moving to Opportunity (MTO) experiment provides the most carefully researched and certainly the most expensive set of studies of the first type. The MTO programme was set up in 1992 to 'assist very low income families with children …. to move out of areas with high concentrations of persons living in poverty to areas with low concentrations…'[7] Because it was so large in scale and has been so extensively studied the MTO is worth examining in some detail (see also chapters in this book by Small and Feldman 2011; Deluca et al. 2011).

The MTO was carried out in five cities: Baltimore, Boston, Chicago, Los Angeles and New York. For the purposes of implementation 'neighbourhoods' were defined as census tracts, so on average they contained around 4,400 people. The issue of what constitutes a neighbourhood is obviously an open one (see for example: Ellen and Turner 2003; Hardman and Ioannides 2004; Durlauf 2004; Bolster et al. 2007; or Krupka 2007) but census tracts, which are designed to be relatively homogeneous in terms of population characteristics, are widely used as approximations in empirical research in the US.

To be eligible for the programme a family had to live in public or assisted housing in a 'poor' neighbourhood – one in which 40% or more of residents were below the poverty line. They also had to have at least one child under 18, not be behind with

[7] Housing and Community Development Act 1992.

the rent, all family members had to be named on their current lease and no member of the family should have a criminal background. Thus, there were already two stages of selection before a family got on to the programme: (1) since only volunteers participated, all sampled households had to want to move into a more affluent neighbourhood; and (2) households had to pass the eligibility criteria. This selection alone would be likely to have increased the chances of finding positive effects of moving poor families to affluent neighbourhoods. The most problematic families were ineligible and, presumably, only those who thought they had a chance of benefiting from such a move, volunteered.[8] There were also other factors, however, which arguably may have reduced the chances of finding positive effects of the move: for example, a significant proportion of the children involved remained in the same school which had served their deprived neighbourhood (see also Deluca et al. 2011 in this volume).

Once in the programme families were randomly assigned to one of three groups. Group 1 received a subsidy only spendable if they moved to a relatively affluent neighbourhood. An affluent neighbourhood was defined as one in which 10% or less of the residents lived below the poverty line. Such families received expert advice to help them find suitable homes. Group 2 received a housing voucher/subsidy spendable in any location and no advice. Group 3 – the control group – got neither help nor money to move although of course free to move using their own resources.

Across the five cities about a quarter of potentially eligible families applied for the programme with about 13% of those applying being excluded because they did not meet the conditions for selection. That still left some 4,600 families – enough for statistical analysis. Families who managed to move were more likely to be enrolled in adult education and drive a car: they tended also to have been more dissatisfied with their existing housing and neighbourhood. Escaping from high neighbourhood crime rates was the most common reason for volunteering for the project.

The early analysis of partial data summarised in Goering and Feins (2003) suggested some positive findings but they found no observable differences in economic outcomes. Incomes and other labour market indicators for families moving to affluent neighbourhoods showed no improvement relative to other groups. But even if the programme was greeted with cautious optimism, as Goering and Feins (2003) went on to note, its apparent modest initial success did not mean that it was a policy success. The impacts were modest while costs were considerable. The advice of the housing experts alone cost $3,000 per family that successfully moved.

Longer Term Follow-up

Subsequent research (Kling and Liebman 2004; Kling et al. 2005, 2007) on the MTO, tracking families over a longer period, destroyed even this cautious optimism: or at least suggested causal processes are considerably more complex and outcomes of moving to an affluent neighbourhood more difficult to anticipate.

[8] It is worth noting that these methodological deficiencies would have disbarred the study had it been a field trial for a new drug or medical procedure.

Kling and colleagues (2005) report on a follow up study analysing changes 4–7 years after families had moved. Their study used sophisticated statistical methods and focused on differences in crime and behaviour of adolescents.

Kling and colleagues longer term follow-up confirms the finding of no improvement in economic indicators for adults who moved but the researchers looked at a wide range of indicators relating to educational achievement, health and welfare and also behaviour for younger people.[9] They focused on the 15–25 age group in which it was most reasonable to look for signs of improvement. It is this age group which in the general population has the highest incidence of behavioural problems and within which educational progress might be most plausibly concentrated. So if moving to a more affluent neighbourhood produced any behavioural benefits these should be easiest to find in this age group. For none of the indicators, however, did they find any significant overall differences between the groups that moved neighbourhoods compared to the control group that was not helped to move. For the age group as a whole some indicators were better and some were worse but, despite the large sample, none of these differences was statistically significant.

Subdividing into males and females did reveal some significant differences, however. Within the set of behavioural indicators were a number relating to criminality. Kling and colleagues (2005) extended the self-reported data set by also tracing administrative arrest records providing two independent sources of data. They found that while for violent crime there continued to be non-significant but – if anything – favourable effects for both the groups which moved, for property arrests there were significant differences for girls compared to boys. For both boys and girls in the first 2 years after moving, property arrests fell, although the reduction was not statistically significant, but for boys it then rose and rose significantly compared to the control group. Overall – for both sexes combined over the whole 4 years – there was no significant reduction in either total arrests or in property arrests because the differences for boys and girls balanced out. For a small sub sample which it was possible to track over a 6 year period, the increase in property crime arrests for boys continued at about the same level. Similar, but non-significant, gender differences are reported, in passing, for mental and physical health, education and substance use. Overall, males in the moving group had scores on the behavioural problem index some 20% worse and arrest rates for property crime some 30% higher than those of the control group of young males who did not move from their poor neighbourhoods.

They carefully sifted the evidence for explanations of these differences in behavioural outcomes. The reduction in girls' arrest rates for property crime suggested the increase in arrests for boys could not be explained by more efficient policing in the

[9] Kling et al. (2005) report briefly on such factors as getting into fights, getting along with teachers, perceptions that school discipline was 'fair', having five or more friends and reported feelings of worthlessness, finding no significant differences on any measure. A wide range of educational, mental and physical health and behavioural indicators was examined in Kling and Liebman 2004. In general they reported some significant beneficial changes for girls but negative and mainly not significant effects for young males.

affluent neighbourhoods. Peer group sorting effects were implausible as an explanation since similar patterns of change were evident for both boys and girls even when they were subdivided into those with or without a history of criminal or behavioural problems before the move. If peer group sorting was the explanation then one would expect those who had had worse behaviour prior to moving would not have improved (if girls) or got worse (if boys) after the move. Differences in coping strategies in relation to the upset caused by moving to a different type of neighbourhood did not seem plausible as an explanation because immediately following the move both boys and girls showed similar reductions in arrests: it was only after two years, when presumably most young people would have got over the disruptive effects of the move, that boys' arrest rates for property crime rose significantly.

Kling and colleagues (2005) come down in favour of what they call a 'comparative advantage in property crime' explanation partly by elimination but also because of the evidence in relation to educational performance. Although the schools which young people went to after moving to more affluent neighbourhoods were better on academic performance indicators for the children attending them, it turned out that moving did not significantly improve the educational performance of the individual children. Thus, children who did not move ended up doing better in school relative to their peers than children who moved. The children who did move now had academically stronger peers against whom they were measured. Moreover moving boys did worse than girls relative to their new peer groups. Boys were also less subject to parental supervision, had more absences from school and lower educational ambitions than girls. The girls who moved had improved expectations for completing college compared to the control group, greater participation in sports, a reduction in school absences and an increased association with peers who engaged in school activities. None of this was true of the boys who moved. Thus, the authors conclude the most plausible explanation is that as boys adjusted to their more affluent neighbourhoods, they found they had a comparatively worse position in educational terms compared to their new peer group but a realm in which they could succeed in their new neighbourhoods was property crime.

Kling and colleagues (2007) largely confirmed the findings of the earlier study by Kling and colleagues (2005). It is methodologically even more rigorous and confirms yet again that moving to an affluent area generated no economic impacts for adults nor any improvements in physical health. They did find improvements in mental health for both adults and young women apparently related to reductions in (fear of) crime. For young women they also found positive educational outcomes, less risky behaviour and improved physical health but again in the change for young females and males together these were offset by significant deteriorations in the same indicators for young males. So there was no net gain for young people overall from moving.

One might comment, moreover, that if the improvements on mental health indicators resulted from reductions in experienced crime then it would be significantly more cost-effective to achieve the reductions in crime in deprived neighbourhoods in the first place by more effective policing. The MTO programme, in so far as it had produced the improvements, did so only for the tiny proportion of the original inhabitants who benefited. Measures to effectively reduce crime within deprived, high crime neighbourhoods would benefit all residents not just the lucky few subsidised to move.

The MTO experiment has been summarised at length because, given the manifold difficulties, it is still the best source of evidence for identifying the effect of moving from a really deprived neighbourhood to a more affluent one on those who make the move; it is equally the best source of evidence for identifying any beneficial effects of constructing mixed neighbourhoods. Other earlier studies and the initial evaluations of the MTO project are summarised in Durlauf (2004). Durlauf concludes that on the basis of studies then available, the balance of the empirical evidence did suggest there was a significant influence of neighbourhood, although he was acutely aware of the difficulties of identification.[10]

This conclusion is overtaken by the longer term follow up studies of the MTO project. These show that moving to a more affluent neighbourhood does not improve an adult's economic situation and outcomes for children who move are complex and causation is uncertain, even when there appear to be significant effects.[11] On balance, there seem to be negative outcomes for boys on a range of indicators and positive outcomes for girls. One of the few indicators showing an improvement for both boys and girls is an important one – arrest rates for violent crime – but so far research does not show this to be statistically significant.

Two other studies of moving are relevant and produced results entirely in line with the long term evaluations of the MTO programme. Edin and colleagues (2003) and Weinhardt (2010) both ingeniously exploit peculiarities of their national social housing systems – respectively Swedish and British – to construct large samples of quasi-exogenous movers. Edin and colleagues (2003) was concerned with impacts of ethnically homogeneous neighbourhoods on labour market outcomes. They exploited the Swedish policy of housing political asylum seekers in non-ethnically concentrated neighbourhoods. They found this was harmful and statistically significantly harmful to the later earnings of refugees. Refugees in more ethnically

[10] Durlauf (2004) compared results of 25 studies published between 1982 and 2003. Outcome measures ranged from marriage rates and teenage pregnancies through school drop out rates to standard labour market measures, such as wages or unemployment. All studies surveyed were econometric in nature and while some found no evidence of neighbourhood effects, the majority did conclude there was an impact of neighbourhood on outcomes for individuals. However, as Durlauf notes, methodological problems are severe and such evidence was unlikely to convince those who were sceptical. He wrote before the methodologically most convincing studies, those of Oreopoulos (2003) and Kling et al. (2005, 2007), were available.

[11] Apart from the increase in boys' arrests for property crime in the longer term, the cause of some of the improvements in health measures are unclear. For example Katz et al. (2001) note that the significant improvement in childhood asthma for both families moving to more affluent neighbourhoods and those moving to locations of their own choice, could be due to characteristics of the structures and particularly the absence of rats – a common asthma trigger – in the new homes and locations: not to classic neighbourhood effects. If reduction in exposure to rats were the cause then getting rid of rats would seem to be a very much more cost-effective policy to achieve the health improvement than mixed communities.

concentrated neighbourhoods gained 13% in earnings for a one standard deviation increase in the ethnic concentration of their neighbourhood.

Weinhardt (2010) uses very high rates of social housing tenure to identify disadvantaged neighbourhoods demonstrating that this is a valid identifier. He then exploits the scarcity of social housing – there are approximately 1.8 million households on the waiting list for social housing and refusal of an offer of a social housing tenancy almost guarantees the household will fail to get a further offer – to identify a substantial (2,047) sample of young people who were exogenous movers into deprived neighbourhoods. The study then estimates the impact of such a move on educational attainment at age 14. Exploiting the timing of the move and comparing outcomes in two public attainment tests: Key Stage 2, taken at age 11, and Key Stage 3, taken at age 14 and comparing outcomes in terms of Key Stage 3 results of different groups moving at different stages in their school careers into the most deprived areas. Failing to control for family and personal characteristics yields the usual result that moving to a 'bad' school worsens a child's performance. However once personal and family characteristics are fully controlled for this finding disappears. Moving to a really badly performing school in a neighbourhood of severe deprivation has no statistically significant effects on educational outcomes – at least over the span of children's lives covered by Weinhardt's analysis.

The Evidence from Other Studies on Neighbourhood Effects

As an alternative methodology to observing the outcome of moves which result from more or less random processes – so abstracting from the effects of selection bias – long term cohort studies offer the best solution for identifying the pure impact of neighbourhood on life chances. Three of the most convincing of these cohort studies, one in Canada and two in Britain, show a similar lack of significant long term effects of neighbourhood on life outcomes. Oreopoulos (2003) tracked individuals assigned as children to public housing locations in Toronto over 30 years. He starts with a sample of children born between 1963 and 1970, living in public housing projects with very different neighbourhood characteristics, and matches the individuals, using an administrative data base, to their labour market characteristics in 1999. The simple relationship between neighbourhood and earnings appeared to be significant but, of course, families have a big influence on the behaviour and choices of children. Once a full range of personal characteristics and the earnings of siblings were added as explanatory variables, the statistical influence of neighbourhood entirely disappeared. The final conclusion was that the characteristics of the neighbourhood in which an individual was born or raised had no statistically significant effect on long term labour market outcomes or subsequent prosperity.

This finding is consistent with the study of Bolster and colleagues (2007). Using a British Household Panel Survey derived cohort dataset and following individuals for 10 years, they find no evidence that original place of residence had any statistically significant influence on subsequent labour market success, whether measured as household incomes or as earnings. Their results may be slightly less persuasive than those of Kling and colleagues or Oreopoulos since their data track individuals over only 10 years and they investigate only economic outcomes. But they explore a range of neighbourhood definitions, concluding that a small unit, of only about 500 people, is the most appropriate measure of neighbourhood, and they use statistically sophisticated techniques. They cannot entirely reject the possibility that the original neighbourhood in which someone lived, influences their future prosperity but they find no statistical evidence that it does. Indeed, although not statistically significant, their result is in fact the unexpected one. After standardising for all the other factors which influence incomes and earnings, coming from a poorer neighbourhood is associated with increased current prosperity!

Finally a very recent study by van Ham and Manley (2010, and see also Manley and van Ham 2011 in this volume), exploiting the Scottish Longitudinal Study, analyses the labour market outcomes of the tenure structure of the neighbourhood of origin. This is of interest both because a high proportion of social tenure is – at least according to Weinhardt's findings – strongly correlated with neighbourhood deprivation but also because policies for mixed communities explicitly strive for tenure mix as well as mixing on the basis of socio-economic or other characteristics. Van Ham and Manley's results are highly consistent with those of Bolster and colleagues (2007). High simple correlations between more social renting in the neighbourhood of origin and labour market outcomes entirely disappear once a full range of personal characteristics is included. The only surviving 'neighbourhood effect' is very small and on labour market outcomes of home owners living in the most deprived neighbourhoods but as the authors note this result is most likely explained by the unobserved personal characteristics of home owners in such neighbourhoods.

Using an extraordinarily large dataset but rather different methods, Galster and colleagues (2008) come to somewhat different conclusions. They acknowledge the problem of selection bias – people choose the neighbourhood in which they live – and the role of personal characteristics in determining income and labour market success. Their approach is essentially that of a cohort study but their 'cohort' is all Swedes of working age living in metropolitan areas in 1991; so it is very large indeed. They track changes in mean annual income from work between 1991/94 and 1996/99 and control for a few observed personal characteristics – age, sex, marital status, children, education and country of birth. They attempt to control for unobserved personal characteristics by using (a) the change in income between the two periods and (b) focusing on individuals who do not move between neighbourhoods over the period. Critical to this strategy for identifying the pure effect of neighbourhood is the assumption that changes in neighbourhood composition (measured in terms of income mix) are exogenous to individuals' unobserved characteristics. Oreopoulos (2003) used sibling earnings to offset for unobserved personal characteristics.

Galster and colleagues (2008) find, as would be expected, that the estimated impact of neighbourhood effects is reduced when differences in earnings are measured rather than levels and are reduced further for non-movers when compared to changes for those who moved neighbourhoods. However for at least one group – males working less than full time in the first period – an apparently significant, if not straightforward, neighbourhood effect remained. Patterns were apparently complex, however: there was not just a simple association between living in neighbourhoods with higher concentrations of low income people in the period 1991–95 being associated with smaller subsequent income gains. The impact of the initial neighbourhood mix only occurred when the shares of both low and high income neighbours were initially below their median values. In such circumstances 'replacing middle income neighbours with either low income *and/or* high income residents' (Galster and colleagues 2008, p. 865) was associated with lower subsequent incomes.

The main concern however must be about the identification strategy. We cannot really conclude anything about the significance or otherwise of the mainly small but complex neighbourhood effects estimated if how a neighbourhood changes in the subsequent time period is associated with the unobserved personal characteristics of those living in it in the present. That in turn would imply people do not form expectations about how neighbourhoods will evolve in selecting neighbourhoods in which to live and equally that their unobserved characteristics do not influence how neighbourhoods change. We have already seen, however, that in 'buying' school quality (Cheshire and Sheppard 2004) or access to open land (Irwin 2002) people appear to buy expected future values of these characteristics not just those observed in the present. Similarly it would not seem unreasonable that adult Swedish males choosing to work less than full time to take part in child care, for example, will not have other unobserved personal characteristics that both lead them to live in particular types of neighbourhood and will also influence how the neighbourhoods in which they live evolve over time. Thus, quite apart from the small impact of neighbourhood effects the Galster and colleagues (2008) study estimates, some caution would seem to be necessary in accepting their characterisation of them as pure neighbourhood effects.

Perhaps the strongest evidence yet found for 'neighbourhood effects' comes from France. It relates to a restricted effect in educational achievement at school, however, rather than to long term outcomes for life chances (Goux and Maurin 2007) but is certainly worth taking into account. Their study exploits a feature of the French Labour Force Survey which samples clusters of neighbours and, using an Instrumental Variable (IV) approach, they show that various educational performance indicators are statistically related to those of immediate peers in the neighbourhood rather than the classroom; for example, the probability of a 15-year old being held back a grade is about 8 percentage points higher (+16% of an SD) if other adolescents in their neighbourhood were born at the beginning of the year (which in itself increases the probability of being held back by 15 percentage points).

Some Social Advantages of 'Specialised' Neighbourhoods

As noted above, the tendency for people to sort into segregated or 'specialised' neighbourhoods is a very strong one.[12] Cities which are socially segregated along income lines are a universally established fact. Authors who have recently addressed such issues include Hårsman (2006), Meen (2006) and Musterd (2006). Meen's work for Britain shows beyond argument that not only is segregation on income and other measures a feature of cities at least since the late nineteenth Century, but the incidence of such segregation is very persistent over time. Many of the same cities with the sharpest incidence of spatially segregated neighbourhoods measured on 1971 data, reappear in 2001 data. Many of the most deprived (and most affluent) neighbourhoods in London in 1881 appear again in much the same positions in 2001.

Hårsman (2006) documents the stability of patterns of both income and ethnic segregation, particularly in Stockholm. In his detailed study of the long term evolution of patterns of ethnic segregation, he shows how its incidence has tended to intensify over the last 20 years and is only partly explained by income differences. His evidence is at least consistent with people from ethnic minorities mainly choosing to live in ethnically specialised neighbourhoods, despite official Swedish policy pushing for ethnic mixing.

Musterd (2006) synthesises his work on three related areas. In particular, he shows that highly skilled workers in different service sectors choose different types of neighbourhood. Workers in Information Communication Technologies, financial services and banking choose to concentrate in the suburbs of Amsterdam while skilled workers in the creative industries are selectively concentrated in central neighbourhoods.

Labour Market Matching

These findings are consistent with those of Bayer and colleagues (2005) for Boston. For a sample of 110,000 employed people, they match the precise location of residence and jobs and find a very strong tendency for people who live in the same

[12] This is not to deny the fact that cross sectional data shows significant income mixing in even small neighbourhoods. Hardman and Ioannides (2004) report some two thirds of micro neighbourhoods (consisting of 10 households) contain at least one household with an income in the poorest one sixth of all households: and a half of micro neighbourhoods contain a family in the richest 20% of the income distribution. Krupka (2008) comments that in most US cities well over half the variance in income came from variations *within* the neighbourhood, as opposed to variation *across* neighbourhoods. But as he also points out this is still consistent with spatial segregation on the basis of income being the equilibrium outcome. Cities as systems are subject to continuous shocks in terms of their size and the distribution of household incomes and adjustment to such shocks may be slow given the costs of moving house.

neighbourhood (defined as a census block) also to work in the same census block. They make an elaborate and convincing effort to eliminate the effects of transport networks and other factors which might explain this finding independently of social interactions with neighbours. They find evidence showing that such interactions between neighbours strongly influence the job locations of neighbours and that such interactions are more influential if neighbours are of a similar level of education, both parties have children and are of similar age. Their conclusion is that social interactions within neighbourhoods between people similar to each other are a significant factor in how urban labour markets work and people find jobs.

This is consistent with earlier US findings, such as those of Blau and Robins (1992), about the importance in job search of informal networks. Blau and Robins found that while this was a frequent – but not the most frequent – method of job search, and particularly important for the less skilled, it was the most successful form of job search from the point of view of both workers and employers. It produced the highest rate of job offers per contact and the highest rate of job offer acceptances. In their recent review of the literature, Ioannides and Loury (2004) report that, in addition, such jobs found through personal contacts lasted longer, so that around half of all jobs were held by people who had found them this way. Ioannides and Loury also report a persistent increase in the use of informal contacts as a means of job search over time – despite the rise in the internet – and that it is more prevalent, the larger a city is: in cities of more than 500,000 more than half of unemployed job searchers relied on friends and neighbours; in cities smaller than 100,000 less than 10% did. Friends and neighbours were a much more important source of jobs for those who were job-seeking while unemployed than for those searching while employed.

These last two observations are particularly important in the present context. They show positive effects of specialised neighbourhoods are relatively more important for lower skilled than for more skilled workers – unemployed job seekers are on average less skilled than employed job seekers but use friends and neighbours more. The fact that the use of friends and neighbours increases with city size is consistent with the idea that specialised neighbourhoods represent a form of agglomeration economy. An advantage of larger cities is that they can support a greater range of specialised neighbourhoods and such neighbourhoods seem to be a fertile source of effective job matching. Of course, neighbourhoods could have such a high proportion of unemployed in them that job search advantages disappeared. The point is that forcing neighbourhoods to be mixed is likely to result in a loss of agglomeration economies and a loss that has a disproportionate impact on the less skilled.

Another less obvious example of the ways in which specialised neighbourhoods may increase productivity is provided by work on ethnic neighbourhoods. The research of Edin and colleagues (2003) showing that more ethnically homogeneous neighbourhoods improved earnings outcomes has already been cited. There have been numerous studies of the role ethnic neighbourhoods play in mediating access to jobs but a recent example is provided by the work of Coniglio (2004). He has a model in which minority non-local language speakers access labour markets via neighbourhood bilinguals who intermediate information within the wider labour market.

Thus, for those who cannot speak the locally dominant language, living in ethnically segregated neighbourhoods does not just provide consumption benefits in terms of access to culturally familiar goods and services but it generates higher productivity and incomes. He shows that such a model is consistent with the formation and stability of ethnic neighbourhoods in Norwegian cities.

Consumption and Welfare Benefits

There seems to be quite persuasive evidence that specialised neighbourhoods have labour market advantages, even for the poor; indeed particularly for the less skilled who rely on personal contacts to a greater extent to find jobs. Even if there are some possible negative neighbourhood effects for poorer groups – and the more meticulously studies have been able to offset for other factors influencing personal outcomes, the less have they found any such effects – still the question also has to be asked: are there also consumption benefits from living in specialised, and so segregated, neighbourhoods? People systematically tend to choose such neighbourhoods. That, itself, suggests there might be benefits which would be lost if we adopt policies to engineer mixed neighbourhoods.

Choice of neighbourhood is constrained by income because, as we saw above, houses in nicer neighbourhoods cost more, but people choose neighbourhoods on the basis of what a neighbourhood offers them which will either yield welfare directly or increase their expected incomes. Specialised neighbourhoods provide direct consumption benefits, and so contribute to welfare, both because they increase the range of choice for people with respect to the types of neighbourhood in which to live; and people and families of similar incomes, tastes or stages in the life cycle tend to consume similar goods and services and require similar amenities. Living in a neighbourhood with a local wholefood supermarket, Montessori school, gastropub or microbrewery commands a premium: neighbourhoods with pawnbrokers, a local Aldi or discount store and a takeaway are cheaper. If you are a recent immigrant and want to continue to speak your language, engage in your native culture or religion, and buy food or other items you have developed a taste for, then there are great advantages in living in neighbourhoods with concentrations of people of similar origin. This is one obvious source of the ethnic neighbourhoods of large American and European cities. A recent study of children in primary schools found 300 different language communities in London (Baker and Eversley 2000) living in linguistically and culturally specialised neighbourhoods. Bowes and colleagues (1997) find a similar pattern in Glasgow.

Such agglomeration economies in consumption are not confined to ethnic groups. Families with young children will find benefits of networks and facilities, and mutual support as well as information, if they live in neighbourhoods with substantial numbers of families at the same stage in life. Young singles who eat out and have a taste for urban entertainment and culture will similarly find agglomeration economies in consumption if they find neighbourhoods in which large numbers of like minded people are concentrated. More educated people, and people working in the liberal professions may prefer to live in neighbourhoods with concentrations of similar types, sharing leisure and cultural pursuits and seeking similar local shops; business people may equally gain consumption benefits from concentrating in neighbourhoods in which other business people live.

Luttmer (2005) explores another possible welfare gain from income segregated neighbourhoods. He analyses the implications of people's welfare depending not just on their own income but on their income relative to other people in their community. He quotes John Stuart Mill to demonstrate the distant roots of this idea: *"…. men do not desire merely to be rich, but richer than other men"* (Luttmer 2005, p. 963). In testing this proposition there are a number of methodological problems – particularly the possibility that welfare is itself a relative concept. However, Luttmer (2005) goes to considerable lengths to eliminate possible biases from his estimates, including testing against absolute measures related to welfare, such as marital conflicts, as well as against reported personal welfare itself. He analyses a sample of about 10,000 individuals from two phases[13] of the National Survey of Families and Households living in a sample of 965 neighbourhoods – or 555 neighbourhoods for the sub sample of neighbourhoods with respondents living in them at both time periods. His findings are striking. Roughly speaking losing $1,000 of own income seems to make people feel about as much worse off as their neighbour gaining $1,000! The estimated impact of a positive change of household income on reported welfare is quantitatively almost the same as a similar, negative, change in neighbourhood mean incomes.

He subdivides the sample into households of single adults, couples living together at both sampling dates (stable couples) and adults living with different partners in the second time period. The strong negative impact of neighbours being richer on peoples' sense of wellbeing estimated for the sample as a whole turns out to be explained mainly by the (large) sub sample of stable couples. Single people do not seem to experience a loss of wellbeing from neighbours' extra income. Moreover, the effects are stronger for individuals who socialise with neighbours and the effect of neighbours' incomes is stronger if the neighbour is more similar to you. If disaggregated measures of reported welfare are analysed then the main effects were in terms of time with family and hours worked. That is, it appears that people living in communities where neighbours have higher incomes relative to their own, compensate by working longer hours and spending less time with their families and in leisure. This causes them to feel worse off and have lower reported welfare. The evidence points to a real impact of relative, as well as own, income on welfare.

Perhaps the main problem with what is a very careful study is the definition of 'neighbourhood'. For reasons of data availability these are the Public Use Microdata Areas which, in the 1990 Census, had a mean population of 144,000 people. They are certainly considerably larger, therefore, than the usual concept of a neighbourhood. On the other hand, given that the findings capture a real effect of relative

[13] 1987–88 and 1992–94.

neighbourhood income which, on the basis of the accumulation of evidence Luttmer (2005) provides, seems plausible, then having data for smaller areas, corresponding more closely to conventional ideas of neighbourhoods, would seem likely to make the impact more significant still. The study finds strong evidence that social interaction with neighbours is a causal factor and presumably social interactions per neighbour are considerably greater with your nearest 500 neighbours than with those living far away and not sharing the same schools, shops or parks. In a neighbourhood of 144,000 there will be few such opportunities to interact with most 'neighbours'; in a smaller neighbourhood of 500 the chance of such interactions increases greatly. So the incomes of nearer neighbours seem more likely to affect one's sense of wellbeing than those of more distant ones and consequently one might expect Luttmer's results to have been even more powerful had he had data for smaller neighbourhoods.

The implications of these findings, therefore, would be that welfare in society is totally unaffected by whether the rich live mixed with the poor or not. The welfare gains of one group would be offset by the losses of the other. It is just that if the poor do live close to the rich they will feel worse of than if they do not. The rich, of course, would feel even better off so mixing communities effects a net welfare transfer from the poor to the rich: plausible but not the intended result of mixed community policy.

Conclusions

This chapter does not argue that 'neighbourhood effects' do not exist. While the evidence is overwhelming that poor people are priced into cheap neighbourhoods because they are poor, living in the most deprived neighbourhoods is almost by definition not a life enhancing experience. Because of peer group and role model effects it certainly seems plausible that the experience would impair the life chances of those who live in them and especially those of children raised in them. Before engaging in significant efforts and spending substantial resources to use policy to force neighbourhoods to be mixed, however, it seems essential to have a clear idea – if such neighbourhood effects do exist – of how large they are and what benefits specialised, homogeneous neighbourhoods may confer on both the richer and poorer households who live in them.

The evidence surveyed here suggests that the benefits of specialised neighbourhoods are significant both in terms of finding suitable jobs and increasing the range of choices available to people and the welfare they derive from living in cities. Specialised neighbourhoods are a significant reason why living in cities enhances welfare. Moreover, if we accept Luttmer's (2005) findings, then having poor people living close to richer people reduces the welfare of the poor because of the effect of relative income.

Despite major research efforts, persuasive evidence of any significant negative effects on poor people of living in deprived neighbourhoods (compared to the fact of poverty and the factors which tend to make someone poor in the first place) is very elusive. The evidence reviewed here, particularly the most recent findings from cohort studies and the MTO project, does not support the conclusion that neighbourhood effects are quantitatively all that important nor that moving the poor to affluent neighbourhoods on balance improves their welfare. However, we do know that the rich can always outbid the poor for nicer neighbourhoods because the desirable attributes of these neighbourhoods are fully reflected in the prices of houses within them. To the extent that this is true, social segregation in cities must largely reflect economic inequality rather than causing it. Forcing neighbourhoods to be mixed in social and economic terms is, therefore, mainly treating the symptoms of inequality not the causes. It may make decent people feel better but it does not address the real problem.

At the same time there seem to be direct welfare benefits from living in specialised neighbourhoods with other complementary and similar households and productivity benefits, too, because of better labour market networking and matching. These seem to apply to poorer, less skilled people more than to the rich and educated. To the extent that these are significant, mixed neighbourhood policies directly destroy a potential source of welfare and a portion of the consumption and productivity benefits cities are capable of delivering. All these possible losses need to be balanced against any benefits of reduced negative 'neighbourhood effects'.

That the disadvantaged are concentrated in poor neighbourhoods does not demonstrate that poor neighbourhoods are a cause of their disadvantage. Given the lack of evidence for any strong neighbourhood effects, the conclusion for policy is to reduce income inequality in society not build 'mixed neighbourhoods'. Mixed neighbourhood policies may divert attention from the need for effective income redistribution. These arguments do not, of course, imply that it is not useful to deliver policies aimed at reducing societal inequality in poor neighbourhoods (e.g. programmes to improve labour force skills or reduce crime). It is in the poorest neighbourhoods that those who most need the help of people-targeted policies are concentrated.

References

Alonso, W. (1964). *Location and land use*. Cambridge: Harvard University Press.

Anderson, S. T., & West, S. E. (2006). Open space, residential property values and the spatial context. *Regional Science and Urban Economics, 36*(6), 773–789.

Baker, P., & Eversley, J. (Eds.). (2000). *Multilingual capital: The languages of London's schoolchildren and their relevance to economic, social, and educational policies*. London: Battlebridge.

Bayer, P., Ross, S. L., & Topa, G. (2005). *Place of work and place of residence: Informal hiring networks and labour market outcomes*. NBER Working Paper, No. 11019.

Black, S. (1999). Do better schools matter? Parental valuation of elementary education. *Quarterly Journal of Economics, 114*(2), 578–599.

Blau, D., & Robins, P. (1992). Job search outcomes for the employed and unemployed. *Journal of Political Economy, 98*(3), 637–655.

Bolster, A., Burgess, S., Johnston, R., Jones, K., Propper, C., & Starker, R. (2007). Neighbourhoods, households and income dynamics: A semi-parametric investigation of neighbourhood effects. *Journal of Economic Geography, 7*(1), 1–38.

Bowes, A., Dar, N., & Sim, D. (1997). Tenure preference and housing strategy: An exploration of Pakistani experiences. *Housing Studies, 12*(1), 63–84.

Brueckner, J. K., Thisse, J. F., & Zenou, Y. (1999). Why is Central Paris rich and Downtown Detroit poor? An amenity-based theory. *European Economic Review, 43*(1), 91–107.

Cheshire, P. C., & Sheppard, S. (1995). On the price of land and the value of amenities. *Economica, 62*(258), 247–267.

Cheshire, P. C., & Sheppard, S. (1998). Estimating demand for housing, land, and neighbourhood characteristics. *Oxford Bulletin of Economics and Statistics, 60*(2), 357–382.

Cheshire, P., & Sheppard, S. (2004). Capitalising the value of free schools: The impact of supply constraints and uncertainty. *Economic Journal*, November, F397–F424.

Cheshire, P. C., & Sheppard, S. (2005). The introduction of price signals into land use planning decision-making: A proposal. *Urban Studies, 42*(4), 647–663.

Coniglio, N. D. (2004). *Migropolis: Migration networks and the formation of ethnic clusters in cities.* 44th ERSA Congress.

Deluca, S., Duncan, G. J., Keels, M., & Mendenhall, R. (2011). The notable and the null: Using mixed methods to understand the diverse impacts of residential mobility programs. In M. van Ham, D. Manley, N. Bailey, L. Simpson, & D. Maclennan (Eds.), *Neighbourhood effects research: New perspectives*. Dordrecht: Springer.

Durlauf, S. (2004). Neighborhood effects. In J. V. Henderson & J. F. Thisse (Eds.), *Handbook of regional and urban economics* (Vol. IV). Amsterdam: Elsevier.

Edin, P. A., Fredriksson, P., & Åslund, O. (2003). Ethnic enclaves and the economic success of immigrants – Evidence from a natural experiment. *Quarterly Journal of Economics, 118*(1), 329–357.

Ellen, I. G., & Turner, M. A. (2003). Do neighborhoods matter and why? In J. Goering & J. D. Feins (Eds.), *Choosing a better life?: Evaluating the moving to opportunity social experiment* (pp. 313–338). Washington, DC: Urban Institute.

Galster, G., Andersson, R., Musterd, S., & Kauppinen, T. M. (2008). Does neighbourhood income mix affect earnings of adults? New evidence from Sweden. *Journal of Urban Economics, 63*(3), 858–870.

Gibbons, S. (2004). The costs of urban property crime. *Economic Journal, 114*(449), F441–463.

Gibbons, S., & Machin, S. (2005). Valuing rail access using transport innovations. *Journal of Urban Economics, 57*(1), 148–169.

Glaeser, E. L. (2001). 'The future of urban research: Nonmarket interactions.' *Brookings-Wharton Papers on Urban Affairs*, 101–149.

Goering, J., & Feins, J. D. (2003). *Choosing a better life: Evaluating the moving to opportunity experiment.* Washington, DC: The Urban Institute Press.

Goodin, R., & Le Grand, J. (1987). *Not only the poor: The middle classes and the welfare state.* London: Allen and Unwin.

Gordon, I. R., & Monastiriotis, V. (2006). Urban size, spatial segregation and inequality in educational outcomes. *Urban Studies, 43*(1), 213–236.

Gordon, I. R., & Monastiriotis, V. (2007). Education, location, education: A spatial analysis of English secondary school public examination results. *Urban Studies, 44*(7), 1203–1228.

Goux, D., & Maurin, E. (2007). Close neighbours matter: Neighbourhood effects on early performance at school. *Economic Journal, 117*(523), 1193–1215.

Hardman, A., & Ioannides, Y. (2004). Neighbors' income distribution: Economic segregation and mixing in US urban neighborhoods. *Journal of Housing Economics, 13*(4), 368–382.

Hårsman, B. (2006). Ethnical diversity and spatial segregation in the Stockholm region. *Urban Studies, 43*(8), 1341–1364.

Haurin, D., & Brasington, D. (1996). School quality and real house prices: Inter – and intrametropolitan effects. *Journal of Housing Economics, 5*(4), 351–68.

Hedman, L., van Ham, M., & Manley, D. (2011). Neighbourhood choice and neighbourhood reproduction. *Environment and Planning A, 43*(6), 1381–1399.

Hilber, C. A. L. (2010). New housing supply and the dilution of social capital. *Journal of Urban Economics, 67*(3), 419–437.

Iaonnides, Y. M., & Loury, L. D. (2004). Job information networks, neighborhood effects and inequality. *Journal of Economic Literature, 42*(4), 1056–1093.

Irwin, E. (2002). The effects of open space on residential property values. *Land Economics, 78*(4), 465–481.

Katz, L. F., Kling, J., & Liebman, J. (2001). Moving to opportunity in Boston: Early results of a randomised mobility experiment. *Quarterly Journal of Economics, 116*(2), 607–54.

Kling, J., & Liebman, J. (2004). *Experimental analysis of neighborhood effects on youth.* IRS Working Paper No 483, Princeton.

Kling, J., Liebman, J. B., & Katz, L. F. (2007). Experimental analysis of neighbourhood effects. *Econometrica, 75*(1), 83–119.

Kling, J., Ludwig, J., & Katz, L. F. (2005). Neighbourhood effects on crime for female and male youth: Evidence from a randomised housing voucher experiment. *Quarterly Journal of Economics, 120*(1), 87–130.

Krupka, D. J. (2007). Are big cities more segregated? Neighbourhood scale and the measurement of segregation. *Urban Studies, 44*(1), 187–197.

Krupka, D. J. (2008). *The stability of mixed income neighbourhoods in America.* IZA Discussion Paper No 3370, Bonn: IZA.

Luttmer, E. F. P. (2005). Neighbors as negatives: Relative earnings and well-being. *Quarterly Journal of Economics, 120*(3), 963–1002.

Manley, D., & van Ham, M. (2011). Neighbourhood effects, housing tenure and individual employment outcomes. In M. van Ham, D. Manley, N. Bailey, L. Simpson, & D. Maclennan (Eds.), *Neighbourhood effects research: New perspectives.* Dordrecht: Springer.

Massey, D. S., & Fischer, M. J. (2003). The geography of inequality in the United States. *Brookings-Wharton Papers on Urban Affairs*, 1–40.

Meen, G. (2006). *Modelling local spatial poverty traps in England.* International Centre for Housing and Urban Economics Discussion Paper No 1, University of Reading.

Meen, G., Meen, J., & Nygaard, C. (2007). *A tale of two Victorian cities in the 21st century*, mimeo. International Centre for Housing and Urban Economics, University of Reading.

Musterd, S. (2006). Segregation, urban space and the resurgent city. *Urban Studies, 43*(8), 1325–1340.

Oates, W. (1969). The effects of property taxes and local public services on property values: An empirical study of tax capitalisation and the Tiebout hypothesis. *Journal of Political Economy, 77*(6), 957–971.

ODPM. (2005). *Sustainable communities: People, places and prosperity, Cm 6425.* London: HMSO.

Oreopoulos, P. (2003). The long-run consequences of living in a poor neighbourhood. *Quarterly Journal of Economics, 118*(4), 1533–1575.

Rosen, S. (1974). Hedonic prices and implicit markets: Product differentiation in pure competition. *Journal of Political Economy, 82*(1), 34–55.

Schelling, T. C. (1969). Models of segregation. *American Economic Review, 59*(2), 488–493.

Small, M. L., & Feldman, J. (2011). Ethnographic evidence, heterogeneity, and neighbourhood effects after moving to opportunity. In M. van Ham, D. Manley, N. Bailey, L. Simpson, & D. Maclennan (Eds.), *Neighbourhood effects research: New perspectives.* Dordrecht: Springer.

Tiebout, C. (1956). A pure theory of local expenditures. *Journal of Political Economy, 64*, 416–424.

Troy, A., & Grove, J. M. (2008). Property values, parks, and crime: A hedonic analysis in Baltimore, M.D. *Landscape and Urban Planning, 87*(3), 233–245.

van Ham, M., & Manley, D. (2010). The effect of neighbourhood housing tenure mix on labour market outcomes: A longitudinal investigation of neighbourhood effects. *Journal of Economic Geography, 10*(2), 257–282.
Weinhardt, F. (2010). *Moving into the projects: Social housing neighbourhoods and school performance in England.* Spatial Economics Research Centre No 22, London School of Economics.
Wilson, W. J. (1987). *The truly disadvantaged: The inner city, the underclass and public policy.* Chicago: Chicago University Press.

Index

DOI 10.1007/978-94-007-2309-2,